PORSCHE DECADES

JAY GILLOTTI

DALTON WATSON FINE BOOKS

PORSCHE DECADES

An Introduction to the Porsche Story

Jay Gillotti

Published 2024
ISBN: 978-1-956309-16-4

All rights reserved. Apart from any fair dealing for the purpose of private study, research, criticism or review, as permitted under the terms of the Copyright, Design and Patents Act of 1988, no part of this book may be reproduced or transmitted in any form or by any means, electronic, electrical, chemical, mechanical, optical including photocopying, recording or by any other means placed in any information storage or retrieval system, without prior permission of the publisher.

This book contains chassis histories based on publicly available information and interviews and discussion with some of the recent or current car owners or their representatives. While the author has used best efforts to ensure that the content of the book was based on the best available information at the time of writing, the author and publisher accept no liability for statements that are inadvertently incorrect or superseded by more accurate information in the future. The author and publisher disclaim any liability to any party for any loss, damage, or disruption caused by errors or omissions, whether such errors or omissions result from accident, negligence, inadvertence or any other cause. This book is intended for entertainment purposes only and should not be used for valuation purposes.

Printed by Interpress, Ltd; Hungary
for the publisher
Dalton Watson Fine Books
Glyn and Jean Morris
Deerfield, IL 60015 USA
www.daltonwatson.com

TIME

Time is one of the few things man cannot influence. It gives each of us a beginning and an end…and this makes us question how we use what comes between. We all have a desire to create something that will show we were here, and did something of value… to create something timeless…

For something to endure, it must be unique, yet so universal anyone can appreciate it. The one thing that is universal is the search for purpose. It is the essence of the human spirit to seek a reason for being. This is why functional designs are so beautiful, so calming.

Of course, timeless design is wasted if it cannot survive. This is why we race and test our cars so hard…

Creating something that time cannot erode, something which ignores time, both physically and in concept… this is the ultimate victory.

The first Porsche, built in 1948, is still with us. Still clearly the inspiration for everything we have done. It will live on in all our cars.

-Ferry Porsche

Table of Contents

Chapter One: 1875 to 1919

- Ferdinand Porsche's Early Life and Career 13
- Lohner 15
  - Lohner-Porsche Vehicles 16
  - Marriage and Children 17
- Austro-Daimler 19
  - Emil Jellinek 19
  - Ferry Porsche's Childhood 21
- Military and Aviation Projects 22

Chapter Two: 1920s

- Austro-Daimler Sascha 25
- Daimler Motoren AG 27
  - Targa Florio 31
  - Nürburgring 33
- Steyr 36

Chapter Three: 1930s

- People's Car 41
  - The Russia Project 41
  - Torsion Bar 42
  - Volkswagen Precursors and Influencers 42
- Auto Union 43
  - Edmund Rumpler and the Swing Axle 44
  - Ferry Porsche's Marriage 47
  - Hans Stuck & Son 50
  - Visits to America 51
- KdF-Wagen 53
  - Auto Union versus Mercedes-Benz 53
  - Politics and Citizenship 54
  - 'Boxer' Engine 56
  - VW Turbo? 57
- Other projects 58
- Mercedes Record Car 58
- Volkswagen 60K10 59

Chapter Four: 1940s

- Kübelwagen 63
  - "Now Hitler's Gone Mad!" 63
- Schwimmwagen 65
- Tank Designs 65
  - Porsches, Father and Son 65
  - Porsche vs. Speer 66
- Move to Austria 68
- Cisitalia Grand Prix 71
  - Piero Dusio and Cisitalia 71
- Porsche 356 75
  - Porsche Synchromesh 75
  - History of 356 Chassis 001 78
  - Erwin Komenda 78
- 356/2 79
  - Formation of Porsche Konstrucktionen GmbH 79

Chapter Five: 1950s

- Le Mans, 1951 ... 87
  - Le Mans ... 88
  - Max Hoffman ... 90
- 356 Developments ... 92
  - Porsche Crest ... 94
- Overseas Assignments ... 95
  - *Christophorus* magazine ... 95
  - Huschke von Hanstein ... 95
  - Anton Piëch ... 96
  - Louise Porsche-Piëch ... 96
  - Ignition Switch Placement ... 97
- Porsche 550 ... 99
  - Ernst Fuhrmann ... 102
  - James Dean ... 104
- Porsche Speedster ... 105
  - Porsche Club of America ... 105
  - Poopers? ... 106
- Porsche 356A ... 108
- Porsche 550A ... 110
  - Porsche Tractors ... 110
- Porsche 718 ... 113
- Porsche 356B ... 116
  - Ferry Porsche on Fashion and the Automobile ... 117

Chapter Six: 1960s

- Early 1960s Carreras ... 119
- Open-Wheelers ... 121
- 356 Development ... 123
  - Reutter and Recaro ... 123
- Porsche 901 ... 124
- Porsche 904 ... 130
  - Weissach ... 130
  - Leopard Tank ... 130
  - Ferdinand Alexander 'Butzi' Porsche ... 131
- 'Plastic' Porsche Racing Cars ... 132
- 1967 ... 136
- 1968 ... 137
- 1969 ... 140
  - Porsche 909 ... 140
  - Ferdinand Piëch ... 141
- 911 Developments ... 142
  - Targa Top ... 142
- 911 In Competition ... 144
  - Fuchs Wheel ... 144
  - Janis Joplin ... 144
  - Trailing Throttle Oversteer ... 146
  - Volkswagen of America ... 146
- Porsche 917 ... 147
  - Erich Strenger ... 147

Chapter Seven: 1970s

Porsche 914 151
911 Developments 154
Porsche 917 155
   Le Mans 155
   Porsche 908/03 157
   Company designations in German 158
   Hans Mezger 158
Porsche 917 Can-Am Turbos 159
   Family Business 159
Porsche 911 RS/RSR Carreras 161
   917/30 Closed Course Record 161
   Safari Rally to Safari Builds 161
   IMSA 161
Impact-bumper 911 165
   Anatole 'Tony' Lapine 168
   2.7 Carrera MFI 168
Porsche 911 Turbo 168
   912E 169
Porsche 934/935 172
   Manfred Jantke 176
Porsche 924 177
   Type 2304/05 Wiesel AWC 177
Porsche 936 179
Porsche 928 183
Porsche 911 SC 188

Chapter Eight: 1980s

Peter Schutz 193
High Performance 924s 194
911 Developments and Cabriolet 195
   1981 24 Hours of Le Mans 195
Porsche 956 199
Porsche 944 203
   TAG Turbo F1 Engine 204
   924S 204
944 Turbo 205
   Helmuth Bott 206
Porsche 959 207
Porsche 962 212
928 in the Eighties 215
   Peter Falk 215
   PDK Transmission 215
   Special Wishes 217
Porsche CART/IndyCar Program 218
   Porsche 989 220
   Porsche Tuners 221

Chapter Nine: 1990s

Porsche 964 224
Porsche 968 227
928 in the Nineties 228
   Mercedes and Audi Outsourcing 232
Porsche 993 232
   1994 Le Mans 233
   Kaizen 238

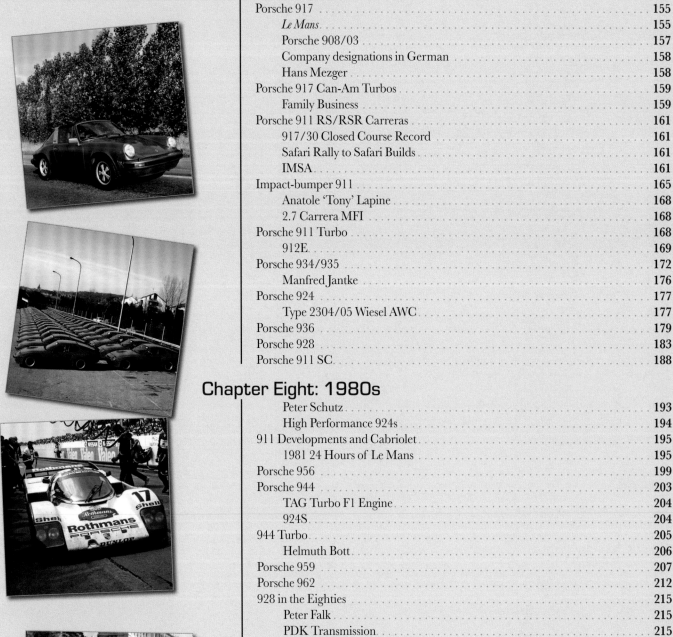

    WSC 95 238
        C88 - China People's Car Project 239
    Porsche Boxster 241
        Porsche Engineering Services 241
    Porsche 911 GT1 246
    Porsche 996 249
        Wendelin Wiedeking 249

Chapter Ten: 2000s

    Porsche Cayenne 257
        Rennsport Reunion 260
    Porsche 996 in the 2000s 261
    Carrera GT 264
    Boxster in the 2000s 266
        M96-M97 Engine Issues 267
    Porsche Cayman 270
        Norbert Singer 272
    RS Spyder 272
    Porsche 997 274
    Porsche 997.2 276
    Panamera 277
    Porsche versus Volkswagen 280
        New Porsche Museum 281

Chapter Eleven: 2010s

    Super 911s 283
        2010 24 Hours of Daytona 284
    987.2, Boxster Spyder and Cayman R 285
        Boxster E 286
    Cayenne E2 and E22 287
    Diesels and Hybrids 290
    Porsche 991 291
        Dieselgate 291
    Porsche 981 293
    Porsche Macan 296
    Porsche 918 297
    Porsche 919 302
        Passing of Butzi Porsche and Ferdinand Piëch ... 305
        Porsche Customs 305
    Porsche 718 Boxster and Cayman 306
    Four-Door Fortitude 309
        Special 911s 310
    Porsche 992 311
        Wolfgang Porsche 311
    Porsche Taycan 313

Chapter Twelve: 2020s 316

End Matter

    Afterword 324
    Appendix 1: Porsche Project Numbers 326
    Appendix 2: Porsches in the Movies 332
    Bibliography and References 336
    Index 344

Foreword *Miles C. Collier* **Founder of Revs Institute**

Porsche automobiles have gone from strength to strength over the three quarters of a century since the first prototype 356 stuttered to life in Gmund, Austria. And it has been quite a journey starting as it did in 1948 with little more than a pre-war, hot-rodded Volkswagen engine and transmission installed in a simple platform chassis covered in a hand-made aluminum body. Seventy-six years later, Porsche is arguably the most successful high-performance car manufacturer on the globe, having produced millions of cars since 1949. Go to one of today's major auctions, especially in the States, and you will observe Porsche after Porsche selling at truly astonishingly high prices. Indeed, the value the market assigns to later model cars from the 1990s, or even the 2010s shows that the originally rather dubious architecture of the overhung rear engine design, despite many premature prognostications of its death, has proven to be successful over the years and over countless drivers, whether skilled racers or novice pilots.

Consider that since 1963, the 911, successor to the original 356, has evolved over more than sixty years and eight generations of development. Try to find another manufacturer who has had such longevity with a single model and done so by remaining, not just desirable and contemporary, but actually going up-market in performance, technology and prestige. If you are going on a cross-country trip staying with blue roads, which car would you pick? Surely not the Aston Martin Vantage, nor the Pagani Zonda. I'd guess because the designers in Stuttgart have built a car that has the performance of the two exotics mentioned, but the reliability and usability of a Toyota. This is no trivial accomplishment.

Such a technological feat reflects the legacy of Ferry Porsche's original "future proof" design dictate for the 901, as the 911 was first numbered in the early days. The simple profile of the body has become universally recognized. Today, despite the iconic form, the 911 still looks fresh and contemporary. Its performance, relying on the original flat six cylinder layout has survived translation from air to water cooling with, not only no lack of performance, but in fact, with more performance. We have only to consider the normally aspirated 911 GT3 RS engine with its 125 horsepower per liter.

Where did that genius for inspired design originate? What was the tradition and experience and, yes, corporate engineering culture that could give rise to such an incredible achievement?

Well, one answer to those questions can be found within the covers of this book. In the end, great enterprises are the product of people, inevitably extraordinary ones, often flawed ones. And the story of such people, their trials, heartbreaks, triumphs, and failures are the elements of the engaging saga we encounter here.

Human stories are at the center of Porsche's history. Those stories pique our interest in everything Porsche. Certainly, while we want to know what clever engineering was encapsulated into each new model that came off the assembly lines, we also want to know the personal stories of those people, and, because for most of its life Porsche was a family business, we want to read the story of the family behind the company. And what a story that is.

That story is filled with triumphs, conflicts, struggles, and loss, good decisions and bad. *Porsche Decades* is devoted to more than just the design and production history of Porsche automobiles. It engages the reader because it goes beyond the objective facts and figures of models and their constituent subassemblies to address the bigger picture: what went on in society at the time, within the family, and among the actors who shaped Porsche's history and the cars themselves. And this is where Jay delivers. We are exposed to expanded accounts, vignettes, and surprising factoids set off in their own boxes. Compared to what we can obtain from other Porsche books, *Porsche Decades* connects the technical history of the cars to the family history. The large number of plates and images also help to amplify the text, especially from the less well known era before the advent of the Porsche automobile. We get glimpses of Dr. Porsche's character. He was brilliant, self-taught, impatient with those who couldn't keep up with his extraordinary abilities, and yet he was otherworldly and naïve to a fault.

For those who want to have a thorough grounding in Porsche lore and history, this book will serve very well. With its emphasis on the backstories and context, it will give the experienced Porsche enthusiast, as well as the new Porsche aficionado, a useful perspective on, and appreciation of the engineering tradition that went into the car they drive. Few consumer products have such a tradition of focus and tenacity. Few consumer products elicit such praise, esteem, and such a following. By coupling the context of the human history to the design history, Jay makes this book one of the Porsche volumes that deserves to be on every Porsche lover's bookshelf.

Miles C. Collier **Founder of Revs Institute**
March 2024

Preface

Why another Porsche book?

In my volunteer role as Historian for the Pacific NW Region of the Porsche Club of America, I began teaching a class on basic Porsche history in 2019. Doing the research and preparation for the class re-energized my own interest in the arc of the Porsche story. This history now dates back nearly 150 years. As we approached the 75th anniversary of Porsche as an automobile manufacturer, I thought it might be a service to the Porsche community to attempt to write an introduction to Porsche history, not unlike a textbook, if there was such a class at the local college. Another part of the inspiration was a sense that in two hours, I could not present the level of detail that I desired for a group of interested Porsche Club members.

As a result, this book was written with club members in mind. However, it is intended for anyone who may be new to Porsche ownership or to learning about the history of the company, its origins and its achievements. In this way, it is the opposite of my first book, *Gulf 917*. That was a deep dive into a very narrow (but exciting) slice of Porsche's racing history. This book is meant as a very broad overview and jumping-off point for further exploration.

For example, readers who are new to Porsche's history may decide to learn much more about diverse topics such as World War II vehicles, Porsche's racing history, personalities like Ferry Porsche, Ferdinand Piëch or Norbert Singer, or specific Porsche production models. The goal of *Decades* is to open the door to a fascinating and ongoing story while also grounding students of Porsche in the basics. Hopefully, even the more experienced and knowledgeable Porsche enthusiasts will find new and interesting aspects of the story as we travel through time. The detail in the bibliography is meant to assist the reader in finding resources for deeper exploration into areas of personal interest.

Decades is an arbitrary division of time and history. It was suggested in part by some significant events in Porsche's history that fell at or near the beginning of a decade. In 1930, Ferdinand Porsche went independent and started his own company. In 1950, production of Porsche sports cars moved from Austria back to Stuttgart. In 1970, Porsche won the 24 Hours of Le Mans overall for the first time. Looking at the decades also allows for a manageable perspective on the Porsche story – a story that is incredibly rich in its full detail.

(Dean Holbrook)

The reader should recognize that in most cases I am vastly oversimplifying complex engineering efforts made by the many thousands of skilled designers and engineers working with Dr. Porsche and for the Porsche company over the decades. As noted above, a large amount of detail is available in the reference sources listed in the bibliography. Much of this material is fascinating if the reader takes an interest in specific eras or vehicles.

The use of sidebars allows the reader to step out or pause the main narrative for important background or details. For example, if the reader is not familiar with great races like the Targa Florio or the 24 Hours of Le Mans, a minimal sense of these events is important to understanding the Porsche story. In addition, please note that the text will generally refer to Ferdinand Porsche (born in 1875) as 'Dr. Porsche' after the period when he was granted an honorary doctorate. His son, 'Ferry' Porsche (whose real name was also Ferdinand, born in 1909) is referred to by his nickname throughout the text (even though he was also granted honorary doctorate status). Ferry's son, Ferdinand (born in 1937) is generally referred to by his nickname, 'Butzi'.

Also, let the internet and a favorite search engine be the reader's friend. In this volume, there is not enough space to explain every automotive term being used. However, nearly any term can easily and quickly be defined on the internet. For those who may be less familiar with automotive systems and mechanics, a simple search on 'two-stage supercharger', 'automotive engine torque', 'pushrod engine' or 'vacuum servo brake assist' will yield good results. The same is true for information about many of the cars, racing drivers and other personalities mentioned in the text.

Regarding the 1930s and 1940s, I had no intent to detail persecution, atrocities and war crimes committed by the Nazi regime and certain individuals mentioned in Chapters Three and Four. As with other people and companies in this period, Porsche's role may remain controversial. In this volume, Dr. Porsche's work in supporting the German war effort is covered as a significant portion of his engineering career. As an amateur historian of World War II, I must say that nothing can overstate or forgive the horrific loss of life and suffering caused by the Axis powers, including Nazi Germany. Interested readers should research this topic further and make their own judgments regarding the actions of Porsche/Piëch family members during that period. This also includes assessing the relevance of the Nazi period to the post-war history of Porsche. Porsche AG has taken significant steps in recent years to research and address the role of Porsche before, during and after World War II. Please see the references in the bibliography relating to this topic.

The large number of sources in the Porsche literature occasionally conflict on small and sometimes significant details. Every effort has been made to resolve these conflicts and I have generally deferred to the most recent published source. I invite feedback on any areas with potential inaccuracy.

I also wish to recognize the important work of so many authors who have written a multitude of Porsche books and articles over the years. In many cases, the writers have done the invaluable service of recording information from people who are no longer with us. They have also cataloged technical information that might be difficult to reassemble today. I am honored to encourage readers who might be new to exploring the library of Porsche information.

All photos courtesy of Porsche AG Corporate Archive unless otherwise noted.

Jay Gillotti
March 2024

Acknowledgments

Special thanks to the following people, without whom this book would not have been completed:

First Readers: Joe Gillotti, Duncan Newell

Photos and research assistance:
Jens Torner and the Porsche AG Corporate Archive

Additionally:
Chris Powell, Mike Piccolo, Ben Przekop, Sherwin Eng, Bill Sargent, Dean Holbrook, John Mueller, Brian Redman, Frank Mandarano, Nick Leader, Paul Hageman, Karl Ludvigsen, Jason Tang, Karl Noakes, John Ficarra, Deanna Johnson, Kurt Oblinger, Norbert Kremsner, Christian Bouchez, Brent Jones, Harry Hurst, Rob Powell, Patrick Kelley, PCA Historian Linda Goodman, Dan O'Connell, Eric Kent, Bruce Canepa.

Sincere thanks to Miles Collier for the Foreword.

Glyn and Jean Morris and especially Jodi Ellis for her design skill, enthusiasm and positive energy.

Finally, I must thank the late Allan Caldwell and Paul Risinger for their inspiration, firsthand stories, experience and perspectives shared as Porsche Club members over many years.

– Jay G.

1875 to 1919

1875	>>	Ferdinand Porsche's Early Life and Career
1898	>>	Lohner
1906	>>	Austro-Daimler
		Emil Jellinek
		Birth of Louise and Ferry Porsche
1910	>>	Prince Henry Competition
1914	>>	World War I
		Ferry Porsche's Childhood and Education
1916	>>	Ferdinand Porsche Managing Director
1917	>>	Ferdinand Porsche Honorary Degree

Looking back, I never knew my father to design any automobile without keeping open the possibility that it might be used for racing. – *Ferry Porsche*

Ferdinand Porsche was born in 1875, the same year that Austria's first gasoline-powered cart was built. The cart was a forerunner to Siegfried Marcus' 1889 *Strassenwagen*, Austria's first internal combustion passenger vehicle. Porsche was born into a world where electricity was unknown to the vast majority of people. As the third child of a demanding, hard-working tinsmith, young Ferdinand was expected to follow in the family business. Instead of going down the prescribed path, he became a self-taught electrician and studied electrical engineering. In a world of rapidly advancing technologies toward the end of the 19th century, the path of electricity led him to automobile design and mechanical engineering.

Ferdinand Porsche went on to work for four prominent automobile companies in the first three decades of the 20th century and he pursued a passion for racing. In 1930, at the age of 55, he founded a company that would eventually become something like a second family for thousands of employees and customers around the world. His talent for design and selecting associates led to the creation of several of the most famous and successful cars in history.

Ferdinand Porsche's Early Life and Career

Ferdinand Porsche's father, Anton, lived in Maffersdorf in Bohemia. This German-speaking region of Austria-Hungary was a remnant of the larger Austro-Hungarian Empire. After the Austro-Prussian War, it was in the Czech part of Austria-Hungary (close to the border with Germany). Porsche's ancestors lived in this region for at least five generations. Today the town is known as Vratislavice nad Nisou, a district of Liberec in the Czech Republic. Much of father Anton's daily work was plumbing and carpentry, but he also held various public offices, including Vice-Mayor of Maffersdorf for a period of time. The relationship of the metalworking trade to automobile construction would become important in Ferdinand Porsche's future.

Anton Porsche and his wife, Anna, had five children. The eldest, also named Anton, was expected to take over the family business but died as the result of an accident in the family workshop. As the next oldest male, third child Ferdinand was then expected to take Anton's place and carry on the family business. However, even at age 11, Ferdinand was underfoot as men worked to electrify the town's largest employer, the Ginzkey carpet factory. By the age of 15, Ferdinand had to spend nearly all of his time as an apprentice, preparing to someday manage

Ferdinand Porsche with his home-built electrical system, circa 1890.

> The passenger vehicle built in Austria by the German Siegfried Marcus, known as the *Strassenwagen*, still exists in the Vienna Technical Museum. The correct date of its creation remains somewhat uncertain. It may have been as early as 1875 (the year of Ferdinand Porsche's birth), 1877 or, most likely, around 1889. See the article in *Automobile Quarterly* Volume 3, number 3.

> Two other giants of the automobile industry, Walter Chrysler and Alfred P. Sloan, were also born in 1875.

> Ancestral forms of the Porsche family name were Borso, Boresch, Porse then Pursche by 1672.

the family business. But the curious and strong-willed boy was bent on experimenting with electricity in whatever spare time he could find. This was in a town where electricity could only be produced from a battery or local generator. Father and son apparently fought mightily, but his mother looked the other way when Ferdinand was working at his hobby. She was more inclined to see her son pursue his obvious passion.

Young Ferdinand was eventually allowed to attend evening classes at the vocational school in nearby Reichenberg. His father's wish was that Ferdinand would have a secure and reliable means of making a living (meaning, the tinsmith's trade). However, Anton got the shock of his life one evening when the family supper was suddenly illuminated by electric light. Ferdinand had secretly electrified the house, setting up a hand-cranked generator, wiring and the lights all by himself. This made the Porsches' home the only electrically lit building in town aside from the carpet factory. The owners of the factory eventually prevailed on Anton Porsche to allow his son to go to Vienna to continue his education, and they paved the way by

Ferdinand Porsche working at VEAG, 1895.

> Like his son, Ferdinand, Anton Porsche rebelled against the traditional family trade (cloth making), becoming a master craftsman in metalwork. Like Ferdinand Porsche, Enzo Ferrari was the second son of a metal craftsman and Enzo also chose not to take up his father's trade.

> Young Ferdinand Porsche was an avid ice skater and even fashioned battery-powered electric lights, one for each skate.

> Ferdinand Porsche's first vehicle, the Egger-Lohner C.2 Phaeton, sometimes referred to as the P1, was unveiled in Vienna on June 26, 1898. 'Phaeton' was used to refer to light, open carriages, then open cars with a front and back seat.

securing a job for Ferdinand at Vereinigte Electricitäts AG/Béla Egger (VEAG). At age 18, Ferdinand was off to study in Vienna and follow his calling. His younger brother, Oskar, carried on the family business. His older sister, Hedwig, married an employee of the Ginzkey carpet factory. Younger sister, Anna, married an innkeeper in Reichenberg.

VEAG was a manufacturer of electrical equipment, so Ferdinand was immediately in his element. He started as a trainee, doing the most mundane jobs like sweeping the floor and oiling machinery. Although he was allowed to sit in on classes at Vienna University of Technology, the workshop became his real schoolroom. His instincts for engineering and problem solving led him to become manager of the testing department and first assistant in the calculating section. Porsche took a keen interest in motorized vehicles and built his own electric-powered bicycle. Through his job at VEAG, Porsche began working with one of their clients, Lohner, on electric vehicles in 1898.

Lohner

Late in 1898, Porsche was recruited by Ludwig Lohner of the famous carriage-building company. Jakob Lohner & Co. built coaches for the Austrian aristocracy, but Ludwig saw the future of transportation coming. He wanted an electrical engineer to help build self-powered vehicles and was already familiar with Ferdinand Porsche's creativity. At the time, methods of propulsion were still in competition. Internal combustion was far from a certain winner versus steam or electric power. Lohner doubted that his noble and elite customers would want to arrive at the opera in a vehicle powered by a hot, noisy, dirty or smoky engine. Plus, until 1912 or later, combustion engines had to be

Lohner C.2 chassis as displayed in the Porsche Museum.

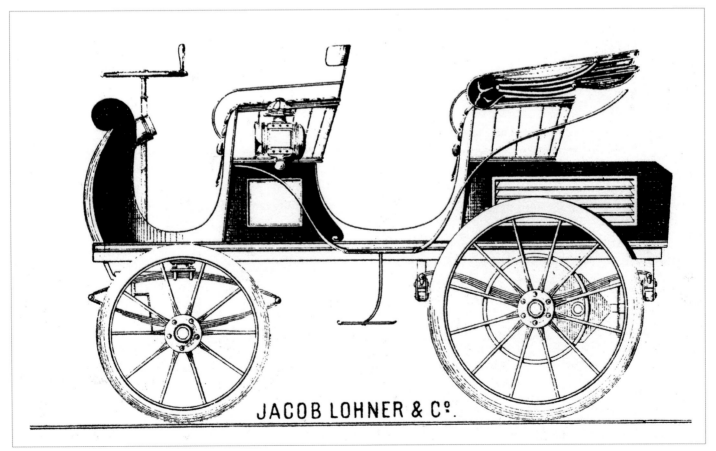

Period sketch of the Lohner C.2.

Lohner-Porsche Vehicles

Lohner's workshop had the experience and skilled craftsmen to produce vehicles of many configurations and Porsche's imagination was fertile. Below are just some of the early vehicles designed by Ferdinand Porsche.

Egger-Lohner – Resembling a horse-drawn open sleigh without its bodywork, the C.2 Phaeton was built in 1898. It is firmly in the 'motor carriage' category. With a single three- to five-horsepower motor it could reach a speed just over 20 mph, with a maximum range about 49 miles. Battery weight was 1,103 pounds. Ferdinand Porsche won an 1899 competition for electric vehicles in Berlin with a C.2. He finished the race portion 18 minutes ahead of the second place vehicle. Sometimes referred to as 'P1' (the first Porsche), a bare chassis C.2 was donated to the Porsche Museum by Wolfgang Porsche in 2014. It should be noted that Ferdinand Porsche did not design the whole vehicle, but rather the motor/drive.

1900 *Voiturette* ('small car') – This was a two or three-passenger carriage with front-wheel drive, each wheel hub motor producing approximately three horsepower for speeds as high as 36 mph. The batteries weighed some 900 pounds with a total vehicle weight of 2,160 pounds.

1900 Type J – Ferdinand Porsche's first purpose-built competition car was a stripped-down EV, with two open seats and the batteries on springs to protect the cells from damage. This vehicle was capable of a record speed at 24.88 mph on the Semmering hill climb.

La Toujours Contente ('always content') – This was the 1900 four-wheel drive racing Phaeton built for English distributor E.W. Hart. The batteries alone weighed 3,790 pounds! Total vehicle weight was more than four tons. It did not achieve great success in part due to the lack of tires capable of supporting the weight.

Semper Vivus ('always alive') – Lohner's first 'mixte' (hybrid) using two single-cylinder de Dion gasoline engines to drive a generator for the front wheel hub motors and to charge the batteries as needed.

Concepts like front-wheel drive, four-wheel drive, hybrid power and electric motor braking were coming off Ferdinand Porsche's drawing boards at the dawn of the 20th century.

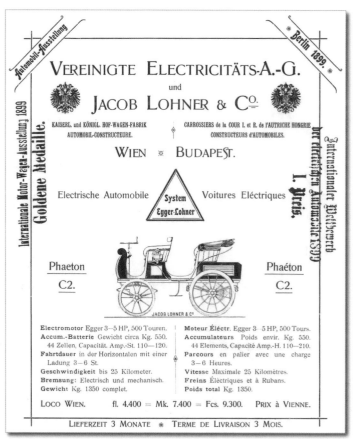
1899 advertisement for the Lohner C.2.

started by hand-cranking. Electric power seemed to be the most dignified solution, and Ferdinand Porsche was chosen to lead the new projects.

By 1900, work had progressed to the point where the System Lohner-Porsche 'Electromobil' (Model 27 *Elektrischer Phaeton*) would be displayed with pride in the Austrian pavilion at the Paris World Exposition. One of the main selling points of the vehicle, which at this stage still resembled something like a horse-drawn carriage, was the elimination of any need for a transmission (less complexity, no loss of power to friction). Further, front-wheel drive with electric motors at the hubs was noted as an advantage for traction on the poor roads of the day. Positioning on the front wheels also helped to keep the motors cool and the rear-wheel friction brakes could be supplemented by braking effect from the front wheel motors. Porsche had applied for patents on his innovative hub-mounted electric motors in 1897 while still working for VEAG/Béla Egger.

Competition and racing are an integral part of the Porsche story and date back to this very early stage of Ferdinand Porsche's career. In September of 1900, he personally drove a competition version of the Lohner-Porsche to establish a 10 kilometer speed record on the Semmering road near Vienna. The vehicle was lightened by removal of unnecessary equipment and even had improved aerodynamics in the form of an angled sheet metal faring or bonnet in front of the two tandem seats. Porsche lowered the record for an electric-powered vehicle by more than eight minutes at a sensational average speed just under 25 mph. In his autobiography, *We At Porsche*, Ferry Porsche wrote: *"Looking back, I never knew my father to design any automobile without keeping open the possibility that it might be used for racing."*

During this same period, Porsche began to work on what he referred to as the *mixt* (in German, or *mixte* in French) power concept. In modern times, this is what one might refer to as a 'hybrid'. At a time when batteries were excessively heavy and highly inefficient by modern standards, charging and range were huge technical challenges. The mixed concept began to solve this problem by using a combustion engine to run a generator, making electricity on board. Excess electricity from the generator could be stored by a small battery, which could run accessories and help start the engine. Gasoline engines were sourced from the Daimler Company at this stage, although the first Semper Vivus used two single-cylinder de Dion-Bouton engines. Porsche also began to work on all-wheel drive by installing electric hub motors on all four wheels of a vehicle. Racing vehicles were built on the mixed concept and military use, another key activity in the future of Ferdinand Porsche, began in this period. In 1902, Porsche personally drove Archduke Franz Ferdinand in a mixed vehicle during military maneuvers in West Hungary. In 1905, Porsche was recognized with the Pötting Prize for Austria's most outstanding automotive engineer.

1900 Lohner Type J, Ferdinand Porsche's first purpose-built competition vehicle.

1900 Lohner La Toujours Contente competition electric vehicle for E.W. Hart. Ferdinand Porsche shown in the passenger seat at the front.

Marriage and Children

While working at VEAG/Béla Egger, Ferdinand Porsche met Aloisia Kaes, known as Louise. Author Karl Ludvigsen wrote in *Professor Porsche's Wars*: *"On a sunny Saturday, 17 October 1903, the two were married in the church at Maffersdorf where Ferdinand had been baptized. True to Porsche's workaholic character, their honeymoon in Austria, Italy and France included several business meetings."*

Their first child, Louise, was born in August 1904. Their son, also named Ferdinand, was born in September 1909. Ferdinand would be known by his nickname, Ferry. Both children went on to play important roles in the future of Porsche.

Initial sales for the mixed cars were slow despite Porsche's constant improvements and innovation. The Lohner cars proved to be too expensive and heavy to achieve significant commercial success. Adjusting for inflation, the Lohner cars' cost was comparable to today's Porsche Taycan. Perhaps Lohner itself was too small to sustain Ferdinand Porsche's ambitious and capital intensive design imagination. The *mixte* concept had better success for fleet vehicles such as fire trucks, taxis and buses. However, by 1906, Ludwig Lohner had decided to phase out production of complete vehicles. Lohner EVs would continue to be built in partnership with Austro-Daimler until 1915. According to author Karl Ludvigsen in *Genesis of Genius*, Lohner built a total of 354 vehicles to Porsche designs with 301 being pure EVs.

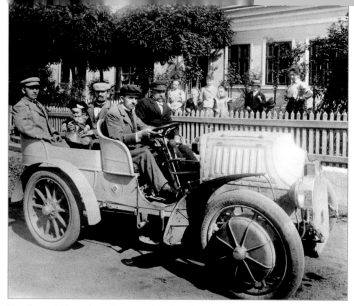

Ferdinand Porsche, shown with family members in 1902, driving a front-wheel drive Lohner in front of the family home.

Ferdinand Porsche in 1903 driving a front-wheel drive Lohner *mixte*.

A transition presented itself in 1906 when Emil Jellinek took an interest in hiring Ferdinand Porsche for Jellinek's expanding automotive businesses.

Austro-Daimler

In 1906, Porsche was hired by Austro-Daimler to be technical director. The company pre-dated the automobile, but had initiated automobile production by licensing designs from Daimler in Stuttgart. Porsche was preceded as head of design by Paul Daimler, son of Gottlieb Daimler. Although early automobile projects under Ferdinand Porsche were not greatly successful, in part due to a slowdown in the economy and the production of military vehicles becoming an important focus for Ferdinand Porsche and Austro-Daimler.

There was a period of transition where Porsche worked on Mercedes-branded vehicles for Emil Jellinek, who had invested the proceeds of his sale of the Mercedes brand (to Daimler) in Austro-Daimler, as well as his Paris-based subsidiaries set up to sell 'Mercedes-Electric' vehicles. During the transition period (and beyond), Lohner continued to manufacture electric and mixed vehicles based on Porsche designs. These included cars as well as commercial vehicles. Lohner also provided coachwork for some Austro-Daimler cars. Jellinek had hired Porsche into a company that would challenge and develop Porsche's engineering, design and management skills over the next sixteen years.

Porsche's first gasoline engine-only project, the Maja, was not a sales success, in part due to a faulty transmission (not designed by Porsche). The Maja was initiated by Emil Jellinek and named for another of his daughters. However, a redesign of the Maja, the Austro-Daimler 28/32, found better success and led to a number of variants.

Emil Jellinek

Emil Jellinek was a diplomat, businessman, investor, sportsman, and automobile entrepreneur. He was also one of the great enthusiasts of the horseless carriage and early automobile age. He is perhaps most famous for commissioning a run of 36 special cars from Germany's Daimler (DMG) to be designed by Wilhelm Maybach and built to Jellinek's exacting specifications. It turned out to be the last Daimler design under the founder, Gottlieb Daimler, who died in March of 1900. It was also one of the first vehicles to be engineered from the ground up as a motor car, rather than a horseless carriage. These innovative vehicles, the '35 HP', were to be named for Jellinek's daughter, Mercédès. They were a great sales success in the south of France, where Jellinek was based as Austrian Consul General and well-connected to the aristocracy. The 35-horsepower cars were also unbeatable in racing during the Nice 'Speed Week' competitions of 1901. The famous Belgian driver, Camille Jenatzy, won the prestigious 1903 Gordon Bennett Cup race in Ireland, driving a 60-horsepower Daimler-Mercedes. Also in 1903, Jellinek went so far as to change his name to Jellinek-Mercedes.

As early as 1901, Jellinek purchased and raced Lohner 'mixte' cars, so he was familiar with Ferdinand Porsche's work. After selling a 60% share of the 'Mercedes' brand to Daimler in 1905, Jellinek became a controlling shareholder in Austro-Daimler. He knew that Porsche was the engineer that Austro-Daimler needed at that stage. Jellinek continued to represent Austria-Hungary as a diplomat and gradually phased out of automotive and business activities by 1910. When Daimler merged with Benz in 1926 it was decided that their automobiles would be branded 'Mercedes-Benz'.

PORSCHE POINTS

- Early 20th century autos were often known by a combination of numbers showing first the taxable horsepower (based on a calculation from cylinder dimensions) and second the approximate actual engine horsepower. Hence the 1908 Austro-Daimler 28/32, derived from the Maja. Its true horsepower was closer to 35.

- The 1909 Prince Henry competition was won by Wilhelm Opel in one of his own cars. Ferdinand Porsche won it in 1910 for Austro-Daimler.

- On the day of Ferry Porsche's birth, Sunday, September 19, 1909, his father was away from home driving in the Semmering Hill Climb, winning his class.

As ever, automobile competition was a passion for Ferdinand Porsche. After a poor showing in 1909, he designed a new car for the 1910 'Prince Henry' competition. Prince Heinrich of Prussia sponsored this highly prestigious competition, which involved a series of trials or stages over several days, running from Berlin to Homberg, near Frankfurt, after passing through Kassel, Nuremburg, Stuttgart, Strasbourg and Metz. The stages included hill climbs and speed tests along the way. The number of competitors could exceed 200 entries and the cars were four-seat touring models required to carry a passenger along with the driver, the customary riding mechanic, and an official observer. Ferdinand's wife, Louise, was an enthusiastic passenger with her husband driving one of the Austro-Daimler entries.

Suffering from the sting of losing, and finishing well back in the order, Porsche started work on the new design during the prize-giving dinner at the 1909 event! The new car was much more powerful, producing 95 horsepower from a 5.7-liter, four-cylinder, overhead camshaft engine closely related to one of his aircraft engines. The car had a new, more aerodynamic shape (referred to as *tulpenform*, or 'tulip-shaped'). It had a top speed of 82 mph, faster than other cars with more powerful engines. According to author Karl Ludvigsen in the book *Professor Porsche's*

Austro-Daimler logo circa 1910.

Ferdinand Porsche driving the winning Austro-Daimler in the 1910 Prince Henry competition.

Austro-Daimler advertisement circa 1911. (Alamy)

Ferry Porsche's Childhood

Ferry was a curious boy for whom his father's factory was a playground. Not enthused by school, he preferred to roam the factory or help with the family garden and farm animals. Luckily for future Porsche owners, Ferry's favorite subject in school was mathematics. During World War I when fuel was in short supply, Ferry would sometimes be driven to school in an electric Austro-Daimler town coupe. A bicycle enthusiast, Ferry was in a potentially fatal crash on his Puch racing bike when he collided with a team of horses pulling a wagon. By age 13, Ferry was already assisting his father with tasks like the careful break-in driving of Austro-Daimler racing cars. His small hands would even help the mechanics with gearbox repair at the Italian Grand Prix of 1922, by reaching into the case to retrieve broken gear teeth. At age 14, Ferry and eventual Mercedes legend, Alfred Neubauer, could have been killed in an accident when Neubauer allowed Ferry to drive, but Neubauer's sudden correction of Ferry's steering spun the car which then flipped over across a ditch.

Ferry admitted to having difficulty with languages and was teased for his Austrian dialect by the Swabian kids when the family moved to Stuttgart in 1923 (Swabia is the region of southwestern Germany where Stuttgart is located). At age 18, Ferry began the courtship with his eventual wife, Dorothea Reitz, affectionately known as 'Dodo', while his sister Louise married Anton Piëch, a lawyer in Vienna. They met while Louise was studying painting at art school in Vienna. Around the same time, Ferry finished school in Stuttgart and began a 12 month apprenticeship at Bosch, the manufacturer of electrical and other components for automobiles. When the family moved back to Austria in 1928, Dr. Porsche insisted that Ferry continue his studies. Ferry's weekends were then spent haunting the Steyr factory to see what his father was working on.

LEFT: Ferry and Louise Porsche portrait from 1915.

Wars: Its mandatory four-passenger bodywork was misleading, for his 1910 'Prince Henry' Austro-Daimler was the first pure racing car from the pen of Ferdinand Porsche. He designed its every part for speed with reliability. ...Shaping the body to reduce drag, said Porsche, 'I have gone so far as to shroud every single nut so the air has nothing to catch on'.

In the 1910 event, Austro-Daimler won all the special tests and nine of the twelve overall prizes. The car driven by Ferdinand Porsche was first overall, with two other Austro-Daimlers finishing second and third. They became known as the 'iron team' and went on to campaign successfully in other competitions. As a touring car, the 22/86 'Prince Henry' became a successful seller, showing that 'win on Sunday, sell on Monday' is an axiom deeply rooted in both automotive and Porsche history. The 22/86 was one of the fastest road cars available in its day.

Military and Aviation Projects

Located close to military barracks, training grounds and an airfield in Wiener-Neustadt, Austro-Daimler became heavily involved in producing military vehicles during the lead up to World War I. One of the main products was a series of artillery 'tugs'. These were massive vehicles designed to tow heavy guns into battle. The series of tugs culminated with the massive M17. This vehicle had a 13.5-liter, four-cylinder engine, weighed ten tons, and rode on wheels 57 inches (!) in diameter. It was appropriately nicknamed 'Goliath'.

Porsche also designed 'land trains', locomotive-like vehicles that could pull a train of carts on land or on rails. These could be used to supply troops on the front lines in battle or to carry component sections of massive artillery pieces. The land trains operated on the 'mixed' system, with giant engines driving generators to produce power for electric motors at each wheel. At bridges, the individual cars could decouple but connect electrically to the 'engine' and cross one at a time. During this period, the Skoda Works, prime producer of heavy weapons for the military, became the controlling shareholder in Austro-Daimler. This made sense on a simplistic level, as Skoda's heavy guns were useless unless they could be positioned correctly for battle. Porsche also designed two-wheeled tractors called the *Pferd* (horse) for towing light artillery.

Aircraft engines were a market segment with commercial potential and Ferdinand Porsche became involved in a most enthusiastic manner. This also related to the military since aircraft were seen as a potential tool for warfare. For both lighter-than-air (airship) engines and early fixed-wing aircraft, Porsche was constantly exploring new ideas. He had at least two harrowing test flights in airships where the outcome could easily have been fatal for Ferdinand and others on board. In one case, he saved the day through his own hands-on, mechanic's talent by repairing a stuck gas valve in flight (when the balloon-construction ship was in danger of rising too high and bursting).

Porsche's first engine for an airplane was built in 1910, when only a handful of airplanes existed in Austria. However, the supply of engines led to rapid development, and by the beginning of the World War, Austria's airplane population was sixth largest in the world. As ever, competitions were a part of the development process and Aero-Daimler engines had success in flying meets and prize challenges from the beginning. Porsche-designed engines powered planes built by his former employer, Lohner, who also manufactured propellers. Austro-Daimler built an inline six-cylinder engine for fighter aircraft that was considered the best airplane engine available to the Central European powers during the war. Among a bewildering array of other projects, Porsche and his team designed a rotary engine, a V12, a W engine with three rows of cylinders, and even experimented with a primitive helicopter design during the war.

By 1916, Ferdinand Porsche had become Managing Director

PORSCHE POINTS

» Ferry Porsche's first nickname was 'Ferdy'. There are two stories about how it changed. In his own book, Ferry Porsche states that his governess didn't like the sound of it and changed to 'Ferry' (Hungarian for Franz). Another explanation is that his parents thought 'Ferdy' sounded like the name of a common coachman. The author favors Ferry's writing on this topic.

» Josip Broz Tito, future communist leader and 'president for life' of post-World War II Yugoslavia, worked under Ferdinand Porsche as an Austro-Daimler mechanic.

» Josef Goldinger became Ferdinand Porsche's personal driver while working at Austro-Daimler. He filled the chauffeur role for most of the rest of Ferdinand Porsche's life.

of Austro-Daimler. He was awarded the Officer's Cross of the Franz Josef Medal by the Emperor for excellence in military technology. Perhaps a more important award came in 1917 when he was given an honorary doctorate by the Technical University of Vienna for his work in automobile and aircraft engineering. With larger responsibilities and greater numbers of projects and products, Dr. Porsche honed a style of leadership with the designers and technicians working for him. His ability to collaborate with a team would serve him well over the years. However, he could be gruff and temperamental, roving from drawing board to drawing board as teams worked on design tasks. In one episode during the war, a frustrated Dr. Porsche offered a spontaneous bonus of 500 Kronen for the best overnight solution to a particularly stubborn problem (having to do with the wheel tread system on the Pferd). The winner of the bonus was young Karl Rabe, who would go on to head the design office at Austro-Daimler. More importantly, Rabe would be Dr. Porsche's choice to be Chief Designer when Porsche started his own firm in 1930.

As World War I ground to a conclusion, Austria-Hungary collapsed amid political and ethnic conflict. In the autumn of 1918, the former constitutional monarchy began splitting into new countries, including Czechoslovakia, Poland, Hungary, Romania, Yugoslavia and Austria. Ferdinand Porsche soon found himself with a domestic market that was a tiny fraction of its pre-war size and a loyal but hungry workforce to whom he personified Austro-Daimler. The company was now effectively owned by investor Camillo Castiglioni through his General Deposit Bank. Castiglioni was known as one of the wealthiest men in Central Europe and had accumulated interests in dozens of companies (including BMW and other aviation pioneers). He had worked with Austro-Daimler and Porsche early on as an owner of a company that supplied rubberized fabric balloon material for airships. Conflict with Castiglioni would be Dr. Porsche's first, but not last, battle with the bankers.

Austro-Daimler M17 'Goliath' towing a mortar cart in 1916. Ferdinand Porsche's office was in the building behind.

1920s

1922	≫ Austro-Daimler Sascha
	1922 Targa Florio
1923	≫ Daimler Motoren AG
1924	≫ 1924 Targa Florio
	History of the Targa Florio
	History of the Nürburgring
1926	≫ Mercedes S-Series
	Daimler and Benz
1929	≫ Steyr

On June 19, 1927, the new 680S Mercedes finished first and second, followed by a 630K in third place, at the very first automobile race ever run on the Nürburgring…

1920s

The post-World War I years were not a happy time for Austro-Daimler nor for Dr. Porsche. The country of Austria was reduced to a small fraction of its pre-war size. The economy was poor, with hyperinflation and labor strife. Porsche struggled to find and keep work for the 6,000 employees of Austro-Daimler. The company would not only have to transition away from military production to passenger cars, but also to find export markets. A bright spot appeared in the form of patronage from Count Alexander 'Sascha' Kolowrat-Krakowski who commissioned a racing car. Built to contest the 1100cc class, the ADS-R had a four-cylinder engine with two bevel-driven camshafts making 50 horsepower. Bevel gear camshaft drive would appear some 30 years later in the Fuhrmann-designed Porsche racing engines of the 1950s. The 'Sascha', as it was nicknamed, weighed 1,700 pounds and had a top speed of 89 mph. It was clear to Porsche that smaller cars were needed if Austro-Daimler were to achieve higher volume production and sales. The new sports car was intended by Dr. Porsche as the basis for just such a car.

Ferry Porsche with his Austro-Daimler toy car in 1921.

PORSCHE POINTS

» At the Targa Florio in 1922, the Austro-Daimler Saschas were painted red to blend in with the Italian cars. They were also adorned with playing card symbols so that each car could be easily identified. The suits were hearts for Count Kolowrat himself, diamonds for Neubauer, spades for 'Fritz' Kuhn (who won the 1.1-liter class) and clubs for Lambert Pöcher. The suit of cars motif was repeated in 1970 and 1971 on the Porsche 908/03s at the Targa. The Saschas also wore only one front fender, on the driver's side. The riding mechanic had no such protection.

» As a teenager, Ferry Porsche scrounged most of the parts needed to build an Austro-Daimler Sascha (without his father knowing), however, the parts had to be abandoned when the family moved to Stuttgart in 1923.

ABOVE: Ferdinand Porsche (behind the hood) with the Sascha driven by Alfred Neubauer at the Targa Florio in 1922.

LEFT & OVERLEAF: 1922 Austro-Daimler Sascha in the Porsche Museum. Note the 1912 Austro-Daimler fire truck in the photograph at left.

The little 'Sascha' racing cars scored first and second place in class at the 1922 Targa Florio in Sicily, although it should be noted that the Saschas were the only entries in the 1.1-liter 'production' category. The Count drove a third 'Sascha' himself but did not finish the race. The fourth Sascha was driven in the event by none other than Alfred Neubauer, who would go on to become the legendary Mercedes-Benz racing manager. At the time, Neubauer was an Austro-Daimler test driver and had impressed Dr. Porsche with his organizational skills. Neubauer finished 19th overall, running in the 'race car' class. Also driving in the epic 1922 Targa was a young Italian, and future legend, Enzo Ferrari. He made his first attempt in 1919 and finished second in the 1920 running of the Targa, driving an Alfa Romeo. The great Sicilian road race would figure prominently in both Ferrari's and Porsche's post-World War II competition success. The Saschas later found success in numerous other competitions, with the Count himself winning his class in the 1922 Semmering hill climb.

During 1922, relations soured between Porsche, Camillo Castiglioni and the Board of Directors at Austro-Daimler. Castiglioni felt that Porsche was spending too much on R&D (including racing), while Porsche doggedly stuck to his principle that racing improves the breed. The 1922 Targa Florio (run at a distance of 268 miles over rough, gravel roads in the Sicilian mountains) and numerous other racing wins proved the point in his mind. The situation worsened when Austro-Daimler's best driver, Fritz Kuhn, was killed in a crash at Monza during practice for the Italian Grand Prix. This further caused the board to question the expense, in terms of money and men, involved with racing. A final, acrimonious board meeting in February of 1923, saw Porsche refuse to carry out massive layoffs at the company that he had come to personify. Some accounts include Dr. Porsche hurling a candlestick in fury. After storming out of the meeting, he was essentially fired the following day. Not long after, Ferdinand Porsche decided to accept a new job at Daimler in Stuttgart as Technical Director. Austro-Daimler carried on during the 1920s under the engineering leadership of 27-year old Karl Rabe.

Daimler Motoren AG

In May of 1923, Ferdinand Porsche officially started employment in Stuttgart (Untertürkheim) for Daimler, although he had been hard at work in Stuttgart as early as February. On the positive side, Daimler was firmly committed to racing and competition. However, overall vehicle sales were in decline with general economic difficulties and hyperinflation plaguing Germany in 1923. Daimler was also lacking in designs for future products. Another challenge was Dr. Porsche's style, which was sometimes at odds with the Daimler team who were accustomed to his predecessor, Paul Daimler (who had also preceded Porsche at Austro-Daimler). Daimler was more orthodox and hierarchical in style, very different from Porsche's direct and hands-on approach.

In Richard von Frankenberg's book, *Porsche – The Man and his Cars*, this is illustrated with the following anecdote: *"Porsche's style was unusual in that he suddenly appeared in the experimental department just after a car in trouble had been taken there. Whilst the engineers in their white coats still stood around and discussed matters, Porsche would be given an oily works overall and would crawl under the*

Ferdinand Porsche, 1923.

PORSCHE POINTS

» Ferdinand Porsche liked Pilsner beer while Ferry did not like beer of any kind.

» Both Ferdinand and Ferry Porsche considered 13 to be a lucky number.

car so as to personally discover the trouble. One day an engineer made the mistake of asking: 'Can you see anything, sir?' Porsche raked the assembled technicians with an annihilating look and yelled at them: 'Why don't you have a look for yourself, you….' (It was) an insult the engineers probably had not ever heard in their own homes! Porsche banged the door shut and did not speak to his collaborators for days."

Porsche's style apparently divided those who were put off by his direct, sometimes rude approach from those who appreciated his wisdom, guidance and decisiveness.

A Daimler-Mercedes had won the Targa Florio overall in 1922, the same race in which Porsche's Austro-Daimler 'Sascha' performed so well. However, that Mercedes was a privately-owned car, painted Italian racing red for its entrant/driver, Giulio Masetti. The factory-entered Mercedes Grand Prix cars could do no better than a sixth place finish. For 1923, Daimler skipped the Targa, but during the winter of 1923-1924, Ferdinand Porsche and the Daimler team prepared a new racing car. They also gained support from the board for a full factory team effort; three cars, mechanics and Dr. Porsche himself as the racing manager. The drivers would be Christian Werner, Christian Lautenschlager and Alfred Neubauer.

Daimler had developed a four-cylinder, 2.0-liter supercharged racing car for the 1923 Indianapolis 500. The cars were a miserable failure at Indy, likely due to lack of development time (one symptom being incorrectly tuned exhaust pipes). Ferdinand Porsche's first important task for Daimler's racing effort was to develop this car and make it a winner. He often drove the race car himself at the test track, acquiring that extra bit of firsthand experience as changes were made. Part of the improvement in performance was for the drivers to learn the proper technique of using the throttle to maintain optimal benefit from the engine-driven supercharger. In typical Porsche fashion, no stone was left unturned. He and the Daimler team revised numerous detailed aspects of the engine including a new supercharger design and larger exhaust valve stems filled with mercury. As author Karl Ludvigsen details in *The Mercedes-Benz Racing Cars*: *"Leaving alone the basic dimensions and valve sizes, (Porsche) enlarged the valve stem diameter to 12 mm to make it easier to provide, in the exhaust valves, a drilling right down to the valve head that could be filled with mercury, which splashed and circulated to help conduct heat upward to the guides. This was perhaps the first use of internal cooling in a racing engine valve."*

The chassis and bodywork were also revised for the Targa entry including an early use of a clear windscreen/rock deflector for the driver (no such protection for the riding mechanic). The cars were painted red instead of German racing white, presumably to blend in with the Italian entries.

In the 1924 Targa Florio, the Mercedes team was opposed by a strong entry from Peugeot, who also brought three factory cars. Other potential winners came from Ballot, Hispano-Suiza and the Italian favorites, Alfa Romeo and Fiat. At the end of the first lap, Giulio Masetti led in his 3.6-liter Alfa ahead of André Dubonnet in the 6.6-liter Hispano, custom built for the Targa. Christian Werner was a close third for Daimler. The top five cars were within 30 seconds of each other. Hot, dry weather took a heavy toll on the drivers and riding mechanics, while the rough and rutted roads shredded the tires. At the end of lap two, Werner was up to second place. Attrition and crashes then delayed or eliminated several of the top competitors. Werner led at the end of lap four, securing the victory in the Targa portion of the race. For 1924, the Targa and Coppa Florio prizes were combined into one event. Werner carried on for the fifth lap, trying to win the Coppa Florio. He then lost the lead to Antonio Ascari (father of Alberto) in an Alfa. However, with only several hundred yards left of the 67-mile lap, the engine in Ascari's Alfa

---

PORSCHE POINTS

» In 1923, because of hyperinflation in Germany, Ferry Porsche had to pay several million marks for the streetcar ride to school. He eventually resorted to walking since he couldn't carry enough money for the ride. At one point the $1 equivalent was 4,200,000,000 marks.

» Anton Piëch's earliest legal effort for Porsche was to represent Dr. Porsche in a long-running case involving the sale and repurchase of patent rights for Porsche designs by Lohner. The dispute began in 1916 and was finally resolved in 1926. In 1924, Daimler (DMG) acquired the rights to all of Dr. Porsche's existing patents.

» The Avus circuit in Berlin was named by acronym, *Automobil Verkehrs und Übungsstrasse* 'automobile and training track'. It opened in 1921, but the daunting north curve was not banked until 1937. Although modified over time, the circuit survived as a racing venue until 1998.

Targa Florio

The Targa Florio was founded in 1906 by the Sicilian Count Vincenzo Florio when he was only 23. This epic race over public roads through the Madonie mountains in northern Sicily pre-dated the 24 Hours of Le Mans, the Mille Miglia and the Indianapolis 500. As a test of speed, endurance and mechanical reliability, the Targa was a highly prestigious competition especially in the period before World War II.

Vincenzo Florio and his older brother, Ignazio, who was the head of the family business, were both automotive enthusiasts at the turn of the century. Their father had left them a large estate and business interests in shipping, railroads and exporting lemons, olive oil and wine. Vincenzo became interested in motorized vehicles and developed a keen desire to race. He won a speed trial at Padua in 1902 and entered many other races, including the Gordon Bennett Cup. Ignazio was very concerned for Vincenzo's safety given the extreme danger of racing in those days and likely was happy when Vincenzo turned his attention to organizing his own racing event in Sicily. He had a special trophy (a solid gold plate, or *targa* in Italian) made in Paris for the winner and even competed in his own race in 1909, finishing second. In the early days, the tough and treacherous gravel roads were a stern test for men and machine.

The length and exact configuration of the circuit changed many times over the years, but generally covered a course of roads with a lap distance from 44 to 91 miles. A few early races even covered a circuit around the entire island. The Targa was always a race against the clock, with competitors starting at timed intervals. This along with the condition of the roads, which were gravel and dirt in the early days, made the race similar to a modern day rally. The Targa Florio became part of the World Sportscar Championship in 1955 and lasted until 1973. It figures prominently in Porsche history from 1922 all the way until 1973, when a Porsche 911 RSR won the last World Championship race in Sicily. In all, Porsches would win the Targa eleven times in the post-World War II era. See Chapters Five, Six and Seven.

---

appeared to seize up and the car spun violently. Ascari and his riding mechanic tried to get the engine going, but Werner soon passed the stricken Alfa to win the 'cup' in addition to the 'plate'. The other Mercedes finished 10th (Lautenschlager) and 15th (Neubauer).

A telegram sent to Daimler in Stuttgart read: *"Overall result Werner wins Targa and Coppa Florio, also Coppa Caltavuturo for shortest time from start to that place, also Coppa Villa Igiea for lap record, also Grand Gold Medal of King of Italy, ditto Motor Club of Sicily, also all prizes awarded by Palermo Merchants' Chamber stop class result Werner first, Lautenschlager second, Neubauer third, Mercedes team wins Coppa Termini for best factory team."* The spectacular triumph of the Daimler-Mercedes team led to Ferdinand Porsche being awarded a second honorary doctorate, this time from the Technical Institute of Stuttgart.

Another project inherited by Ferdinand Porsche at DMG was an inline eight-cylinder Grand Prix car initiated by Paul Daimler. Redesigned by Dr. Porsche, the supercharged 2.0-liter entry had a tragic beginning when Count Zborowski was killed at Monza in the 1924 Italian Grand Prix and the Mercedes team was embarrassed by the P2 Alfa Romeos taking the top four places in the race. The tricky handling car had a few hill climb/sprint successes driven by Christian Werner, who was allowed to build his own modified version. For the first German Grand Prix, held in 1926 at the Avus circuit near Berlin, two of the cars were modified to run as 'sports' four-seaters by tacking on rear bodywork barely big enough for the two additional seats. This was possible in part due to Dr. Porsche's location of the fuel tanks under the midsection of the cars to lower the center of gravity and reduce changes in handling character as the fuel was used. The circuit suited the Mercedes '8' since it was made up of two 6-mile long, straight sections of highway connected by loops at either end.

In the race, the young German, Rudolf Caracciola, stalled his Mercedes at the start (a common fault with the high-revving engine) and had to be pushed by the riding mechanic to get going. Mercedes teammate, Adolf Rosenberger led the field in front of 200,000 spectators (Neubauer's article in *Automobile Quarterly* puts the crowd at 500,000!). On lap five, the weather changed from dry to pelting rain. On the seventh lap, Rosenberger crashed and hit a scorer's shed in the North turn, killing three occupants. Caracciola drove heroically as the track changed again from wet to dry but had to stop to change spark plugs, losing two minutes. He was also handicapped by not knowing his relative place in the field. Overcoming fatigue and a sense of futility, thinking he was behind the leader, Caracciola found that he had won the race, bringing glory for himself and Mercedes.

It was the first of his six victories in the German Grand Prix. This 1926 race led Daimler-Benz employee, Alfred Neubauer, to begin developing a signal system to report information from the pits to the drivers. The pit signaling concept had been mentioned in the DMG report from the Indianapolis 500 race of 1923 noting the advantage it gave to the American teams. Neubauer went on to personify the role of racing team manager from the 1930s to the 1950s.

Although Germany was prohibited by treaty from producing most types of military vehicles and arms, there was an allowance for defensive and policing vehicles that could be produced by

Ferdinand Porsche with Alfred Neubauer, at the wheel of the Mercedes, at the 1924 Targa Florio.

PORSCHE POINTS

>> In 1926, Louise Porsche was entered in the Kartellfahrt Rally in a Mercedes and recorded a faster time in a hill climb section than the factory drivers Adolf Rosenberger and Rudolf Caracciola. Ferry Porsche wrote that this was due to Louise over-revving the engine while the hot shoes kept to the prescribed rev limit.

>> Adolf Hitler first met Dr. Porsche on July 11, 1926 at the German Grand Prix at Avus. It was a brief congratulatory moment between the two Austrians after Caracciola had won the race for Mercedes in front of 200,000 spectators.

>> In the late 1920s, the Mercedes racing cars were known as 'White Elephants' due to their size and the German national racing color being white. In 1927, red bands on the hoods were used to make the works-entered cars easier for fans to spot. In the 1930s, silver famously became the color for German grand prix cars, hence the 'Silver Arrows'.

Nürburgring

It would be difficult to appreciate any story concerning the German automobile industry without knowing a bit about the Nürburgring. While it is one of the most famous race tracks in the world, it is also a popular and demanding test track used by many of the world's automobile companies right up to the present day.

The track was conceived as an economic development project for the area near the town of Adenau. Road racing had been popular in the area near Bonn and Cologne but the authorities decided to build a proper, dedicated facility. Construction of the track began in 1925 and was completed in 1927 at a cost of approximately $50 million in today's money. The original layout consisted of four separate configurations; the North loop, South loop, combined North and South (*Gesamtstrecke*), and a small warm-up loop around the pit area (*Betonschleife* or concrete loop). The North loop, *Nordschleife*, circles around the ancient and foreboding castle in the town of Nürburg. The nature of the terrain in the Eifel Mountains made the Nürburgring similar to the Targa Florio course in Sicily. The 1927 German Grand Prix was held at the new track, and from the beginning, it was open to the public as a one-way toll road when not in use for racing or other activities.

In the 1930s, Grand Prix races were held on the Nordschleife and the 1935 edition is considered a classic. Tazio Nuvolari in an outdated Alfa Romeo defeated the might of Germany's Mercedes-Benz and Auto Union racers. After World War II, the Nürburgring hosted the German Grand Prix as well as the ADAC 1000 KM race for sports cars. Often wet, lined with trees and hedges and consisting of many irregular turns as well as jumps where the fastest cars could become airborne, the 14-mile long Nordschleife was widely considered to be the most difficult race track in the world. World Champion Jackie Stewart nicknamed it the 'Green Hell'.

In the 1970s, safety concerns and ever faster racing cars became issues for the facility. Armco barriers were added and the track was modified in some sections to improve safety. However, Nikki Lauda's near fatal accident in the 1976 German Grand Prix ended the use of the Nürburgring for Formula One. 1983 was the last year for World Championship sports car/endurance racing at the Nürburgring and Stefan Bellof in a Porsche 956 set the lap record at 6 minutes, 11 seconds, an average speed of 125 mph. The record stood until 2018 when Porsche broke it with the Evolution version of the gasoline-electric hybrid 919 driven by Timo Bernhard at 5:19.546.

In 1984, the largely disused *Südschleife* (South loop) was replaced by a modern Grand Prix track which has hosted Formula One and World Championship sports car racing. The Nordschleife is still used for racing, most notably the 24 Hour race for production-based sports cars as well as the World Touring Car Championship.

Daimler (and others). Daimler's DZVR armored car pre-dated Porsche but was further developed during his tenure. Dr. Porsche also worked on a design for a six-wheel car-like military vehicle that could be used to carry officers and soldiers in the field. In a competition between Daimler, Benz, Horch and others, Porsche's design lost out to Paul Daimler's design for Horch. A much more ambitious design for the military came in the form of an eight-wheel, amphibious reconnaissance vehicle.

Perhaps the most famous project and products for Daimler-Benz under Ferdinand Porsche were the cars known as the 'S Type' and their variants. They were based initially on the earlier, large engine 24/100/140K which itself was a shorter wheelbase version of the most prestigious Mercedes touring car. The 24/100/140 had proved its potential as a competition car as early as 1925. Renamed the '630', this platform was also used with large cabriolet and limousine bodies including some built for royalty and heads of state. However the Model K, as it eventually was named, was more sporting in its intent and Otto Merz won his class in one at a 12 hour endurance race in San Sebastian, Spain during 1926. The Mercedes team lost the race overall to cars with much smaller engines, in part due to significant tire problems. Rudolf Caracciola won the prestigious Klausen and Semmering hill climb events using the Model K.

For 1927, Porsche and his design team focused their racing efforts on the inline six-cylinder engine from the Model K (*kurzer radstand* or 'short wheelbase') with its capacity increased to 6.8 liters. In typical Porsche fashion, the engine and the chassis were extensively improved with a plethora of innovations and technical developments. Since the Benz side of the board was opposed to spending on purpose-built racing cars, Porsche and

his men made do with adapting this high-performance road car for racing. The sporting new 'S' model was built on a much lower chassis with only six inches of ground clearance. The rear of the frame rails curved to a higher position to accommodate a revised rear suspension and axle layout. The hood line was lowered to almost sit on top of the engine which itself was moved back in the chassis to improve weight distribution. Aside from the basic configuration, the engine was totally redesigned and in supercharged form it produced 180 horsepower.

On June 19, 1927, the new 680S Mercedes finished first and second, followed by a 630K in third place, at the very first automobile race ever run on the Nürburgring. Caracciola won the race before a crowd of 500,000 spectators. Just a month later, Otto Merz won the 1927 German Grand Prix, also held at the Nürburgring, in his Mercedes S when Caracciola's car broke down. Christian Werner and Willi Walb were second and third for Mercedes. Numerous other racing successes in 1927 swiftly built the reputation of the Type S.

The 'SS', for Super Sport, version was mainly intended as a touring car, now with a 7.1-liter engine. Again, the internals of the engine were significantly revised with bigger valves, new pistons, new connecting rods, a redesigned crankshaft and changes to the supercharger drive system. As with earlier supercharged Mercedes racers and road cars, the supercharger was activated by the driver using an extra push of the accelerator pedal toward the floor (which could also activate changes to the air ducting and fuel pressure to the carburetor). The newer engine could be adapted to the proven S chassis for racing and Rudolf Caracciola and Christian Werner won the epic, broiling hot 1928 German Grand Prix in an 'SS' (an S chassis with the newer engine). No less than 16 Bugattis lined up to challenge five Mercedes cars along with entries from Bentley, Talbot and others.

Caracciola led the first half of the race, until tire failure led to a two minute pit stop. Werner dislocated his shoulder trying to hold the steering wheel and turned his Mercedes over to Willy Walb on the ninth lap. However, Werner was persuaded by Neubauer to take over for Caracciola, who was suffering from heat exhaustion, a few laps later. Caracciola recovered sufficiently to take a second stint before handing back to Werner. At that time, a lap of the Nürburgring took about 15 minutes. In something more like a five hour endurance race than a modern Grand Prix, Werner passed Otto Merz' Mercedes on the last

Otto Merz in the winning Mercedes-Benz Type S, German Grand Prix at the Nürburgring in 1927.

lap when Merz had a tire shred. The Mercedes SS cars finished first, second, third and fifth, with Werner credited in both the winning and the third place car.

The SS led to the 'SSK' derivative, a two-seat short-chassis car with the 7.1-liter engine. This was the last competition car developed under Dr. Porsche at Daimler-Benz. The 19-inch shorter wheelbase improved cornering for racing on tighter circuits and especially for hill climbs. It also reduced weight; always helpful for performance. Caracciola won with the SSK first time out at the Gabelbach hill climb, setting a new record. He also broke the record at Semmering. The SSK and SSKL (lightweight) served the company well even after Dr. Porsche had moved on from Daimler-Benz and were competitive up to 1932. The S-series cars, especially the SS touring cars and SSKs, are now highly valuable and prized by collectors.

Interestingly, an article in *Motor Sport* magazine from 1928 stated that Dr. Porsche was dissatisfied with the state of the *"present racing rules, on the grounds that even with a weight limit, engines of unlimited capacity can be made to develop so much power that the cars would be too fast to be held on the road. Under these circumstances the engines would never be stressed to their limit, and consequently no lessons would be learnt by manufacturers."*

It may be worth noting that the 1.5-liter Grand Prix formula for 1926/27 was unpopular with many manufacturers and entrants. As a result, many 'Grand Prix' races were run with sports car rules or with a 'formula libre' (run what you brung) format.

Daimler and Benz & Cie had signed a cooperation agreement as early as 1924 and began coordinating product lines. This agreement and the formal merger in 1926 (which Dr. Porsche opposed) caused several problems, not the least of which was trying to integrate the projects and employees of two highly competitive former rivals. The new Managing Director, Wilhelm Kissel, came from the Benz side, and Dr. Porsche did not get along well with him. Porsche also had to contend with another Chief Engineer, Hans Nibel, from the Benz side although Porsche was given overall engineering responsibility. In addition to company culture differences, there were now five separate factories in five different geographical regions to coordinate. Also around this time, Dr. Porsche initiated a new, simplified project numbering system using the now-famous 'W' (for Wagen) prefix followed by sequential numbers.

During the mid-to-late 1920s, further military projects included a tank design, and not surprisingly, several aircraft engines. The first airplane engine fully developed on Porsche's watch was the interestingly named 'F1', a three-cylinder, 1.5-liter radial producing 34 horsepower and weighing only 126 pounds. On the other end of the size spectrum, Daimler designed and built the 'F2', a 54-liter (!) supercharged, 800-horsepower V12 for the German navy. Although this engine never made it into an airplane, per se, a further version was adapted to power a fast attack gunboat (three engines working together including one with its crank rotation reversed to balance torque effect). After Porsche's departure, Daimler-Benz produced a modified diesel-powered version both for marine use and for airships.

In a situation similar to his departure from Austro-Daimler, Dr. Porsche left Daimler-Benz at the end of 1928. The board was reluctant to increase investment in new models. Porsche felt his engineering and design staff was overworked and underpaid. Although passenger car sales and productivity were trending positively, the board elected not to renew Porsche's contract. As with Austro-Daimler in the 1920s, Daimler-Benz carried on with projects and products initiated under Ferdinand Porsche well into the 1930s.

Steyr

Daimler-Benz offered Dr. Porsche the chance to stay on as an advisor and consulting engineer. Instead, he decided to accept an offer to return to Austria at the beginning of 1929 to work as Chief Engineer and a member of the board of Steyr. The potential of Steyr and Porsche caused a great deal of enthusiasm in the Austrian press. Steyr was mainly an arms manufacturer until the end of World War I but started to manufacture automobiles in 1920. They had good success and by 1929 were the largest competitor to Austro-Daimler, possessing a larger factory. A potential advantage for Porsche was that Steyr could produce nearly any component in-house using their own foundry, forge and milling capacity.

Dr. Porsche set to work developing two new products for Steyr based on their existing models. The first was powered by a 2.1-liter engine of the type Porsche had tried unsuccessfully to introduce at Daimler-Benz. Known initially as the XXX, the Type 30 would carry on successfully for most of the 1930s. This innovative design used cross-bracing of the frame and allowed the bodywork to be 'unstressed', having no subframe. In Ferry Porsche's book, *We At Porsche*, he pointed out that this basic construction method continued in many modern cars, including the current Porsches at the time of his writing in the mid-1970s. He also notes this car as the first use of 'chill' casting of an aluminum crankcase. The temperature-controlled casting process offset the higher cost of aluminum by reducing the production time and cost for the finished product.

1929 photograph at the Steyr factory showing the elegant Austria (right) with the Type 30 (left) and Type 20 (middle).

The engine was noted for its durability and influence on other German engine designs of the 1930s.

Dr. Porsche's other project was the Steyr Austria, an elegant five seat cabriolet with a 5.3-liter straight-eight. He drove the car himself to the Paris Salon in the fall of 1929 where the Steyr products were well-received.

However, the stock market crash and the start of the Great Depression stopped the Steyr/Porsche efforts almost immediately. The bank that supported Steyr, Bodenkreditanstalt, collapsed while Dr. Porsche was in Paris and was taken over by the bank supporting Austro-Daimler-Puch, Kreditanstalt, controlled by Camillo Castiglioni. Given this and the severe economic hardship to come, it was inevitable that Steyr would be absorbed into Austro-Daimler-Puch. Large cars like the Austria would likely be produced under the Austro-Daimler side of the newly combined company and Dr. Porsche had no interest in working under Castiglioni. His three-year contract with Steyr was terminated after only one year. The combination of Austro-Daimler-Puch and Steyr would eventually be known as Steyr-Daimler-Puch AG, later the maker of the Pinzgauer. They also came to build the Mercedes G-Class vehicles and to develop four-wheel drive technology and components. The current company is based in Graz, Austria, and is known as Magna Steyr. It develops and assembles vehicles for other manufacturers under contract. Following his time at Steyr, Dr. Porsche's next move would be the most consequential of his career.

1930s

1930	»	Porsche Design Office
1932	»	Small Car Projects
		The Russia Project
		Torsion Bar
		Volkswagen Influencers
1933	»	Volkswagen Project Beginnings
1934	»	Auto Union Grand Prix
		Edmund Rumpler
		Porsche Visits to America
1935	»	KdF-Wagen
1938	»	Mercedes T80 Land Speed Record Design
1939	»	Volkswagen 60K10

You are actually *building* such a car in your garage? It's hard to believe my eyes!

When Ferdinand Porsche's duties with Steyr officially ended in April of 1930, he had already made the momentous decision to set up his own design office and be independent. There were other options. Ferry Porsche wrote that his father had an offer from Skoda, the large Czech manufacturer of machine tools, weapons, vehicles of many types and power generation equipment. Working for Opel was also a possibility. However, it is clear that Dr. Porsche was tired of working for other people at this point in his career and hoped to pursue engineering excellence without the internal politics of a large company. Much of 1930 was spent recruiting the staff, and actual design work started as early as April. As to the location of the new business, Dr. Porsche decided that the commercial prospects and supplier infrastructure were better in Stuttgart than in Austria. This would give the family a chance to eventually move back into the beautiful villa on Feuerbacher Way, constructed when Dr. Porsche first went to Daimler.

Dr. Porsche selected a small group of key employees, many of whom would go on to have long and significant involvement with Porsche designs and eventually, Porsche and Volkswagen cars. Interestingly, all of the first nine designers were Austrians, including the Chief Engineer, Karl Rabe who had worked under Dr. Porsche at Austro-Daimler. Erwin Komenda, who eventually styled the Porsche 356, came from Daimler-Benz (in November 1931). Josef Mickl, who had a long career at Porsche, was a mathematician, aircraft designer and expert in metal, fluid and gas dynamics. He became Porsche's in-house master of calculations. Josef Kales (engines), Josef Zahradnik and Karl Fröhlich (transmissions) followed Dr. Porsche from Steyr to the new company. Ferry Porsche, just turning 21 was also one of the original nine, and Dr. Porsche's young nephew, Ghislaine Kaes, who started as Dr. Porsche's secretary at Steyr continued in the role of personal assistant.

For commercial management, Dr. Porsche called on one of his former racing drivers, Adolf Rosenberger. Rosenberger was a wealthy iron and steel dealer who also became an investor in the new Porsche enterprise. He worked with Porsche directly until 1933, when Hitler came to power. Rosenberger fled Germany

Ferdinand Porsche at work in the design office, 1937.

for the first time in 1933, living in France and Switzerland. Persecuted for being Jewish, he was forced to leave Germany for good after being imprisoned in a concentration camp in 1935. He represented Porsche in France and England until 1938 before moving to America and changing his name to Alan Arthur Robert. The value of his 10% stake in the original Porsche design company became a controversial restitution issue after World War II.

Dr. Porsche's son-in-law, Anton Piëch, handled the company's legal affairs. Piëch and Rosenberger recruited the Austrian enthusiast and racer Hans von Veyder-Malberg to take over Rosenberger's role as commercial director and to (nominally) buy out Rosenberger's share in Porsche in January 1933. The actual disposition of those shares became part of the post-war controversy. Importantly, in 2022 Porsche AG announced funding and cooperation for an independent study, in conjunction with the non-profit Adolf Rosenberger company, meant to achieve a more complete understanding of Rosenberger's life, role with Porsche, and post-1933 relationship with Porsche.

The design office was located at Kronenstrasse 24 starting on December 1, 1930. The initial name was *Dr. ing. h.c. Ferdinand Porsche GmbH, Konstruktionsbüro für Motoren und Fahrzeugbau*. The first part stands for 'Doctor Engineer *honoris causa*' referring to Ferdinand Porsche's honorary degrees. GmbH is the abbreviation for '*Gesellschaft mit Beschrankter Haftung*'. This translates to 'limited liability company'. The final part translates to 'design office for engine and motor vehicle construction'. An alternate variation of the name adds *Luftfahrzeug und Wasserfahrzeug*, or 'aircraft and watercraft'. The formal registration documents for the Porsche company were completed on April 25, 1931.

The first car designed by Porsche was a new medium-size vehicle for Wanderer (the W21/22). It was powered by a Porsche-designed inline six-cylinder engine eventually available in two sizes. The design work initiated Porsche's sequential 'Type' or project numbering system. The ledger was famously started with the Wanderer project as design number 7. The generally-accepted story is that the Porsche team did not want their client to think that they had done no previous work, so numbers 1 through 6 were left vacant. The numbering system would eventually yield many famous Porsche model numbers starting with the 356. The Wanderer design was completed very quickly, and prototypes were running as early as the summer of 1931. However, the struggle to maintain financial stability (and simply pay the employees) when working on a project-to-project basis was a constant worry for Porsche in the first few years of the company.

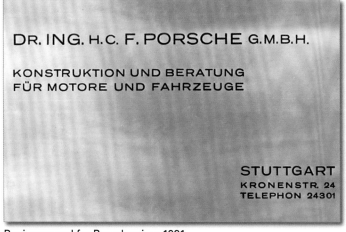

Business card for Porsche circa 1931.

Wanderer W22, design Type 7, in the Porsche Museum.

PORSCHE POINTS

» In 1931, Mercedes-Benz became the first German car to win Italy's great road race, the Mille Miglia. Rudolf Caracciola won in an SSKL, the basic design having originated during Dr. Porsche's tenure.

» The second design from the original Porsche team, under Type numbers 8 and 9, was a larger, more luxurious car for Wanderer. Two prototypes were built before Wanderer merged into Auto Union where Horch was assigned to build the larger cars. Dr. Porsche kept the supercharged Type 9 prototype as his personal car for several years during the 1930s.

People's Car

The idea for a small, affordable car had been with Dr. Porsche for many years, dating back to his time with Austro-Daimler. At Daimler-Benz, he was again unable to obtain support for such a vehicle from the board, although testing had been conducted on prototype small cars. One was based on a 1.4-liter six-cylinder engine and 30 of these cars were built for evaluation. The other was powered by a 1.2-liter, horizontally-opposed, air-cooled, rear-mounted four-cylinder engine. It had independent suspension and semi-monocoque 'shell' construction. After Dr. Porsche left Daimler-Benz, his designs were rejected in favor of a small car powered by a 1.3-liter liquid-cooled, rear-engine design (the 130 H, for *Heckmotor*, introduced in 1934). The conservatively-styled 130 and its rear-engine derivatives never found great success in the market. Perhaps customers had a hard time accepting the idea of a lower-priced and rear-engine Mercedes. The later Mercedes 170 H (W28) is quite Beetle-like in its appearance, and survivors are now rare, sought-after collectibles.

However, in the early 1930s, two projects continued Porsche's work on a 'people's car' and both involved motorcycle manufacturers. First came Zündapp, looking to boost production at a time when motorcycle sales were faltering. Zündapp owner, Fritz Neumeyer, had started to study the

The Russia Project

A strange but pivotal episode came in 1932, when Dr. Porsche was invited to tour Russia and review various industrial facilities. He and his associates thought that there might be an offer to consult on vehicle or other design projects. With the Soviets determined to modernize their economy, including agriculture, perhaps there would be an opportunity to design a 'people's tractor' for Russia. Dr. Porsche was taken to numerous cities and shown factories and products, including some that were considered secret. The tour lasted several weeks, including a period of bed rest when Dr. Porsche suffered a leg injury. As the tour progressed, he began to think the Stalin regime might want something more than the design for a tractor. At the end of the tour, Dr. Porsche was stunned to be offered a job with the title 'State Designer of Russia'. It appeared that the Soviets were willing to offer unlimited resources for research and design work to help accelerate the modernization of Russian industrial capabilities and production.

Although Dr. Porsche would be allowed to bring his family to Russia, his contacts with Germany would be cut off and his personal liberties limited. At this stage of his career, he still had three great design ambitions. One was something like a people's tractor (a rugged, affordable, mass-produced tractor available to nearly any farmer). Second was a 'people's car' (also rugged and affordable to the masses as personal transportation). Last was the creation of a truly great racing car (Porsche's love of competition and racing was still burning brightly). In Russia, Dr. Porsche thought the people's car would never be a priority and the racing car likely unthinkable. Although more advanced than outsiders may have realized, the Soviets needed a systematic industrial overhaul. There was also concern about overcoming the language barrier (significant when working on technical matters). In the end, Dr. Porsche turned down the offer, unable to reconcile leaving his homeland and with the thought that Germany held the best prospects for fulfilling his treasured design ambitions. He also told his son Ferry he felt too old to start a new career along the lines envisioned by the Soviets. One shudders to think what might have ultimately happened to Dr. Porsche and his family had he accepted the Soviet offer. It seems very unlikely anyone would be driving Porsche cars today.

PORSCHE POINTS

> Some sources indicate there was conflict between Dr. Porsche and Herr Neumeyer of Zündapp on the use of a five-cylinder, radial engine, with Dr. Porsche preferring an air-cooled four. Ferry Porsche states this was not true. The surviving Zündapp prototypes (Porsche Type 12) were destroyed by Allied bombing in World War II.

> The second set of 'people's car' prototypes, built for NSU, included two with imitation leather bodywork over wood frames. The third had a steel body constructed by Reutter in Stuttgart. Reutter had also constructed bodies for Wanderer and would play a significant role in the history of Porsche production cars.

> ## Torsion Bar
>
> The torsion bar suspension was a significant example of development work from the Porsche design office. Typically a long metal bar or bundle of smaller bars is attached to the chassis of the car on one side and a lever at the suspension side. Movement of the suspension is moderated or dampened by the twisting action of the torsion bar.
>
> The concept was invented in Norway and patented for wagons in 1878 by Anton Lövstad. Further use and patents for automobiles preceded Porsche's work. According to author Karl Ludvigsen in *Battle for the Beetle*: *"Porsche's accomplishment was the practical realization of torsion bar springing through the mastery of stress calculation for durability and the provision and patenting of suitable suspension linkages."*
>
> Early adopters of the torsion bar system included Hanomag, Morris, Citroen, Volvo and Alfa Romeo. Torsion bars also were used on the Grand Prix cars from Auto Union and the British E.R.A. The Volkswagen and the Porsche 356 made use of torsion bar designs for suspension springing along with the Porsche 911 which employed them right through the 1980s.

Volksauto or *Auto für Jedermann*, a 'car for everyman' concept as early as 1925. Porsche's design for Zündapp was similar in many respects to the Volkswagen except for the use of a water-cooled, five-cylinder radial engine behind the rear wheels. Three five-cylinder prototypes were built and tested in 1932, but Neumeyer elected not to invest in producing the car as motorcycle sales improved.

The next people's car design project came from NSU. This design was much closer to the eventual Volkswagens built in the late 1930s. Although the prototypes were larger and more boxy, the familiar VW shape is evident. The engine was a 1.4-liter, air-cooled flat-four, with the engine still behind the rear axle line, transmission forward of the rear axle. A significant development was the appearance of a torsion bar suspension. This innovation not only found its way into production Volkswagens, but also the Porsche 356 and 911. Manufacturers from many countries soon licensed the torsion bar design, providing a stream of income to the company. As with Zündapp, only three NSU prototypes were built before the company ran out of capacity due to increasing demand for motorcycles.

In the late summer of 1933, the Volkswagen project got its real start when Dr. Porsche was mysteriously invited to Berlin to meet with Jakob Werlin, Hitler's unofficial automotive advisor. It turned out to be a discussion about Hitler's interest in a small car. The surprise was that Chancellor Hitler himself suddenly appeared and joined the meeting. Hitler outlined his vision

> ## Volkswagen Precursors and Influencers
>
> ◆ Béla Barényi was an automotive engineer, designer and inventor who conceptualized something similar to Dr. Porsche's 'people's car' as a student in 1925. He went on to work for Austro-Daimler, Steyr and for many years at Daimler-Benz. He is best known for his post-war safety innovations, including the crumple zone concept and deformable steering column. Berenyi successfully sued Volkswagen in the 1950s and received legal recognition for his contribution as an intellectual father of the Volkswagen.
>
> ◆ Josef Ganz was a mechanical engineer and automotive journalist who championed the small car concept. He was an editor of *Motor-Kritik*, a leading German automotive publication between 1928 and 1934. He designed a small car concept proposed to Zündapp in the late 1920s, but it was never built. He was then involved with prototypes built by Ardie, Adler and the Standard Superior between 1930 and 1932. Persecuted as a Jew, Ganz left Germany in 1934. As with Edmund Rumpler (see page 44), the Nazis' attempted to remove all evidence of Ganz' contributions to the automobile from history. He continued to work on small car projects in Switzerland (the Erfiag) and France, then eventually worked for Holden after moving to Australia in 1951. Volkswagen discussed possible employment with Ganz as early as 1961, however, Ganz fell into ill health. Volkswagen assisted with some medical costs, but Ganz never received a pension from VW as had been proposed. Ganz passed away in July 1967.
>
> ◆ Tatra, the Czech auto manufacturer with Austrian engineer, Hans Ledwinka, produced a small car bearing some significant resemblance to the Volkswagen. Ledwinka was acquainted with Dr. Porsche and Adolf Hitler and those relationships likely influenced the people's car concept. The Tatra 97 from 1936 was a fairly small four-seat car with a rear-mounted, air-cooled, flat four-cylinder engine. In 1965, after a lengthy legal battle, Volkswagen paid Tatra a settlement of one million marks for pre-World War II patent infringement.

for the people's car and basic technical specifications (largely discarded by Porsche). Hitler dictated that the price should be no more than 1,000 marks and requested a formal proposal from Porsche. In the winter of 1933 into 1934, the proposal was developed. Dr. Porsche insisted that it be a 'real' car as opposed to something like a cyclecar or microcar. He set the engine size at 1250cc to make 26 horsepower and achieve 36 mpg. The car would have independent suspension for all four wheels and a top speed of 62 mph. However, Porsche's study indicated that it would be difficult to build and sell the car for a thousand marks (equivalent to about $250 at that time). That price certainly left no room for dealer profit and was far less than what Porsche had projected for the NSU design. The Porsche proposal price came to 1,550 marks. However, the June 1934 contract signed with the Society of German Automobile Manufacturers (or RDA for *Reichsverband der deutschen Automobilindustrie*) called for a production cost of only 990 marks with a volume target of 50,000 cars. It also called for a royalty of one mark per car for Porsche.

Porsche received a monthly fee to continue work on the design and development of the people's car. The demand for a running prototype within ten months was certainly challenging and probably unrealistic. Porsche had no workshop, per se, so the garage at the Porsche family home was used by a small crew of workmen to construct the first three cars. Hitler's first public mention of the 'Volkswagen' came in March of 1934 at the Berlin Motor Show where he pointed out that Germany only had one automobile for every 100 inhabitants (compared to one automobile for every six people in America). He called for a car that would be affordable to anyone who could buy a motorcycle and exhorted the German auto industry to design cars that would attract buyers in the millions.

In late 1935, the first two test vehicles, known as V1 and V2 were ready to begin testing. The V1 was enclosed while the V2 was an open car. Testing of different engine designs was one of the main priorities so that a production spec could be finalized. As it happened, the first three formal prototypes, referred to as the VW '3' series, would not run until 1936.

Auto Union

Auto Union was the company formed by the 1932 merger of four separate German auto makers: Wanderer, DKW, Audi and Horch. The four interlocking rings of the Auto Union logo live on in the present as the Audi badge. The financial conditions during the Great Depression led to a merger in which the individual brands could still exist but product planning, design and development would be centralized. The wealthy Adolf Rosenberger suggested that Auto Union use a Porsche design for racing in 1934. As a member of the Berlin Rotary Club, Rosenberger discussed the racing car idea with fellow Rotarian and Wanderer sales manager, Klaus von Oertzen, as early as 1932.

In fact, Porsche had started design work before any agreement with Auto Union. A separate company was even set up by Porsche to build and possibly race the car. It was named *Hochleistungsfahrzeugbau GmbH* ('high-performance vehicle construction'). When representatives from Auto Union formally approached Porsche about designing a Grand Prix racer, Dr. Porsche was able to say he already had it done. In the end, the combination of Porsche and Auto Union proved to be complimentary and highly competitive against Mercedes-Benz.

Auto Union poster showing the individual brand logos.

A meeting was arranged for March 10, 1933 by Baron von Oertzen (who was now Chairman of the Board at Auto Union) between Adolf Hitler, Dr. Porsche, racing driver Hans Stuck and von Oertzen. This was before any discussion of the people's car idea as discussed above. Porsche showed his design ideas and rather forcefully argued that it would be best for two German companies to compete for the honor of Germany's greatest racing car. Although the Chancellor was initially non-committal, the meeting led to a state-sponsored competition. It is thought that Hitler's personal affinity for Dr. Porsche carried significant weight in the decision to split the racing subsidy between Auto Union and Mercedes. This was annoying to Dr. Porsche's old friends at Daimler-Benz who had already designed a car for the 1934 racing formula. The new formula limited the cars to a *maximum* weight of 750 kilograms. This rule is interesting in a modern context where racing formulae usually specify a *minimum* weight.

The Auto Union Grand Prix car was a typically creative Porsche design with one radical element for a racing car. The engine was behind the driver and in front of the rear axle in what is today commonly referred to as the mid-engine configuration. Dr. Porsche had seen the layout before, in the mid-1920s when Daimler and Benz began their cooperation. Benz had some limited success developing and racing a mid-engine *Tropfenwagen* ('teardrop-shaped car'), officially known as the *Rennwagen Heckmotor* ('racing car, rear engine'). Benz' competition cars were then de-emphasized in favor of Daimler's Mercedes racers, as discussed in Chapter Two. However, Ferry Porsche states in the book, *We At Porsche*, that the mid-engine placement for the Auto Union was his idea mainly as a weight-saving measure, reducing transmission weight in addition to improving overall weight distribution.

The silver bodywork was distinctly tear-drop shaped, resembling an aircraft fuselage or even a Zeppelin-style airship. With its driver ahead of the engine, the seat could be positioned quite low (no driveshaft to clear). Compared to its front-engine Grand Prix competition from Mercedes-Benz, Bugatti, and Alfa Romeo, the Auto Union looked radical and aggressive.

The first engine design was a supercharged 16-cylinder arranged in a 45-degree V configuration with a capacity of 4.4-liters. The light alloy engine was designed with increasing capacity in mind, and over the five years from 1933 through 1937, the engine would increase in size from 4.4 to 6.3-liters. Horsepower increased from 295 to 545 over those years. To save weight, the engine was designed with a single camshaft and its narrow-angle V helped reduce the width of the 'fuselage' for better aerodynamics. Unlike other Grand Prix competitors, Porsche chose to design a lower-revving, torquey, less-stressed engine for better reliability and steady power delivery. The transmission was a five-speed, and along with the engine, was angled slightly to position the differential correctly relative to the rear wheels.

The suspension used torsion bar springing (in front and later at the rear as well) combined with friction type shock absorbers. Trailing arms were used at the front suspension with swing axles at the rear, driven wheels. Independent suspension for all four wheels is something we take for granted in all cars today, but this was a new and innovative approach for 1930s racing cars. Porsche was first to call for round chassis tubes in a racing car, a practice that would carry on for much of the next 50 years. The large, longitudinal chrome molybdenum chassis tubes were used to conduct coolant between the engine and radiator although this was abandoned after 1934 when hairline cracks led to coolant leakage. The central location of the fuel tank (between driver and engine) maintained an even weight distribution as

Edmund Rumpler and the Swing Axle

Edmund Rumpler was an innovative Austrian automotive and aircraft engineer who worked for the Adler company and eventually designed a rear-engine car that was the 'star' of the 1921 Berlin Motor Show. He ultimately produced 100 of his teardrop-shaped *Tropfen-Auto* street cars. The mid-engine Benz RH race car was based on the Rumpler chassis concept. The swing axle was patented by Rumpler in 1903. It allowed for independent rear suspension, where each rear wheel could follow an uneven road surface. This not only eliminated camber change on the opposite wheel (as could happen with a solid axle), but also reduced unsprung weight with the differential mounted to the chassis of the car. Swing axle designs would be used on the Volkswagen as well as the Porsche 356 and the Chevrolet Corvair, among others. Modern independent suspensions use constant velocity joints to further improve on the swing axle concept.

Rumpler is one of a small number of automotive engineers who preceded Dr. Porsche as an independent designer/engineer/consultant. Rumpler was persecuted as a Jew by the Nazis and briefly imprisoned in 1933. His business was ruined and he died in 1940 with the Nazis having destroyed all of his records in an attempt to erase his contributions to the automobile industry.

Porsche design office at Kronenstrasse 24 in Stuttgart, circa 1935. A young Ferry Porsche is seen middle left.

fuel was consumed. The mid-engine configuration served to reduce the polar moment of inertia. This allowed the car to rotate more easily on its vertical axis by concentrating the weight between the front and rear axle lines. Drum brakes were hydraulic from Lockheed, and center-lock wire wheels were sourced from Rudge Whitworth. One of the main limitations to performance in this era were the tires, which required a delicate balance between temperature and durability.

The cars were built at the Horch factory in Zwickau. Ferry Porsche did some early test-driving of the Grand Prix car but was soon halted by his father. Dr. Porsche, who also tested the car at least once, explained that he had many good, professional drivers, but only one son! The projection was for a top speed of 180 mph, although it took a good deal of development before that level of speed could be realized. Porsche would be responsible for design work on the Type A, B and C versions of the Auto Union Grand Prix car, taking the program through the 1937 season. In the early years, Dr. Porsche usually attended the races and supervised the team particularly on technical matters. Hands-on as always, Dr. Porsche stepped in to personally assist with repairing a leaking radiator just before the start of a race at the Nürburgring.

The first appearance of the car was in 1934 for a speed record attempt with Hans Stuck driving at Avus and achieving an average speed of 134.9 mph, for one hour. He also set speed records for 100 miles and 200 kilometers. The 1934 German Grand Prix at the Nürburgring was a classic battle between Rudolf Caracciola in the Mercedes W25 and Stuck in the Auto Union. Much to the delight of the spectators, the two aces battled for 12 laps with Stuck clinging to the lead. On lap 13 of 25, Caracciola finally shot past on the pit straight. However, Caracciola retired on the next lap with engine failure. With four laps remaining, Stuck signaled his pit with a coolant temperature indicating 100 degrees Centigrade. Dr. Porsche told the team to signal Stuck to carry on, thinking that the radiator could not have suddenly failed, but rather the gauge itself must be the problem. Stuck took the victory for Auto Union, and the thermometer was found to be faulty.

Hans Stuck won three Grand Prix races and had two second

Hans Stuck in white coveralls with Dr. Porsche at Masaryk, Brno in Czechoslovakia. Victory for Stuck and Auto Union on September 30, 1934.

PORSCHE POINTS

» The first victory for an Auto Union Grand Prix racer was on June 10, 1934 at the Feisberg hill climb near Saarbrücken. King of the Mountains, Hans Stuck, won at an average speed over 90 mph.

» In 1935, Hans Stuck's wedding gift to Ferry and Dorothea Porsche was a terrier named Jackie who became a faithful guardian to the Porsche children.

» The Auto Union Grand Prix cars were nicknamed 'P-Wagen' in recognition of Porsche's design work. By 1935, Auto Union discouraged the use of the nickname so as not to detract from the public relations value of their racing efforts.

speed of attempts in both directions to account for wind and road conditions. This would be prodigious speed, even 30 years later.

The Type B version of the Grand Prix car enjoyed a large increase in horsepower, to 375 from the original 295. The rear suspension now employed torsion bars in place of the original transverse leaf spring. Auto Union won several important races in 1935, but Mercedes won nine of the ten events counting toward the European Championship. It was thought that Mercedes had the greater driving talent, so it is important to note that during 1935 Auto Union took on Bernd Rosemeyer as a new driver. Rosemeyer had been a successful motorcycle racer and was recruited internally from the Auto Union motorcycle team. He became a most capable teammate for Hans Stuck. Perhaps due to his lack of automobile racing experience, he adapted well to the handling character of the Auto Union cars and nearly beat Caracciola in his first race for Auto Union, at the Nürburgring *Eifelrennen*. The race was run in changeable weather conditions and Rosemeyer had to cope with the loss of his windscreen and two cylinders for part of the race. During the season, Rosemeyer began to develop a close relationship with Dr. Porsche, whom he called 'Uncle Doctor'.

The 1936 season was a return to form for Auto Union with Rosemeyer triumphing at the German, Swiss and Italian Grands Prix. The Type C engine grew to a 6.0-liter with 530 horsepower. Handling and power application were further improved by the use of a ZF limited-slip differential, developed during 1935 at Ferry Porsche's suggestion (after a test at the Nürburgring where he noticed that in fast corners the inside wheel would give off clouds of rubber smoke). At the German Grand Prix, Rosemeyer was the first to break the 10-minute barrier for the Nordschleife lap time. According to author Richard von Frankenberg, this was akin to the 4-minute mile

place finishes during 1934. Had there been a European Championship for drivers in 1934, Stuck might have won it. Late in 1934 and early in 1935, he set additional speed records, including the standing-start kilometer and mile. With the Grand Prix car covered in a more streamlined body, a flying-start mile record was set at 199.01 mph. Stuck did average over 200 mph in one direction during this attempt on the Autostrada between Altopascio and Lucca. Record-setting required averaging the

Ferry Porsche's Marriage

In 1935, after knowing each other for seven years, Ferry Porsche married Dorothea Reitz on January 10. 'Dodo's' father graciously granted approval of the long-expected union. Ferry's parents were a bit less sure of the 25-year old, feeling somehow that he might still be too young for marriage, but gave in with less opposition than Ferry expected. Their honeymoon was spent skiing in Oberstdorf, in the Bavarian mountains. On December 11, 1937 their first son, Ferdinand Alexander 'Butzi' Porsche, future designer of the Porsche 911, was born.

DALTON WATSON FINE BOOKS
NEWEST RELEASES

Dalton Watson Fine Books
Glyn and Jean Morris
www.daltonwatson.com
+1 847 274 5874
info@daltonwatson.com

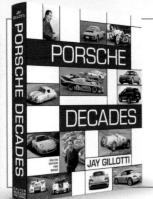

PORSCHE DECADES: AN INTRODUCTION TO THE PORSCHE STORY
Jay Gillotti   356 pages • 312 illustrations $135

Porsche Decades is an introduction to Porsche history from 1875 to 2023. It is written and designed for new Porsche owners, new Porsche club members, or those new to learning about the brand. Each chapter touches on the major events and projects important to Porsche's history during that specific period. The main topics are meant to spark further interest and provide a jumping off point for future exploration of this fascinating automotive and engineering history. Plentiful sidebars add general knowledge and enrich the narrative for new and experienced Porsche enthusiasts alike. **Porsche Decades** includes appendices for Porsche's type number history and prominent appearances by Porsches in Hollywood films. An extensive bibliography details books, magazine and internet articles for readers to explore for more information on topics of interest. Travel across the decades of automotive history and through Porsche's travails and triumphs.

NASH-HEALEY: A GRAND ALLIANCE
John Nikas, with Hervé Chevalier   2 volumes in slipcase • 800 pages • 1,192 illustrations $250

Nash-Healey – A Grand Alliance examines in exquisite and exacting detail the story behind America's first postwar sports car and the unique Anglo-American partnership between Nash and Healey that gave it life, which became an international triumvirate with the later involvement of famed Italian coachbuilder Pinin Farina. Focusing on the lives and careers of the men behind these fantastic machines, this book dives into their prior accomplishments, before reviewing the design and development of the Nash-Healey roadsters and coupes. It also explores the marque's incredible competition record at iconic races like the Mille Miglia, 24 Hours of Le Mans and Alpine and Monte Carlo rallies. Illustrated with more than 1,100 images, most never previously published, including many from Donald Healey's personal collection, this book is essential for all Nash-Healey enthusiasts and fans of sports cars from the breed's golden era.

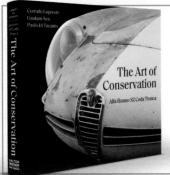

THE ART OF CONSERVATION: ALFA ROMEO SZ CODA TRONCA
Corrado Lopresto • Gautam Sen • Paolo Di Taranto   216 pages • 263 illustrations $110

Alfa Romeo SZ Coda Tronca presents an unforgettable story of an abandoned yet influential vehicle's rediscovery and restoration. Hidden away in a barn for the better part of five decades, a single Alfa Romeo SZ Coda Tronca – a rare and highly significant car in the company's design development – was found in a nearly perfect state of preservation. Drawing on art and archaeological techniques and taking painstaking care to save as much as possible of the original body, prominent Italian collector Corrado Lopresto decided to clean only half the car, leaving the other half frozen in time. In the uncleaned half, Lopresto preserved everything (including the dust) under a thin layer of transparent lacquer. The cleaned part has not been fully restored either, but has been saved by retouching in such a way that the original aspects are not affected. This book captures the rediscovery of a historic vehicle and the way it has been preserved—a fascinating tale of art meeting automobile.

FAST FASTER FASTEST: THE BILL SADLER STORY
John R. Wright   384 pages • 713 illustrations $110

This biography of Bill Sadler tells the story of an innovator who set the racing world astir with his boldly inventive cars. His original creations – including the Sadler Formula Libre and Mark V – brought him racing victories at tracks such as Watkins Glen and the Brighton Speed Trials throughout the 1950s and early '60s. Sadler's career took him from his first shop in Hamilton, Ontario, to the notorious Nevada government facility known as Area 51, where he worked on prototype aircraft and other classified projects during his decades-long hiatus from the racing world.

Written with the full cooperation of Bill Sadler before he passed away in early 2022, **Fast, Faster, Fastest** contains previously unpublished photographs and rarely-heard stories from a brilliant engineering mind.

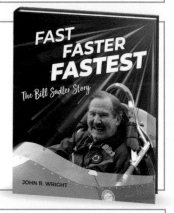

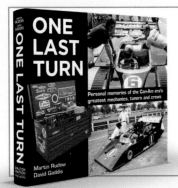

ONE LAST TURN
PERSONAL MEMORIES OF THE CAN-AM ERA'S GREATEST MECHANICS, TUNERS AND CREWS

Martin Rudow and David Gaddis   400 pages • 428 illustrations $125

Can-Am: the words are still magic to a generation of road racing fans to whom the Canadian-American Challenge Cup series represents the pinnacle of the sport they love. Taking over from the USRRC (United States Road Racing Championship) in 1966 as the feature sports car circuit in North America, this competition combined the world's best drivers and racing teams with the lawless spirit of the Can-Am series. Anything went, as long as it had two seats and enclosed wheels. This glorious free-for-all attitude set the stage for years of ground- and pulse-pounding cars powered by daredevil drivers with increasingly bigger engines. Illustrated with many previously unpublished photographs of the cars and people who made Can-Am great, **One Last Turn** is the book that fans of this renegade racing series have been waiting for.

Dalton Watson MyRewards When you buy a book from **daltonwatson.com**, you will receive points for the product(s) you purchase. Not all items accrue points. Each product indicates how many points you will earn below the price in the description. It's a great way to save on your next purchase, whether it's for you or someone special and is our way of saying "thank you" to our returning customers.

COACHBUILDING/DESIGN

Gaston Grümmer: The Art of Carrosserie *(2-Volume Set)*	Philippe-Gaston Grümmer and Laurent Friry	Regular: $295
Marcel Pourtout: Carrossier	Jon Pressnell	$150
Paolo Martin: Visions in Design	Paolo Martin with Gautam Sen	$125
Park Ward: The Innovative Coachbuilder 1919-1939 *(3-Volume Set)*	Malcolm Tucker	Regular: $375; Custom Leather: $1600
Raymond Henri Dietrich: Automotive Architect of the Classic Era & Beyond	Necah S. Furman	$225
The Bertone Collection	Gautam Sen and Michael Robinson	$95
The Kellner Affair: Matters of Life and Death *(3-Volume Set)*	Peter Larsen with Ben Erickson	Regular: $445
Tom Tjaarda: Master of Proportions	Gautam Sen	$150

GENERAL AUTOMOTIVE/RACING

Audi RS: History • Models • Technology	Constantin Bergander	$79
Augie Pabst: Behind the Wheel	Robert Birmingham	Regular: $79; Signed/Numbered: $99
The Automotive Alchemist	Andy Saunders	$115
Bahamas Speed Weeks	Terry O'Neil	$155
Bugatti: The Italian Decade	Gautam Sen	$150
Cobra Pilote: The Ed Hugus Story	Robert D. Walker	$89
Concours d'Elegance: Dream Cars and Lovely Ladies	Patrick Lesueur, Translated by David Burgess-Wise	$69
Cunningham: The Passion, The Cars, The Legacy *(2-Volume Set)*	Richard Harman	Regular: $350; Leather: $1200
Fast, Faster, Fastest: The Bill Sadler Story	John R. Wright	$110
Ferrari 333 SP: A Pictorial History 1993-2003	Terry O'Neil	$150
Fit for a King: The Royal Garage of the Shahs of Iran	Borzou Sepasi	$150
Formula 1	Peter Nygaard	$89
Imagine too!	Patrick Kelley	Regular: $150; Signed/Numbered: $200
KIM: A Biography of M.G. Founder Cecil Kimber	Jon Pressnell	$150
Lime Rock Park: The Early Years 1955-1975	Terry O'Neil	$225
Meister Bräuser: Harry Heuer's Championship Racing Team	Tom Schultz	Regular: $95; Signed/Numbered: $125
Mid-Atlantic Sports Car Races 1953-1962	Terry O'Neil	Signed/Numbered: $155
One Last Turn	Martin Rudow and David Gaddis	$125
QPRS: F1 Grand Prix Racing by the Numbers, 1950-2019	Clyde P. Berryman	$95
Shelby Cobras: CSX 2001 - CSX 2125 *(2-Volume Set)*	Robert D. Walker	$250
Sports Car Racing in the South: Vol. I 1957-1958, Vol. II 1959-1960, Vol. III 1961-1962	Willem Oosthoek	Vol I: $125; Vol II: $155; Vol III (Signed/Numbered): $155
The Golden Days of Thompson Speedway and Raceway 1945-1977	Terry O'Neil	Signed/Numbered: $195
The Straight Eight Engine: Powering Premium Automobiles	Keith Ray	$95
Watkins Glen: The Street Years 1948-1952	Philippe Defechereux	$49

BRITISH CARS

Allard Motor Company: Beyond the Records *(2-Volume Set)*	Gavin Allard	$175
Bentley Motors: On the Road	Bernard L. King	$165
Bentley: Fifty Years of the Marque	Johnnie Green/Hageman, King, Bennett	$92
Making a Marque: Rolls-Royce Motor Car Promotion 1904-1940	Peter Moss and Richard Roberts	$125
Nash-Healey: A Grand Alliance *(2-Volume Set)*	John Nikas, with Hervé Chevalier	$250
Rolls-Royce: Silver Wraith	Martin Bennett	$125
Rolls-Royce: The Classic Elegance	Lawrence Dalton/Bernard L. King	$85
The Rolls-Royce Phantom II Continental	André Blaize	Regular: $395; Leather: $1750
The Silver Ghost: A Supernatural Car	Jonathan Harley	$69
Why Not? The Story of The Honourable Charles Stuart Rolls	David Baines	$89
Jaguar E-Type Six-Cylinder Originality Guide	Dr. Thomas F. Haddock & Dr. Michael C. Mueller	$125
Vintage Jaguar Keyrings, 1955-1980	Morrill 'Bud' Marston	Regular: $95; Signed/Numbered: $135

FRENCH CARS

Crossing the Sands: The Sahara Desert Trek to Timbuktu	Ariane Audouin-Debreuil/Ingrid MacGill	$65
Eighty Years of Citroën in the UK	John Reynolds	Regular: $70; Special Edition: $450
Figoni on Delahaye	Richard Adatto with Diana Meredith	$250

GERMAN CARS

Forty Six: The Birth of Porsche Motorsport	Multiple	Regular: $150; Signed/Numbered: $285
Gulf 917	Jay Gillotti	Regular: $150; Leather (2-vol): $1500
Porsche by Mailander	Karl Ludvigsen	$150
Porsche Decades	Jay Gillotti	$135
Rudolf Uhlenhaut: Engineer and Gentleman	Wolfgang Scheller and Thomas Pollak	$89

ITALIAN CARS

The Art of Conservation: Alfa Romeo SZ Coda Tronca	Corrado Lopresto, Gautam Sen, Paolo Di Taranto	$110
Lamborghini: At the Cutting Edge of Design *(2-Volume Set)*	Gautam Sen	$250
Maserati 300S *(Revised, 2-volume set)*	Walter Bäumer	Regular: $270
Maserati 450S	Walter Bäumer and Jean-François Blachette	$195
Maserati A6GCS	Walter Bäumer and Jean-François Blachette	$175
Maserati A6G 2000 Frua • Pininfarina • Vignale • Allemano	Walter Bäumer	$125
Maserati Tipo 63, 64, 65: Birdcage to Supercage	Willem Oosthoek	Regular: $140; Special Edition: $550

ICON / GENERAL INTEREST

Steve McQueen: In His Own Words	Marshall Terrill	$95
Steve McQueen: The Last Mile Revisited	Barbara McQueen and Marshall Terrill	$49

ORDER FROM: Dalton Watson Fine Books / www.daltonwatson.com / info@daltonwatson.com / +1 847 274 5874

LEFT: Auto Union Grand Prix team in 1936 at Bremgarten.

TOP: 1937 Auto Union Type C at the Nürburgring.

ABOVE: Hans Stuck in practice for the 1937 German Grand Prix at the Nürburgring.

Auto Union Grand Prix car in 1936 at the Nürburgring, Ernst von Delius driving.

Hans Stuck & Son

Born in 1900, the Swiss/German Hans Stuck began racing in 1922 and achieved good success in hill climbs and as a private entrant in Austro-Daimlers. His acquaintance with Hitler helped lead to Auto Union's Grand Prix program, for which Stuck initially was the lead driver. Although successful as a Grand Prix racer, Stuck was best known for his hill climb victories and was nicknamed *Bergkönig*, for 'King of the Mountains'. He did most of his racing, and gained over 70 victories, in Porsche-designed cars. Stuck continued racing after World War II and was German hill climb champion at age 60.

Stuck's son by his third marriage, Hans-Joachim Stuck, was born in 1951. Hans Sr. became an instructor at the Nürburgring and taught his son how to drive there. Hans-Joachim Stuck went on to race in touring cars, Formula One and sports cars as a works driver for BMW and Porsche. He won the 24 Hours of Le Mans as a Porsche factory driver in 1986 and 1987 and has been an enthusiastic ambassador for Porsche over the years.

barrier for runners at the time. Rosemeyer was awarded the European Championship for drivers. Meanwhile, Hans Stuck again set several speed records including long-distance marks ranging from 50 kilometers to 100 miles.

For 1937, the Auto Union engine grew again, now to 6.3 liters providing 545 horsepower for speed record attempts. Even with the standard race engine, the cars were capable of speed in excess of 200 mph on long straights. This was tremendously fast considering the construction of the cars and the track conditions in that era. However, the Mercedes W125 had a significant power advantage with something over 600 horsepower available by mid-season. The 1937 Mercedes also benefited from major chassis and suspension improvements under Rudolf Uhlenhaut, their brilliant new head of the racing department. Mercedes fielded a superb team of drivers, including Hermann Lang and Dick Seaman with stalwarts Manfred von Brauchitsch and Rudolf Caracciola.

Dr. Porsche and his son, Ferry, made a visit to the United States during 1937 that coincided with the running of the revived Vanderbilt Cup race at Roosevelt Raceway, on Long Island, New York. The early-century road racing version had been

Visits to America

Hitler encouraged Dr. Porsche to visit the United States in 1936 to learn from Ford, GM and Chrysler with an eye toward design and tooling for mass production in the Volkswagen factory. In today's context it seems a bit surprising that the Big Three were so helpful. Ferry Porsche wrote that his father returned from that trip both '*optimistic and confident*'.

The 1937 trip was a much more comprehensive visit and included an entourage with Ferry Porsche, Ghislaine Kaes (Dr. Porsche's nephew and secretary who was fluent in English), Jakob Werlin of Daimler-Benz, Dr. Bodo Lafferentz co-director of the VW factory project, plus Bernd Rosemeyer and the Auto Union team members. Incidents during the sea voyage on the *Bremen* gave Ferry Porsche exposure to the disturbing threat of Nazi anti-Jewish fervor. The visit to America was much more pleasant for young Ferry who developed an admiration for the American way of life. Their party also developed great respect for the mass-market Ford V8, which they used for transport during much of their travels. A small number of talented German immigrants were recruited to move back to Germany to assist with the ongoing Volkswagen project.

The trip included visits to companies like Fisher Body, Cincinnati Milling and Budd (interesting as a pioneer in metal stamping and for its use of spot-welding). Once again, the access granted and cooperation from the American companies seem quite surprising from a perspective more than 80 years later. The Porsches visited Ford Research in Dearborn and had at least two discussions with Henry Ford himself. Dr. Porsche invited Mr. Ford to visit in Germany but Ford politely declined, citing the 'unsettled' state of the world and the likelihood of war on the horizon. This was a shock to the completely apolitical and ultra-logical Dr. Porsche, to whom the idea of war was unthinkable.

Ferdinand Porsche in New York harbor with the Statue of Liberty, 1936.

discontinued for safety reasons, but the Cup race returned in 1936 using a permanent circuit.

The great 'silver arrows' grand prix cars from Mercedes and Auto Union made their first and only appearance in America, and it proved quite a spectacle. The primary competition came from Scuderia Ferrari and their Alfa Romeos. Scheduled for Saturday, July 3, rain delayed the race until Monday. Still, some 75,000 spectators came out for the 300-mile race. Rosemeyer and Ernst Von Delius drove for Auto Union, lining up against Caracciola and the Englishman, Dick Seaman, for Mercedes. Guiseppe Farina and the great Tazio Nuvolari drove for Ferrari in the 12C Alfas. Prominent American driver, Rex Mays, drove a privately-entered 8C Alfa and there were various American

1961 *Christophorus* calendar photograph shows Dr. Porsche supervising an Auto Union record run with Bernd Rosemeyer driving.

PORSCHE POINTS

» On a transatlantic crossing aboard the Queen Mary, Dr. Porsche was given the VIP treatment and allowed to tour the engine room. Caught in a heavy storm, Porsche inquired about what degree of roll the ship could survive before turning over!

» In 1936, Auto Union and Dr. Eberan von Eberhorst pioneered an early form of telemetry – a device that could measure and track onboard information. Developed with the Kienzle instrument company, it could record rpm, gear changes, acceleration, speed and braking on a paper disc. Nicknamed 'Isidore', author Ian Bamsey stated that the drivers hated it.

» Bernd Rosemeyer was married to the famous aviatrix, Ellie Beinhorn, who lived to be 100 years old, passing away in 2007. The couple met on the evening of Rosemeyer's first Grand Prix victory, the 1935 Czechoslovakian race at Brno. The press nicknamed them 'the fastest couple in the world'. Their son, also named Bernd, was born only ten weeks before his father was killed. He went on to become a doctor. The memorial at the site of Rosemeyer's crash is still in place alongside the A5 autobahn.

roadsters entered, driven by the likes of Mauri Rose, Jimmy Snyder and Wilbur Shaw.

Caracciola was a little over one mph faster than Rosemeyer in qualifying. Mays was third fastest in the Alfa. Rosemeyer led the race early until being passed by Caracciola on lap three. Rosemeyer retook the lead on lap 11 as the two German aces eased away from the field. Nuvolari retired with engine failure and on lap 22, Caracciola fell victim to supercharger failure and also retired. Rosemeyer and Dick Seaman exchanged the lead during the fuel stops and the Flying Mantuan, Nuvolari, took over Farina's Alfa on their fuel stop. Nuvolari was in great form. He passed Von Delius in the Auto Union and was challenging Rex Mays for third when he suffered his second engine failure of the day on lap 50. Toward the end, Seaman was catching Rosemeyer, but ran out of fuel and had to make an unscheduled stop. Rosemeyer took the win and the $20,000 first prize (about $350,000 adjusting for inflation). Mays finished third in his Alfa Romeo.

The following weekend, the other halves of the great Grand Prix teams staged another titanic battle at Spa-Francorchamps in the Belgian Grand Prix. Rudolf Hasse and Hans Stuck finished first and second for Auto Union over Hermann Lang and Christian Kautz for Mercedes, with Raymond Sommer fifth for Ferrari/Alfa Romeo. The grand era of unlimited engine 1930s Grand Prix machines, which had been dominated by the state-sponsored German entries, was winding down as was Porsche's involvement with Auto Union. Caracciola won the championship for Mercedes, while Auto Union won only five major races against seven for Mercedes. Author Ian Bamsey, in *Auto Union V16 Supercharged*, notes that Rosemeyer could have been less aggressive and more calculating in the German, Swiss and Czech Grand Prix races. The outcome of the drivers' championship would have been much closer if not for Rosemeyer's off-course excursions in those races.

In October of 1937, Rosemeyer set a record for the flying-start five kilometer distance at a speed over 251 mph on the autobahn between Frankfurt and Darmstadt. This was achieved by another special version of the Auto Union Grand Prix car with a fully streamlined, aerodynamic body. Rosemeyer also used the standard open-wheel race car to set new records for the standing-start kilometer and mile. Those marks stood until the late 1950s.

In January of 1938, Mercedes received special permission to challenge the speed record outside of the normal 'record week' period. Caracciola achieved an average speed just under 270 mph in a Mercedes streamliner. Auto Union naturally wanted to compete with Mercedes and produced a further streamlined version of their record car, however, this was designed without direct involvement from Porsche. The new shape had a smaller frontal area but was more vulnerable to crosswinds. The car was capable of matching or exceeding the Mercedes' speed, but after a 267 mph warm-up run, Bernd Rosemeyer was killed making his record attempt. Although the cause could not be determined for certain, it is likely that a strong crosswind swept the car off the road and resulted in a huge crash. When told of the accident, Dr. Porsche commented that had he been present he probably would not have allowed the run given the weather conditions.

The 3.0-liter, 12-cylinder Auto Union racers entered for races in 1938 were designed and prepared without involvement from Porsche. Dr. Porsche's friend and colleague, Auto Union's Dr. Robert Eberan von Eberhorst was responsible for the Type D which did carry on with the mid-engine configuration. The contract between Porsche and Auto Union concluded at the end of 1937 and the Porsche team began a new three-year association with Daimler-Benz.

Auto Union versus Mercedes-Benz

Over the four years of classic competition between Auto Union and Mercedes-Benz, each side had numerous successes. Competitors from France, Italy and England rarely had a chance to win, especially at the high-speed tracks. Mercedes generally had the edge on horsepower, race management (including efficient pit stops) and driving talent. The Mercedes cars were also more traditional and predictable in handling for the drivers of that era. Daimler-Benz is assumed to have enjoyed greater financial resources, government subsidies aside. The Auto Union had a supple, more comfortable ride that helped the drivers in the relatively long-distance grand prix races of the time. The drivers liked the smooth power delivery and acceleration from the V16. Auto Union had less trouble staying below the 750 kilogram weight limit. Although both cars' suspension designs favored the power-sliding oversteer that was typical, the Auto Unions were likely trickier to handle at the limit, especially for drivers who were used to the traditional front-engine layout. A review of the races for 1934 through 1937, including both European Championship and major non-Championship races shows Auto Union with 18 victories compared to Mercedes at 25.

ABOVE: Volkswagen prototype at the Porsche home in Stuttgart.

Politics and Citizenship

Ferdinand Porsche had been born an Austrian, although in the Czech region of Austria-Hungary. After the First World War, he was considered a Czech citizen since his birthplace was now located in the new country of Czechoslovakia. This was in spite of the fact that he lived and worked in the small, new country known simply as Austria. After moving to Germany in the 1920s, then returning to Austria in 1929, and finally settling back in Germany, the Porsche family members were still considered Czech citizens. Ferry Porsche points out in his book that there were certain envious 'enemies' of Dr. Porsche who advised Hitler that the Auto Union Grand Prix cars were being designed and built by a Czech (to the detriment of any propaganda value that might come from racing successes). According to Ferry Porsche, Hitler simply decreed German citizenship in the Third Reich upon the family in 1934. Other accounts state that Dr. Porsche received a letter strongly suggesting that the application for citizenship be completed. Ferry Porsche wrote that his father, a most apolitical man, took a typically resigned stance, saying to his secretary, Ghislaine Kaes: *"I really don't see what we can do about it. We haven't been given any choice have we?"*

KdF-Wagen

In Ferry Porsche's book, *We At Porsche*, he recounts a 1935 visit from the RDA during assembly of the Volkswagen prototypes in the garage of the Porsche family home. Otto Koehler, an engineer from Daimler-Benz who had worked under Dr. Porsche at Austro-Daimler, was part of the evaluation team. After taking a long look at the work being done, he exclaimed: *"You are actually building such a car in your garage? It's hard to believe my eyes! …I must say that I am very much impressed by what you are doing. It's a striking example of what can be done with limited resources but great determination."*

One of the main hardships endured by the Porsche team was a shortage of material resources, due partly to Germany's restrictions on 'hard' currency for foreign exchange. This affected the availability of raw materials, such as rubber and copper. It meant designing and building around the material shortages. Further, the cost of foreign currency meant making choices such as using mechanical brakes on the VW rather than licensing hydraulic brake technology from Teves (which would require a royalty payment to Lockheed in Great Britain).

On October 12, 1936, the 'V3' Volkswagen prototypes were finally complete and ready for formal testing. With the earlier V1 and V2 making a five-car group, testing ran day and night (except on Sunday) on two set loops, taking in sections of Autobahn as well as back roads and twisty, mountainous terrain. The cars were driven by Porsche engineers accompanied by an employee from the RDA whose job was to note any problems along with speed and weather data. The cars covered more than 30,000 miles prior to Christmas of 1936. The RDA observers compiled their data into a report that was overall quite positive on the performance of the cars, allowing for the primitive conditions under which they had been built. No fundamental design flaws were found, but the development process was painfully slow for Dr. Porsche, as any failure led to redesign and hand-building new components. According to Ferry Porsche, his father complained: *"It's easier to build a brand new race car than it is to produce a Volkswagen!"*

The next step was to construct a batch of 30 pre-production prototypes and this work was contracted to Daimler-Benz. The cars were finished by the spring of 1937 and began a much more intensive period of testing. These cars covered approximately two million kilometers with some individual cars exceeding 100,000 kilometers. The driving was handled day and night by men from the S.S., or *Schutzstaffel* (Hitler's protection squadron), usually supervised by Ferry Porsche. This method meant that the cars were driven by people who approximated the skills of ordinary drivers but who could keep the progress of the testing secret. A further 30 cars were assembled for exhibition purposes, and a company was formed to begin the process of constructing a factory for the Volkswagen.

Another series of cars, known as the Type 38 in Volkswagen terms, were made in 1938. They were built to the planned production specification with an engine of 985cc capacity. The Porsche team working on the Volkswagen project included many familiar names in addition to Dr. Porsche himself and his son, Ferry. Karl Rabe led the project along with Erwin Komenda, Josef Kales, Karl Fröhlich, Josef Mickl, Josef Zahradnik and Franz Xaver Reimspiess (who is also one of the people with a claim to designing the famous circular VW logo). Reimspiess was responsible for the final 1936 design of the air-cooled, flat four-cylinder, four-stroke boxer engine that would go on to power millions of Beetles. Reimspiess won over Dr. Porsche by producing a design in just 48 hours that would eliminate problems encountered with two-stroke engines. Rabe was critical to the success of the project overall. In *Battle For The Beetle*, author Karl Ludvigsen presents the following quote from Wilhelm Vorwig of the RDA: *"It was Rabe's extraordinary ability that made the bold Porsche concept come into reality. Yet he was always in the background, never getting credit for his work. Without Rabe, there would be no Volkswagen."*

1936 Alpine test drive for two of the three 'V3' Volkswagens. In the photograph, from left to right are Dr. Porsche, Ferry Porsche, Rudolf Ringel (master mechanic) and Josef Goldinger, Dr. Porsche's chauffeur.

Ferdinand Porsche with a Volkswagen W30 prototype in 1937.

BMW's Franz Josef Popp was primary among the auto industry executives who devised a plan for production of the Volkswagen in a way that would limit the impact to sales and availability of scarce material resources for the other German car makers. His idea was to make the Volkswagen essentially a non-profit and tax-exempt entity. The elimination of tax-related costs would help to achieve Hitler's very low mandated price. He also proposed limiting the purchase of cars to members of the German Labor Front (DAF), which had absorbed the outlawed, pre-Nazi trade unions. The millions of workers in the DAF were Hitler's target market anyway. The Labor Front would be responsible for any marketing or advertising and would make its significant financial reserves (some of which being money effectively stolen from the previously existing unions) available to fund the construction of a factory without assistance from other auto companies or traditional banks. The scheme was similar to other plans for things like refrigerators and even home building. In the end, interested buyers were not limited, and only 5% were DAF members. Only 25% of VW factory workers ultimately signed up to buy the car they were meant to build.

On May 26, 1938, the cornerstone was laid for the VW factory at the Fallersleben site near Brunswick (*Braunwschweig*). The plan was for interested buyers to enter a savings or subscription program under the Nazi leisure organization known as 'strength

'Boxer' Engine

The vast majority of 'flat' or horizontally-opposed Volkswagen and Porsche engines have cylinders that appear to 'box'. Opposing pistons move in the same direction relative to the crankshaft when the engine is running and have their own 'journal' or connection to the crankshaft. In other words, opposing pistons will either be moving toward each other or away from each other. In a 'V' engine, opposing cylinders generally move in the opposite direction relative to the crankshaft (one piston moving away from the crank while the other is moving toward the crank). The Porsche 917 was a rare exception to the Porsche 'boxer' architecture (see Chapter Six).

through joy' (*Kraft durch Freude*) or KdF. This organization was part of the DAF. The KdF sought to promote leisure activities and tourism in Germany as well as the benefits of National Socialism. The scheme was for the peoples' savings or subscription payments to be tracked in a booklet, redeemable for a car when payments were completed. A similar program had been in place for several years to fund workers' vacation travel. Hitler dictated that the car itself would be named for the KdF, not Volkswagen. As such the car became known as the *KdF-Wagen*. Both Dr. Porsche and his son Ferry were unhappy with the name as it would not be meaningful for export markets in the future. Given Germany's invasion of Poland in 1939 and the start of World War II, no KdF subscribers received a car under the original plan. Ferry Porsche stated in *We At Porsche* that some 200,000 Germans had paid in full for their KdF-Wagens by the start of the war. The number of future buyers reached 275,000 by the end of 1939, but that growth declined rapidly as the war went on. By 1944, the number was just over 336,000. It took until 1961 to resolve the post-World War II legal claims of the KdF subscribers. Those with a completed book were granted a choice of 100 Deutschmarks in cash or 600 credit against the purchase of a new Volkswagen.

At the end of 1937, Porsche had changed its status to 'KG' (*Kommanditgesellschaft*) for limited partnership. In June of 1938, Ferry Porsche's third son, Gerhard Porsche, was born. Also in June of 1938, Porsche was able to move into their own office building and workshops, in the Zuffenhausen neighborhood, northwest of Stuttgart. This building became known as *Werk I*. Porsche later paid restitution to an heir of the Wolf family since the property had been acquired at below market value from this Jewish family.

Two production-ready KdF-Wagens were shown at the 1939 Berlin Motor Show. Initial advertising called for the car to be priced at 990 marks, or 1,050 for a sliding canvas sunroof version. Cars would only be available in one color, a blue-gray tone (not Henry Ford's Model-T black). Top speed would be about 65 mph from 24 horsepower and with a vehicle weight of 1,430 pounds. The price of the car had been a constant worry for Dr. Porsche and his son during the previous five years. However, Hitler's plan had always been to eliminate any sales cost (ultimately shifted to the DAF), pointing out that demand

VW Turbo?

In *We At Porsche*, Ferry Porsche mentions that in 1939 Porsche worked on development of an exhaust gas turbocharging system for the Volkswagen. One of the goals was to increase horsepower for high-altitude (mountain) running in military versions. The problems were the small volume of exhaust to work with and, on the intake side, the small size of the manifold and valves. Still, a projected increase of 60% in horsepower was theoretically possible (and confirmed during the war by experiments with supercharging). Turbo development was a casualty of war and it would take more than 30 years before Porsche returned to the concept of turbocharging for an automobile.

PORSCHE POINTS

- According to Richard von Frankenberg, Hitler's minions were horrified that Dr. Porsche addressed the Führer as 'Herr Hitler', but the Führer himself did not seem to mind.

- In May of 1937, Dr. Porsche and his son-in-law, attorney Anton Piëch formed the 'Company for Preparation of the German People's Car' (*Gesellschaft zur Vorbereitung des Deutschen Volkswagens m.b.H*). The abbreviation, GEZUVOR, translates loosely to 'go ahead' or 'in front'.

- Dr. Porsche was awarded the National Prize of the Third Reich in 1937 along with airplane makers Ernst Heinkel and Willy Messerschmitt. The award to Dr. Porsche was for outstanding work on the Auto Union and Volkswagen.

- As a result of publicity from being photographed driving Hitler from the VW factory cornerstone ceremony to the train station (a job he felt "stuck" with), Ferry Porsche received admiring letters and marriage proposals from girls as far away as the United States!

- The town built for the VW factory, eventually known as Wolfsburg, might have been named *Porschestadt* or *Volkswagenstadt*, but Hitler chose *Stadt-des-KdF-Wagens bei Fallersleben* or *KdF-Wagenstadt*.

was not the problem. The only real problem was building the cars. Actual production of cars in the newly completed factory was scheduled for October 1939 and a small number of cars were built (and provided to Nazi Party officials). World War II mooted the issue of price as the factory would be converted for wartime production, in part for military vehicles based on the Volkswagen design (see Chapter Four).

Other projects

Porsche's 1934 Volkswagen 'exposé' (proposal document) mentioned adaptability for military vehicles. Discussions between Porsche and the German military about adapting the Volkswagen design for military use also started during 1934. In 1937, Porsche built a prototype Volkswagen for off-road use and this kept some interest alive with the German military.

In the mid-1930s, Porsche worked on aircraft engine design proposals such as the Type 70, an X arrangement (like upper and lower V16s on a central crankshaft) with 32 supercharged cylinders at 17.9 liters of capacity. The Type 72 was an inverted V16 at 19.7 liters supercharged. Neither of these 900-plus horsepower designs made it to production.

The 1937 Type 110 (with development prototypes through Type 113) 'people's tractor' (*Volksschlepper* or *Volkspflug* 'people's plow') designs first ran in 1938. Powered by an air-cooled, two-cylinder 'V' engine, the design was a forerunner to the Porsche tractors of the 1950s. The multi-speed capability of the tractors made for a versatile vehicle suited to many different jobs on the farm. Prototype testing took place at the Porsche family property in Zell am See, Austria. World War II halted plans for a vast factory near Cologne intended to build 300,000 tractors per year.

Mercedes Record Car

As the contract with Auto Union ended, Porsche began a three-year consulting agreement with Daimler-Benz in 1938. Porsche did have some consulting involvement with the W154 Mercedes Grand Prix cars of 1938 and 1939, including the design of a two-stage supercharger system for 1939. The Porsche-designed Mercedes T80 land speed record car is a fascinating 'what if' vehicle from that era.

It started as a passion project for Hans Stuck who tried at first to interest his own race team, Auto Union, but they declined on the basis of cost. Discussions and design work with Porsche dated back to 1936, before there was a German engine available with enough power to challenge the existing records set by Englishmen like Malcolm Campbell and George Eyston. The

Mercedes-Benz T80 record car bodywork on display at the Mercedes Museum. (Sean Cridland)

Volkswagen 60K10 in 1939.

project was promoted to Hitler by Hans Stuck, who hoped to drive the car to speed records for the flying kilometer and flying mile. Hitler, for his part, appreciated the propaganda value of record-setting. Daimler-Benz eventually agreed to take on the project when Stuck committed to using the new DB 601 engines. Design work accelerated during 1937 as Daimler-Benz was also assisting Porsche with construction and testing of the Volkswagen prototypes. Further land speed record-setting at the Bonneville Salt Flats in Utah forced Porsche to redesign the T80 during 1938. Eyston and John Cobb raised the record above 350 mph, and this exceeded Porsche's original design speed.

The T80 as built was over 26 feet long with a narrow, detachable body of light alloy no more than one millimeter in thickness. Design of the fanciful, futuristic shape was assisted by aerodynamicist and Porsche theoretician, Josef Mickl. Even with its inverted side wings, fins and bulging front fenders, the drag coefficient was a remarkably low 0.18. The mid-mounted engine was a prototype of the newly-developed DB 603. It was a 44.5-liter inverted V12 aircraft engine potentially capable of 3,000 horsepower. There were two driven axles at the rear, making for a six-wheeled vehicle (wheels fully enclosed).

Continental designed special tires to survive at over 400 mph as the vehicle had a theoretical top speed of 466 mph. The six tires would also have to support a total weight of 6,380 pounds.

The target date for a record attempt was January of 1940, but World War II stopped all such activities. The vehicle's engine was removed during the war but the T80 was put in storage and survives as an exhibit in the Mercedes-Benz Museum in Stuttgart. It wasn't until 2001 that a wheel-driven land speed record car achieved the projected speed of the T80.

Volkswagen 60K10

From the time of the earliest Volkswagen prototypes there were thoughts about a sports car version. Ferry Porsche noted that exploring a sport or racing version was standard practice for nearly all the cars his father had designed. There had been a proposal for a road-going Auto Union sports car with a three-seat, center steering configuration (like the McLaren F1). Such a car could have been eligible to compete in 24-hour races at Spa and Le Mans. Beginning in 1937, Porsche considered trying to build their own sports car; a 1.5-liter, V10, mid-engine design formalized as Type 114. There was also an earlier Type 64 proposal for a 1.5-liter sports version of the four-cylinder Volkswagen that was turned down by the KdF leadership.

Although the KdF did not support the idea of a production sports car, there was an important related project. A time-speed-distance rally competition was planned for 1939 between the Axis capitals of Berlin and Rome. Sponsorship came from Germany's motorsport governing body, the ONS, and the KdF organization for Porsche to build a car for the Berlin-to-Rome race. Porsche used their Stuttgart prototype shop to build a lightweight, streamlined version of the Volkswagen. With 40-plus horsepower from a revised version of the air-cooled flat-four, the car had a top speed of about 90 mph. The cars featured a smooth, futuristic body shape designed by Erwin Komenda. Similar to the Type 114 design, but with enclosed wheels, it also predicted the basic shape of post-war Porsche sports cars. The Duralumin bodies were built by Reutter of Stuttgart and the cars were about 150 pounds lighter than a standard Type 60. In Volkswagen terminology, the car was known as the 60K10 (tenth iteration of bodywork on the Type 60 platform).

A Berlin-to-Rome competition had been scheduled for 1938 but was canceled as a result of the crisis over possession of the Sudetenland (the German-speaking area of Czechoslovakia, including Dr. Porsche's birthplace). The event was re-scheduled for September of 1939. With the invasion of Poland by Germany on September 1, it was canceled again.

It appears that three Type 60K10 prototypes were planned, but only two were fully completed. The first car, Chassis 38/41, was completed in August of 1939. It was damaged in an accident in 1939 while being used by Volkswagen Director, Bodo Lafferentz. The second car (38/42) was protected by the Porsche family and sent to Austria during World War II. It was used and abused by Allied soldiers after the war. When the weather got warm, the soldiers cut off the narrow cabin greenhouse. The engine eventually seized and the car was parted-out and scrapped. Some parts were eventually used to create a replica made by the Prototype Museum in Hamburg (that car is painted black). The surviving prototype (Chassis 38/41 with engine 38/43), still exists. Current thinking is that the third body (38/43) and engine may have been used to repair the first car built.

According to the 2019 auction information, in 1947, 38/41 was sent to Pininfarina in Italy for restoration work and came back painted silver. It is possible that Ferry Porsche added Porsche letters on the nose after the restoration, but period photos suggest the letters were added after the car was sold. It is likely that Porsche made some use of the 60K10 as a benchmark during development of the 356. It was sold by Porsche to the Austrian Otto Mathé in 1949 and participated in numerous racing events. His prizes in the 1950 Alpine Rally could be considered the first competition successes for a 'Porsche' sports car. The car was painted a turquoise color during a 1951 refresh and later returned to silver (but the turquoise now shows through). It was retired from competition after a very successful 1952 season of racing in Austria.

In 2019, the ex-Mathé 60K10 was offered for sale at the RM Sotheby's Monterey auction. In a controversial outcome, the car was bid to $17 million, but did not meet its reserve. The Porsche Museum has a replica 60K10 body which is shown in bare metal.

PORSCHE POINTS

≫ England and France declared war on Germany on September 3, 1939, Dr. Porsche's 64th birthday. The Porsche family were on holiday at their home in Wörthersee, Austria when Germany began the invasion of Poland on September 1.

≫ Porsche investor and first commercial director, Adolf Rosenberger, married Anne Mezger when she emigrated to the United States in 1939. They had met when she worked at Porsche as a secretary.

Nr. 14   MOTOR-KRITIK   Seite 315

KdF.-Sportlimousine zu Studienzwecken.

Eine Vorbemerkung für unsere Leser! Beachten Sie bitte die Ueberschrift! In der Entwicklung eines Kraftfahrzeugtyps werden oft verschiedene Einzelausführungen gemacht, die lediglich Studienzwecken dienen. Es ist also zwecklos, wegen der gezeigten hochinteressanten Ausführung jetzt den Hersteller bzw. die Versuchsabteilung mit Rückfragen wegen Preis usw. zu überschütten.

Verwendet ist das KdF.-Fahrgestell mit 1100-ccm-Motor, der bei n = 3600 ca. 36 PS leistet. Der normale Motor des KdF.-Wagens ist also im Hubraum etwas vergrößert, auch sonst auf Höchstleistung getunt, was bereits in der höheren Drehzahl zum Ausdruck kommt. Das Vergleichsbild gegenüber dem KdF.-Wagen zeigt ohne weiteres die ausgezeichnete strömungsgünstige Formgebung des Sportzweisitzers, durch welche der Luftstrom zügig angehoben wird, den Wagenaufsatz mit der geteilten Windschutzscheibe fast widerstandslos umfließt und nach hinten sanft abgleitet. Der dem Luftstrom entgegengeführte Wagenquerschnitt ist möglichst klein gehalten, dementsprechend ist auch der Luftwiderstandsbeiwert sehr niedrig.

Wir hören von einem Wagengewicht von ca. 550 kg (betriebsfertig), also nur rund 15,2 kg/PS, einer Spitzengeschwindigkeit von 140 km/Std. und einem Brennstoffverbrauch unter 5 Liter bei 105—110 km/Std. Dabei ist zu berücksichtigen, daß der Motor ohne Ueberladung arbeitet.

Der Vergleich der Sonderausführung gegenüber dem Normalwagen zeigt die Unterschiede der dem Luftstrom entgegengeführten Flächen.   Phot.: Kreubel.

Page from the *Motor-Kritik* magazine review of the Volkswagen 60K10.

1940s

1940 ›› Volkswagen Factory at War
       Porsche Tank Designs
1942 ›› Kübelwagen
       Schwimmwagen
       Porsches Father and Son
       Porsche Versus Speer
1944 ›› Move to Austria
1945 ›› Porsches Detained
       Piero Dusio
1946 ›› Cisitalia
       Porsche Synchromesh
1948 ›› Porsche 356/1
       Erwin Komenda
       Porsche 356/2

It was perfectly clear to me at that moment that the whole affair would end badly.
– *Ferry Porsche*

At the beginning of 1940, the course of the war and its eventual impact on Germany and its population were far from clear. Some people in Germany may have thought it would be a limited war, and not last very long. Applying logic in place of the madness of Hitler and the Nazi regime, there were those who thought that peace might be negotiated with England, France and Russia. For his part, Ferry Porsche was pessimistic based on his observations of the Nazi Party and the German military. In his books, he wrote about Germany's relative lack of natural, industrial, organizational and human resources in relation to Germany's (eventual) adversaries.

The *KdF-Werk* 'Volkswagen factory' was completed in August of 1939, however, very few of the cars it was intended for could be built while war production took priority. Only 630 'Beetles', as the car later came to be known, were built in the new factory during the war. Those cars were mainly used by Nazi Party officials or by Porsche for development purposes. The relative suitability of the factory for military use had been a source of discussion as far back as the beginning of 1938. Although commandeered by Hermann Goering for the Luftwaffe in September 1939, actual military work was slow to come about. This may have been related to the impression noted above among some people that peace could be negotiated with the British and the French. Porsche even worked on civilian car and truck designs for Steyr during 1940. The truck design would serve numerous functions for the military, with more than 21,000 built during the war.

"Now Hitler's Gone Mad!"

That was Ferry Porsche's reaction to Germany starting what became World War II. His description of his father's reaction, in *Cars Are My Life*, is telling:

"This remark gave rise to a sharp reprimand from my father. His straightforward character and his sense of decency made it impossible for him to imagine that a German government, with responsibility for the welfare of the people, would deliberately follow a policy that could only end in war. Of course, the opportunities for work offered to him in the Third Reich had impressed him, but that was not the decisive factor. It was his uprightness that made it seem inconceivable to him that everything that had been achieved hitherto was to be put at risk by going to war. As an engineer, he was accustomed to thinking logically and clearly, and he could see no sense in what had now happened.

It was perfectly clear to me at that moment that the whole affair would end badly."

In the course of the war, the Volkswagen factory would not only build military versions of the Volkswagen, designed and engineered by Porsche, but its giant size and capacity were also used to build wooden fuel tanks for aircraft, torpedo hulls, portable furnaces, flotation kits for tanks, bombs, mines, and the single shot hand-held anti-tank weapon known as the *Panzerfaust* 'Panzer fist'. The factory was also used to repair Junkers aircraft, building replacement wings, tails and eventually producing most of the major components needed. Another task for the KdF-Werk was production of parts for the V-1 flying bomb. Hitler's 'vengeance' weapon was a crude pulse-jet missile with an 850 kilogram bomb aimed indiscriminately at London. The factory produced fuselages and engine housings for the majority of V-1s. Final assembly generally took place at the launch site. Unfortunately for the factory, involvement with V-1 production made it a high priority target for Allied bombing, especially in 1944.

In general, design and prototype work on military projects was a way for Porsche to remain in business during the war. As reviewed in Chapter One, Dr. Porsche had been prolific in this area and some of the company's design work in World War II would draw directly on experience gained in World War I. In September 1940, Dr. Porsche became an honorary professor at Stuttgart's Technical University, and in October, Ferry's third son, Hans-Peter Porsche, was born. In May 1941, with Dr. Porsche ever more busy on military design work, Anton Piëch took over as the commercial director and general manager of the KdF-Werk, living in the 'Hut', a small wooden house built near the factory as lodging for Dr. Porsche during his frequent visits.

Kübelwagen

The first military version of the Volkswagen became known as the *Kübelwagen*, for 'bucket seat car'. Small German military vehicles of the period did not have doors, so made use of bucket seats to keep the soldiers from falling out. Ironically, the production version of the Kübelwagen did have doors, flat front seats and a rear bench seat. The bench seat allowed room for at least one extra soldier, but the 'bucket' name stuck.

The first prototype, Porsche Type 62, dated back to 1936. It was envisioned as the cross-country or off-road version of the Volkswagen. The production version was Type 82 and featured the familiar, angular shape of pressed body panels that would reappear in similar form during the late 1960s as the VW 'Thing'. The bodywork was produced by Ambi-Budd of Berlin, a partner to the Edward Budd Company of Philadelphia, who pioneered mass production of stamped-steel automobile bodies in the mid-1910s. Unlike the Thing, the Kübelwagens had rounded, Beetle-like fenders. With the cloth top down, the vehicle itself looked a bit like a metal bucket. Among many design innovations, the Kübelwagen benefited from its extremely light weight at around 1,600 pounds ready for battle, plus its ZF limited-slip differential to compensate for lack of four-wheel drive, and reduced idle speed that allowed the vehicle to proceed at a soldier's walking pace.

In March 1942, Hitler declared a monopoly for the Kübelwagen as the only light military vehicle for the German army. In all, more than 50,000 were made during the war including variations such as those with tread drive or skis for winter use plus a steel wheel version for driving on railroad tracks. In *Professor Porsche's Wars*, author Karl Ludvigsen points out: *"Substantial though they were by German standards, these Beetle numbers were an order of magnitude smaller than America's production of 650,000 Jeeps for use in the Second World War."*

The four-wheel drive (Type 87) version of the Kübelwagen was adapted for amphibious use (Type 128). After experimenting

Kübelwagen in 1941 at the Porsche factory in Stuttgart.

PORSCHE POINTS

» A design for a human-piloted version of the V-1 was oddly named the *Reichenberg*, after the largest town near Dr. Porsche's birthplace. Although pilots might have had a chance to bail out, this was basically a suicide weapon and was deemed inappropriate under the German military tradition in March 1945.

» A competitor for Porsche's Type 128 amphibious vehicle was an Opel-powered design from Hans Trippel. Trippel took over management of the Bugatti works at Molsheim during the war and like Dr. Porsche, he was imprisoned by the French after the war. In the 1950s, Trippel was credited with the design of the gullwing doors for the Mercedes 300 SL and he also designed the famous Amphicar which debuted in 1961.

Porsches, Father and Son

Ferry Porsche's professional relationship with his father is interesting and revealing. Dr. Porsche favored rapid innovation and advances in technology. His son was more inclined to cautious, incremental improvement. There was also a great difference in Dr. Porsche's openness to his son's ideas depending on the setting. When others were present, the father's authoritarian or didactic streak was likely to take over as a way to 'save face'. In private, one-on-one discussions, Dr. Porsche was much more open to persuasion. In *We At Porsche*, Ferry wrote: *"On a long car trip, he would be more amenable and would listen to everything I had to say. Since I always drove, it was then possible for him to give me his entire attention; and he seemed genuinely interested in looking for the good points in what I told him. Still, this attitude was not possible even if only my wife or mother were present. The reason was simple enough. My father was a very authoritarian man. He could not stand any contradiction or argument before other people, even the closest to us in the family."*

Ferry Porsche's incremental approach would become a defining element of Porsche's manufacturing and engineering culture.

by welding the doors shut on a standard vehicle, amphibious bodies were then engineered and built specifically for flotation. Another variation was the Type 287 'Command Car', a four-wheel drive, hardtop Beetle. Its three-seat interior was set up to provide workspace for the most senior military commanders in the field. In 1943, all of the Volkswagen-based military vehicles benefited from a larger 1130cc engine with increased torque.

Schwimmwagen

The Type 166 was developed for the Waffen-SS based on an initial request for a motorcycle with a driven sidecar wheel that could "go anywhere" according to Ferry Porsche. Porsche's design was for a four-wheel drive, smaller chassis version of the amphibious Kübelwagen. The shorter wheelbase allowed for easier entry and exit from the water. The body (or hull) design was specific for water use and the engine-driven propeller was designed to fold up at the rear of the vehicle when driving on land. More than 14,000 Schwimmwagens were built between 1942 and 1944 when Allied bombing of the Volkswagen factory halted production. These vehicles were highly valued in the field by the German army and post-war analysis by the Allies was greatly complimentary of Porsche's design.

As for Hitler and his affection for the Volkswagen, author Karl Ludvigsen presented this prophetic, if ironic, 1942 quote from Hitler in *Professor Porsche's Wars*: *"…After the war, when all the modifications dictated by war experience have been incorporated into it, the Volkswagen will become the car par excellence for the whole of Europe… I should not be surprised to see the annual output reach anything from a million to a million and a half."*

The basic Volkswagen engine was adapted to numerous other uses during the war. These applications included electric generators, compressors, barrage-balloon winches, auxiliary power and tank engine starters, a pump for the V-2 rocket launch pad, powering small attack boats, and even powering a Horten 'Flying Wing' powered glider.

Tank Designs

Dr. Porsche's primary role during World War II was to be chairman of the Panzer Commission of the Armaments Ministry. In March 1940, he became Chief Engineer of the Nibelung tank factory in Austria. From mid-1941 until the end of 1943, Dr. Porsche had overall charge of tank design and procurement strategy. However, he and the Porsche design office were also directly involved in tank designs.

Schwimmwagen climbing out of the water, circa 1942.

> ### Porsche vs. Speer
>
> Fritz Todt had been Inspector General for German roadways, in charge of Autobahn construction, then became Reich Minister for Armaments and Munitions. He and Dr. Porsche got along well, and Dr. Porsche respected Todt's ability to grasp technical information. When Todt was killed in a mysterious plane crash, in February 1942, he was succeeded by Albert Speer. Speer was famous as 'Hitler's architect', conceiving grandiose building designs to project Nazi power. The relationship between Speer and Dr. Porsche was distant and strained for the rest of the war.
>
> On a simplistic level, Speer believed in the vitality and ideas of younger men at a time when Dr. Porsche was over 65. In addition to this ageism, Speer also tended to disagree with Porsche's inclination toward complex technological solutions when Germany needed to increase production. Relative to Speer's concerns, it can be noted that medium and heavy tank production increased by a factor of six from 1941 to 1944. Dr. Porsche was suspicious of Speer's architect background and ability to understand matters of engineering. Perhaps Speer was jealous of Dr. Porsche's relationship with Hitler. There was also some stressful wrangling over the terms of Porsche's relocation to Austria in 1944. Over time, Speer edged Dr. Porsche out of favor in spite of Hitler's ongoing enthusiasm for Porsche designs. At the end of 1943 Speer moved Dr. Porsche to the honorary post of *Sachverständig für Rüstungswesen* ('Armaments Expert for the Reich').
>
> After the war, Albert Speer was helpful to the Porsche family by confirming for Allied investigators that Dr. Porsche was probably the most unpolitical person Germany. Speer also confirmed that the Porsches never desired to join any Nazi organization (although they had joined). In *We At Porsche*, Ferry Porsche wrote: *"I certainly don't think he meant it as a form of praise, but merely wanted to state a fact which he knew to be true."*

Porsche's first proposal was the Type 100 (Type 101 engine) nicknamed the 'Leopard'. Powered by two 11.4-liter, gasoline-powered V10s, the design called for an electric motor drive. Dr. Porsche's hybrid concept reappeared as used successfully in World War I. A larger design, the Type 102 was nicknamed the 'Tiger'. It was placed in competition with a heavy tank design from established builders, Henschel & Sohn. There was a rather lengthy period of testing and political maneuvering. The Porsche design ran into difficulty with oil leaks and the shortage of copper needed in great quantities for the electric motors. Oil leaks were significant since the air-cooled engines derived a significant cooling effect from oil circulation. The Henschel design ultimately won the competition and went into production as the familiar Tiger I.

However, the Tiger project was not a complete loss for Porsche. Their design work and chassis already built, perhaps prematurely, for the Porsche version of the Tiger tank were repurposed for the 'Ferdinand' (later referred to as the *Elefant*). The Porsche Type 130 was conceived as a self-propelled anti-tank gun or tank destroyer (*Panzerjäger*). It was powered by two 11.9-liter Maybach V12s coupled to Siemens generators and electric drive motors. Unlike a regular tank, the turret was fixed in place but could accommodate a more powerful and precise, long-range cannon. The first combat duty for the Ferdinand was at the Battle of Kursk in July 1943. This was the largest tank battle in history, and although it became a major defeat and the last German offensive on the Eastern Front, the 88 Ferdinands performed reasonably well. The estimated 10 to 1 kill-to-loss ratio caused the Russians to overestimate the number of Ferdinands present in battle.

Porsche projects included a number of other tank, vehicle and engine proposals during the war including a competitor to Henschel's Tiger II (King Tiger), but Porsche's design again lost out. A most intriguing proposal was a twin-turbocharged, X16 diesel tank engine capable of 700 horsepower. Also interesting in a modern context was Josef Mickl's work on a wind turbine design for power generation (Type 136) that could be used by German settlers farming conquered Russian land.

The last major tank design from Porsche during the war was the *Maus*. The 'mouse' codename was meant to disguise its massive size. Another of Hitler's secret 'wonder' weapons, the Maus, was the largest tank ever built. Eight days before Pearl Harbor, Hitler brought up the topic of a super-heavy tank in discussions with Dr. Porsche. Although Dr. Porsche remained unconvinced of its tactical usefulness, the design evolved over the next two years and two running prototypes were eventually built by 1944. Production was hampered by Allied bombing of the Krupp factory during 1943 and the Maus was opposed by many officials given that its material and production requirements were equivalent to six or seven Tiger tanks.

1940s

Ferdinand tank being loaded onto a rail car, June 9, 1943.

The Porsche Type 205 had a potential battle weight of *207 tons* and required a specially built rail car for transport. Its armor plating was eight to ten inches thick and the tracks were almost fully enclosed. It was powered by either a gasoline or diesel vehicle version of the Daimler-Benz 603 aircraft engine (familiar to Porsche from use in the T80 record car). With Dr. Porsche's mixed drive, the 1,200-horsepower, inverted V12 powered two generators feeding two 1,200-horsepower electric drive motors. The Maus carried two cannons in its giant turret. One was a 128 mm gun that could fire 60-inch shells over a range of two miles. The second gun was a conventional 75 mm. The Maus required tracks of 43 inches in width and a track length more than twice as long as the vehicle width. This ratio violated German tank standards for maneuverability. However, the Maus eventually proved itself quite agile.

On its first running on Christmas 1943, the Maus achieved a seemingly excellent 26-foot turning circle. Upon hearing this,

PORSCHE POINTS

» In November 1941, the German military and Dr. Porsche evaluated a captured Russian T-34 tank. Dr. Porsche agreed with the suggestion that the Germans should simply copy the impressive, less complex T-34. This course was not pursued for a variety of technical, material and political reasons.

» The official name for the Type 130 was *Sturmgeschütz mit 8.8cm PaK 43/2 (Sd. Kfz. 184)*. It's no surprise that one of the Porsche engineers nicknamed the tank destroyer 'Ferdinand'. The name was approved by Hitler in 1943 and although officially renamed *Elefant* in February 1944, the new name was widely disregarded.

Maus tank shown testing, circa 1944.

Dr. Porsche dispatched electrical engineer Otto Zadnick (who was sick in bed) to investigate. After a small adjustment to the tracks, the Maus was able to literally turn in place, defying those who thought it would be impossible for the vehicle to go around corners. During the development and testing process, Dr. Porsche himself was not shy about driving the Maus and personally guided it to its intended top speed of 12 mph. Because the Maus was too heavy to cross most bridges, it was designed to wade or even submerge when crossing a river, using a power connection to another Maus and an air hose for the crew while submerged.

Only two completed Maus tanks were built during the war and there was only one turret. The critics of the project were right that it demanded resources far beyond Germany's capability, especially late in the war. Whether it was a practical battle weapon was never proven either way, but Dr. Porsche and his team deserved credit beyond just designing the Maus. Coordinating the highly complex completion of one of Hitler's impractical fantasies, while Germany was under constant air attack, was also a significant achievement. Captured by the Russians, the two Maus prototypes were consolidated into one complete example which now resides in the Kubinka Tank Museum.

Move to Austria

On May 10, 1943, the last of Ferry Porsche's children, Wolfgang Porsche, was born. This must have been a bright spot in an increasingly dismal time for the family. Stuttgart became an important target for Allied bombing in 1943 and 1944. During 1943, Albert Speer began to push for the relocation of Porsche's people and intellectual assets to a safer location. Dr. Porsche resisted at first, seeking to remain in close touch with his suppliers. However, as the bombing intensified, Ferry Porsche took on the role of searching for evacuation sites. He turned down one in Czechoslovakia, thinking the family would be unwelcome there post-war. Approaching the authorities in Salzburg on his own led to a fortunate offer of the glider flying school at Zell am See. This was adjacent to the family property, a substantial investment of more than 800 acres purchased by Dr. Porsche in 1939. Additionally, the family owned a large lakefront house and property near Klagenfurt, also purchased in 1939. Dr. Porsche wisely sought land over currency as a hedge against war for the family fortune. Both properties were suitable for housing a substantial number of people, including the Porsche and Piëch families plus Porsche employees as needed.

The third property was a sawmill at Karnerau, northwest of Gmünd, in the state of Carinthia. This location was difficult to access but reasonably safe from bombing due to its remoteness. As Allied air attacks on Stuttgart increased during the summer of 1944, the decision was made to move production for prototypes to Gmünd while the property at Zell am See would be used mainly for storage. After some significant construction at the Gmünd buildings, the designers' drawings and workspaces were also located there. A small number of new houses were built for the employees; some with materials moved from the Volkswagen factory location. The locations, being in Austria, were well suited to the Austrian engineers working for Porsche. The engineers preferred not to be in Germany at the end of the war. The workers, equipment and archives could preserve a nucleus of the Porsche company in Austria, while some of the German employees soldiered on in Stuttgart.

Dr. Porsche moved his base to Zell am See in January 1945, and by April, as the war came to an end, the Porsche and Piëch families gathered there. On May 13, Allied representatives arrived to interrogate Dr. Porsche, his son and other employees. Fortunately, the American representative, Major Franzen, had met Dr. Porsche at Chrysler in 1936. The British Colonel Reeves was an enthusiast well acquainted with the Auto Union racing cars. With complete cooperation, the Allies carefully investigated Porsche's designs and technical work during the war. On May 30, Dr. Porsche was formally detained and sent to Kransberg Castle. Code named 'Dustbin', it was a relatively friendly prison setting for many of Germany's most important finance, industrial and engineering figures (including Wernher von Braun) along with Albert Speer himself. Porsche continued to cooperate with Allied investigators and was allowed to return home to Austria on September 11, just after his 70th birthday. Luckily for Porsche, the families and the employees, Gmünd fell in the British occupation zone while Zell am See was in the American zone.

Unfortunately for Ferry Porsche, Anton Piëch and several of the other Porsche engineers, they were arrested on July 30, 1945. This was the result of a vengeful ex-criminal who had been held in a concentration camp but then talked his way into becoming the chief of police in Wolfsburg. When two unidentified dead bodies were discovered in the former Porsche hut at Wolfsburg, this individual convinced the American military police that the Porsche people were suspected of murder. This preposterous and completely unfounded charge caused Ferry Porsche and the others to be detained for three months in miserable prison conditions. Dr. Porsche was eventually able to mount a successful argument for their release.

Later in 1945, Porsche KG received approval from the British military government to complete design and testing work on a farm tractor. The gasoline and diesel versions became Porsche Types 312 and 313. Logically, the British commander, Major Andrews, determined that tractors would be critical to food production and restarting the regional economy. The meager circumstances for Porsche were also helped by doing repair work on VW-based vehicles which were badly needed by the occupation forces for transportation. Also in late 1945, discussions began with French authorities about a French version of the Volkswagen, to be manufactured with equipment taken from the KdF-Wagen factory as war reparations. Dr. Porsche was enthusiastic about the prospects, even though the representations were being made on behalf of the French Communist Party.

Circumstances turned dire when the French auto industry, led by Jean-Pierre Peugeot, became threatened by the notion of a socialized version of the people's car. Charges were filed accusing the Porsches of war crimes relating to the removal of equipment and materials from the Peugeot factory during the war as well as the deportation of French citizens forced to work in Germany. Dr. Porsche, Ferry Porsche and Anton Piëch were arrested by the French on December 16, 1945. Since Ferry Porsche had no role in the management of the KdF-Wagen factory during the war, he was released by the French in July 1946 after considerable additional intrigue.

In May 1946, Dr. Porsche and Anton Piëch were transferred to Paris. During the course of detention in Paris, Dr. Porsche was forced to consult with Renault on their design for the 4CV, meant to be an affordable and small rear-engine car for the post-war market. The circumstances were strained for obvious reasons and nothing very productive came from the nine months that Dr. Porsche spent in Paris. He did offer some design suggestions while at the same time pressing for a fair trial. In February 1947, Dr. Porsche and Anton Piëch were transferred to a prison in Dijon. The experience was miserable with the men in separate and unheated cells. Some people believed that the conditions of their imprisonment affected the health of Dr. Porsche and Anton Piëch, likely shortening their lives.

On May 31, 1947, in a hearing of sorts, official testimony from two Peugeot employees was in Dr. Porsche's favor, including statements that he *reversed* deportation of French workers. The former Gestapo chief responsible for the Peugeot factory also cleared Dr. Porsche of any wrongdoing. Further, testimony implicated Albert Speer's representative, Alfred Nauck, as the prime mover behind the removal of machinery from the Peugeot factory for use at the KdF-Werk. In the end, Jean-Pierre Peugeot did not testify, claiming illness. Although evidence

---

PORSCHE POINTS

›› The official report on Dr. Porsche's 1947 French war crimes hearing was sealed for 100 years. Only in 2047, when the documents are hopefully unsealed, will it be known exactly what happened.

›› The German word for 'Beetle' is *Käfer*.

›› The Cisitalia Grand Prix project involved a complex series of relationships, starting with correspondence from Carlo Abarth to Louise Piëch in 1946. Former Porsche employee, Rudolf Hruska, and Abarth, acting as Porsche's agents in Italy, recruited the great Tazio Nuvolari, who had driven for Auto Union in 1938. The chain of interest, and search for financing led to photographer Corrado Millanta, then to Count Giovanni Lurani and finally to the ambitious Piero Dusio. Millanta later represented Porsche in licensing synchromesh gearbox technology to Alfa Romeo.

›› Porsche friend and partner, Carlo (Karl) Abarth, married Anton Piëch's secretary as his first wife in the 1930s.

Porsche Type 360, Cisitalia Grand Prix racing car.

The first Porsche, photographed at Gmünd with a Beetle in the background. From the left, Erwin Komenda with Ferry Porsche and Ferdinand Porsche.

Porsche 356 Beutler Cabriolet and Gmünd coupe shown at the 1949 Geneva Salon. Ferry Porsche is third from the left, sister Louise Piëch second from the right. Bernhard Blank is in the middle.

panels (including panels that served as the cabin floors). The boxes were relatively easy to make and served as passages for control cables, wiring and support for the doors. Integration of body panels with the chassis made for a 'unit-body' construction familiar to most modern cars. Suspension and steering followed standard Volkswagen practice, but with improved shock absorbers and hydraulic brakes from Lockheed that appeared later in the run of cars produced in Austria. A further important decision was to mount the engine behind the rear wheels, again in standard Volkswagen form, as a way to create a bit of additional space for luggage or occasional rear seating.

The smooth and elegant body shape was styled by Erwin Komenda with guidance from Ferry Porsche and followed styling elements dating to the Volkswagen 60K10 as well as the original Type 60 Volkswagen. An important element for Ferry Porsche was the styling of the front fenders and visibility from the driver's seat. He wanted to assist the driver in knowing where the front wheels were positioned in a turn. This, along with the sloping hood line diving below the tops of the front fenders, remains a part of Porsche's design language to the present day. The same can be said of the sloping 'fastback' rear body section which makes a gentle and continuous curve from the roofline to the engine cover and down to the rear bumper. Aside from the deeply inset wheels, partially covered at the rear, the basic shape is recognizable to anyone with a passing familiarity with the modern 911.

For the first series of cars, made in Austria, the bodies were crafted in aluminum alloy. They were shaped and hammered by hand over a wooden 'buck' or form. The aluminum body and the small size of the car made for a featherweight package by modern standards. Performance was reasonably good even with its Volkswagen-based engine producing only 40 horsepower with dual carburetors. Dr. Eberan von Eberhorst assisted with work on valve drive and combustion chamber design in the effort to upgrade performance of the Volkswagen engine. Larger valves with dual springs and increased compression ratio were part of the program. In spite of the risk presented by generally poor-quality fuel available at the time Porsche elected to proceed with a modest increase in compression. The first coupe-style 356/2 was completed in June 1948 after the bare chassis had been tested extensively on the mountain roads near Gmünd. The first sales took place in the early part of 1949.

Switzerland played a key role in the start-up of Porsche sports car production. Finance in the form of an initial sales channel came through advertising man Rupprecht von Senger in Switzerland who represented an order for five cars. He had also worked with Porsche on an earlier sports car proposal (Type 352) that did not come to fruition. In reality, von Senger was financed by Zurich hotelier and garage owner Bernhard Blank (unbeknownst to Porsche). Once the deception was discovered, Blank took over from von Senger and showed two Porsches at the Geneva Auto Show in 1949. Finally, an import agreement was formed with AMAG, the Swiss Volkswagen representative, when Porsche production moved to Stuttgart. The Swiss relationships provided a source of hard currency which was desperately needed to acquire parts for 356 production. Prior to the implementation of the 1948 agreement with Volkswagen in Wolfsburg, the Swiss connection provided a vital lifeline for necessary components and materials. Some parts, such as spark plugs, were only available in Germany and had to be smuggled across the border into Austria. Currency, licensing, travel and Allied occupation business activity restrictions were all obstacles challenging the Porsche team during this period.

By the time production in Gmünd ended in 1951, Porsche had built just over fifty 356 coupes and cabriolets, with a small number of coupes eventually converted to SLs. The SL 'sport light' 356s were mainly used for competition. Barely breaking even on the sale of Gmünd 356s meant that series production

December 1949 photograph of Herbert Linge discussing a 356 cylinder head with Dr. Porsche in Stuttgart. Porsche was just beginning production processes at the Reutter plant.

of Porsche cars was far from a foregone conclusion at this point in the company's history. The fact that most of the sales were to royals or members of wealthy families was an encouraging sign relative to the appeal of the Porsche car.

During 1949 Ferry Porsche worked actively on a plan for the company to re-establish itself in Stuttgart. Although it was clear that Gmünd could never be a suitable site for volume production of Porsche cars, there was also a concern about offsetting the income in Germany that began to flow from the new Volkswagen contract as well as the rent resulting from US Army occupation of the factory building. Manufacturing in Germany could begin to mitigate the tax consequences of this income. The first step was setting up a small office in the family home, where Ferry's boyhood friend, Albert Prinzing, took charge of day-to-day business. The garage on Feuerbacher Way, where the first Volkswagen prototypes were built, housed the rebirth of Porsche in Germany. Ferry Porsche also began a relationship with the new mayor of Stuttgart, Dr. Arnulf Klett, who was to become an important Porsche supporter. They began the process of recovering the Porsche factory building, which was being used as a repair facility for the US Army motor pool. In the end, this process took several years (including a delay resulting from the Korean War), so Porsche had to explore other options for office and manufacturing space.

1948 photographs showing Ferry Porsche (in car) with nephews Ferdinand Piëch (left) and Michel Piëch (right). One of Ghislaine Kaes' sons in the middle.

1950s

1950	❯❯	Return to Stuttgart
		Porsche 356 Development
1951	❯❯	Le Mans 1951
		History of Le Mans
		Max Hoffman
1952	❯❯	International Business
		Anton and Louise Piëch
1953	❯❯	Porsche 550
		Ernst Fuhrmann
1954	❯❯	Porsche Speedster
1955	❯❯	Porsche 356A
		Porsche Tractors
1956	❯❯	Porsche 550A
1958	❯❯	Porsche 718
1959	❯❯	Porsche 356B

We want to produce an automobile which stands up to the most critical technical tests.
– Ferry Porsche

1950s

The 1950s was the most consequential decade for Porsche relative to the company as it is known today. The decade began with the relocation of design and production activities back to Stuttgart from Austria. In the early years of the decade, the company established a true, series produced vehicle which found success in the form of steadily increasing sales. Design work for Volkswagen added to the foundation of an ongoing business serving other clients as well. The decade also found Porsche becoming a successful constructor, seller and factory entrant of racing cars. By the end of the decade, Porsche was selling thousands of cars each year and exporting to numerous countries including the increasingly important American market. They had also regained their original 1930s production facility and expanded with a new building in the Zuffenhausen district of Stuttgart.

The return to Stuttgart was well underway in 1950. However, with the original Porsche factory still occupied by the US Army, alternative space had to be found for assembly of the 356. Porsche received proposals from three potential partners for production of 356 bodies and the chosen supplier, Reutter, was conveniently able to make a small, 5000 square foot area in one of their factory spaces available for an assembly line. Porsche had worked with Reutter before, on projects such as the production of Volkswagen prototypes, and Dr. Porsche had confidence in their workforce. The decision had been made to produce the 356 with steel for the body, rather than aluminum, mainly to reduce costs. Working in steel was a job that Reutter was prepared to handle and the initial order for 500 bodies was placed in 1949.

However, Reutter still had to satisfy Dr. Porsche. He arrived one day early in 1950 with Ferry Porsche and Erwin Komenda to accept delivery of the first Reutter-built 356 body. Although slowed by age and failing health, the great hands-on engineer still possessed a keen eye. This seminal moment was described by Ferry Porsche in *Cars Are My Life*: "*When we arrived, my father*

Ferdinand Porsche was still active in the Stuttgart drawing office, 1950.

Max Hoffman

Max Hoffman was the man most responsible for the post-war launch of European automobile brands in the United States. In addition to Porsche and Volkswagen, Hoffman also initiated the import of cars from Jaguar, Mercedes, BMW, Fiat, Lancia, Alfa Romeo and Facel Vega among others. By the mid-1950s, his Frank Lloyd Wright-designed showroom on Park Avenue in Manhattan was ground zero for the European car culture that continues today in America. He created the beginnings of dealer networks for several manufacturers.

Max Hoffman was born in Vienna in 1904, so was a contemporary of Ferry Porsche. His father was a manufacturer of sewing machines and later bicycles. Hoffman became a racing driver and dealer for Amilcar. He also imported American cars to Europe during the 1930s while representing other prestige brands in Austria. His Jewish ancestry caused him to move first to Paris and to emigrate to the US in June 1941. With the automobile business shut down during the war, Hoffman dealt in metalized plastic costume jewelry. He started importing European cars to the US after the war and became reacquainted with Ferry Porsche in 1950, around the time Hoffman imported the first Volkswagen Beetles to America. He brought the first Porsches to America that same year.

Hoffman not only commissioned Porsches like the Speedster, he also convinced Mercedes that wealthy Americans would love a car patterned after the 1952 Le Mans winner. The road version became the 300 SL 'Gullwing' although Hoffman himself preferred the smaller 190 SL. The BMW 507 was suggested by Hoffman as a competitor to the Mercedes 300 SL roadster. Hoffman was a fine art collector who took a 'fine art' approach to selling his cars. This was necessary when the price of a 356 was similar to a Cadillac convertible in 1951.

As time went on, the manufacturers bought Hoffman out of his distribution agreements and set up their own North American operations. Hoffman continued to build his art collection and passed away in 1981. A portion of the family wealth continues today in a Connecticut-based foundation that supports education, medicine and the arts.

Ferry Porsche with Max Hoffman in New York City, December 1951.

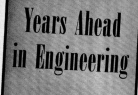

Blanche (the infamous 'white house' corner). The driver was injured and the car could not be repaired.

This left just a lone Porsche to start the race. It was driven by experienced Frenchmen Auguste Veuillet and Edmond Mouche. Porsche's luck improved and the car had a relatively trouble-free run to win the class and finish 20th overall in Porsche's first Le Mans race. The connection with Veuillet's company, Sonauto, would last for decades and numerous Le Mans entries. Ferry Porsche nervously hovered in the Porsche pit area for most of the race but the result quickly reinforced his belief in the advertising benefit of good results in racing. Unlike traditional advertising, competition also fostered direct engineering benefits for the cars and valuable experience for Porsche employees. Ferry Porsche's only regret was that his father did not live to see the success at Le Mans.

356 Developments

The 500th Porsche was built in March 1951. Although Porsche had yet to make their first Le Mans entry, and a 1.3-liter engine was nearing production, thoughts had turned to an even larger and more powerful option. During the summer of 1950, design work began on a 1.5-liter engine. 1500cc was a very popular racing class, so such an engine could be both a key to further competition success as well as a further upgrade for the street-going 356 line. Innovations from Porsche suppliers Mahle (aluminum cylinders) and Hirth (roller bearing, modular crankshafts) allowed for the larger capacity engines to be built within the limits of the existing Volkswagen architecture. The 1.5-liter engine achieved numerous competition and speed record successes for Porsche. By 1953, both the 1.3 and 1.5-liter engines were available in higher horsepower 'Super' versions making use of new valve timing and a special camshaft design from Porsche's rising star, Ernst Fuhrmann. A downside for

Ceremony for the 500th Porsche 356 built, at the Reutter factory in 1951.

Porsche was the relatively short service life and occasionally spectacular failures of the Hirth crankshaft. In April 1952, Porsche changed from the earlier split-window to a more conventional but slightly v-shaped, one-piece windshield.

The 356 itself was constantly being improved and developed in all of its mechanical and cosmetic features. The 1500 Super engine powered an important offshoot from the 356 Cabriolet, known as the America Roadster. Body builder Eric Heuer's firm was the fabricator for this small series of stripped-down, roadster-style aluminum body 356s. Heuer also built some standard Cabriolet bodies but they were unable to build the bodies to the required price. The cars were very expensive at $4,600 and only sixteen America Roadsters were built with eleven thought to survive. Intended for competition rather than daily use, several raced successfully in the United States.

RIGHT: Ferry Porsche with a group of 356s in 1954.

BOTTOM: Cutaway drawing shows details of the engine placement, drivetrain and chassis of the 356 (later body style).

PORSCHE POINTS
» The first Porsche club was founded in 1952 near Essen. The chapter is now known as Porsche Club Westfalen (aka Westphalia). There are now over 700 clubs around the world, in 86 countries and with nearly 250,000 members.

PORSCHE DECADES | 1950s

Porsche Crest

The Porsche crest as we know it today was spurred by the December 1951 order from Max Hoffman for 356s to be exported to the USA. Among the other specifications for the cars, Hoffman formally requested a Porsche-Stuttgart badge. The crest was created by engineer Franz Xaver Reimspiess working with Ferry Porsche and advertising manager Herrmann Lapper. Reimspiess is one of the men credited with designing the circular VW logo and returned to Porsche in 1951 after working for Steyr.

The search for a crest was prompted by Hoffman's visit to the 1950 Paris Salon. Hoffman knew that all great marques need to have a distinctive symbol. Work started in 1951 when Ferry Porsche's friend and customer, Dr. Domnick (a true Renaissance man and purchaser of the first 356 built in Stuttgart), organized a competition among art students from academies all over Germany. Prizes were paid for the top three entries, but Porsche elected not to use them. Since internal work was already underway, Ferry Porsche was famously able to sketch the basic idea on a restaurant napkin during his 1951 visit with Hoffman in New York.

The crest blends the coats-of-arms from the city, Stuttgart, and the state of Württemberg-Hohenzollern. Stuttgart's own black stallion harkens back to the city's horse breeding origins. The

black and red bands in quadrants opposite the antlers on gold are borrowed directly from the state emblem. The exact colors, shape and manufacture of the Porsche crest have evolved slightly over the decades, but remain true to the original sketch. The main differences from the drawing are the uniform gold background and the curved rather than straight top. The crest first appeared in October 1952 on the 356 steering wheel hub.

TOP: The first official drawing of the Porsche crest, completed by Franz Xaver Reimspiess.

LEFT: Porsche reproduction napkin showing Ferry Porsche's sketch of the Porsche crest.

They predicted the concept and success of the now legendary 356 Speedsters.

Overseas Assignments

Porsche's activities expanded rapidly in the early Fifties, on the production car line, in racing and in consulting and design. In addition to numerous proposals for new versions of the Volkswagen, none of which made it to production, Porsche also worked on potential projects for the Americas and even far-away India (a potential tank design that would have been built by the Tata company).

Ferry Porsche's visit to America in December 1951 was mainly to discuss the possibility of engine supply for a very light military vehicle that could be dropped with paratroops. The Mighty Might was designed by Mid-American Research of Kansas City. It eventually reached production, but not with Porsche power. Porsche also worked with the Fletcher company of California on a small jeep project for the US military as well as a helicopter engine adaptation for Gyrodyne on Long Island. Neither of

Christophorus magazine

The official publication of Porsche was founded in 1952. The name is a tribute to Saint Christopher, patron saint of travelers. The first editor was Richard von Frankenberg and the magazine is thought to be amongst the oldest automotive customer publications. It traditionally has covered a wide range of topics relating to Porsche, from new models to company history, motorsport to lifestyle and travel.

Circulation has grown to more than 500,000 around the world and in eleven languages. Since 2013, content has been available online in addition to the print version. Total editions have now exceeded 400.

Huschke von Hanstein

Before World War II, Fritz Huschke von Hanstein was a Prussian baron who had success in motorcycle and car racing as well as hill climbs. In 1940, he won Italy's great road race, the Mille Miglia, with Walter Bäumer in a BMW 328 (that race being held on a short, 104-mile course lapped nine times rather than the traditional 1,000-mile Brescia-Rome-Brescia route). As a result of the war, he lost his family land and fortune in East Germany, including the castle that had been in the family for over a thousand years.

Von Hanstein worked for Vespa before joining Porsche in 1951 where he would stay until 1974. He finished second for Porsche as a driver in the 1951 Liège-Rome-Liège Rally and still had success as a driver as late as 1958. Along with Richard von Frankenberg, he staffed the early Porsche marketing department, handling customer service, PR, press relations and advertising. Von Hanstein also served as Porsche's racing manager until 1968, overseeing the factory's entries in all forms of competition. In addition to recruiting many excellent drivers, he became known for his diplomacy and skill in dealing with customers and race officials. 'Working the refs' at the CSI (sporting commission within the FIA) was an important part of the job. The 'Racing Baron' was an ideal and enthusiastic promoter for Porsche on the global stage of motorsport and in place of the publicity-shy Ferry Porsche. After leaving Porsche, he led the ONS, the auto sporting authority in Germany.

Huschke von Hanstein in racing helmet and gloves, 1955.

these projects got beyond the prototype stage, but they showed the potential reach of Porsche and the air-cooled four-cylinder engine.

The most important client project for Porsche in America came in the form of design work for Studebaker. Thanks to the Max Hoffman connection, in 1952 Ferry Porsche and a small team visited South Bend, Indiana and signed a contract to design a car for the US market. The eventual designs included both air-cooled and partially liquid-cooled V6 engines. Construction was specified as modern-style unibody. Ferry Porsche also proposed a more compact, rear-engine car that coincided favorably with Studebaker market research. The smaller car proposal (Type 633) in some ways predicted the Chevrolet Corvair. Given Studebaker's limited resources, the only completed prototype was for the conventional-size car (Type 542). The prototype built in Stuttgart and tested by Porsche was shipped to America in 1954. With Studebaker's fortunes declining further, they were forced to merge with Packard and the Porsche relationship ended.

Porsche had business prospects in Argentina, initially resulting from the Dusio/Cisitalia relationship. The first project was a design (Type 372) in 1949 for a car that would be built in

Anton Piëch

Dr. Porsche's son-in-law was a lawyer and the son of a lawyer. He was a critical presence, mostly behind the scenes, looking after the Porsche family's finance and legal affairs. He also managed the Volkswagen factory for much of World War II (with attendant controversy over the use of forced labor). His eldest son, Ernst, eventually took an active role in the management of Porsche's Austrian arm and his second son Ferdinand Piëch played a critical engineering role at Porsche in the 1960s.

In August 1952, as Ferry and Dorothea Porsche were about to embark on a voyage to New York, they noticed that Anton was not looking well and he complained of breathing problems and fatigue. He made an odd remark to Dorothea, saying: "If I go quickly, that will be best for me." Two days later, he died from a heart attack when the Porsches were already onboard the Queen Elizabeth. It is possible that, like Dr. Porsche, the conditions of his imprisonment by the French had impacted the elder Piëch's health. The ever-strong and resilient Louise Piëch was left to finish raising their four children and manage her side of the family's growing business interests.

Louise Porsche-Piëch

Louise Porsche was almost five years older than her brother, Ferry. As far as personality is concerned, she seemed to take after her famous father. She was intelligent, resourceful, hard-working and an aggressive, talented driver. She could also be fiercely determined, autocratic and quick-tempered. In the immediate post-war period, Louise had to hold the family and the business together when her husband, father and brother were all detained by Allied authorities. She worked with Karl Rabe to keep some form of business going at Gmünd, taking on work such as repairing Volkswagens and other vehicles. The use of her nationality as owner of Porsche's Austrian company prevented nationalization of Porsche as a German entity in Austria during the immediate post-war period.

It is notable that Dr. Porsche divided his estate equally between his two children. Louise went on to build Porsche-Salzburg as a very successful automobile distributor, first in Austria and then expanded to other countries. By 2000, her company had reached $25 billion in sales over time. Louise was quoted stating proudly that she had never driven cars that were not made by her father, brother or son. Porsche stylist Tony Lapine stated that Ferry would not make an important decision without consulting his sister. Always a force behind the scenes at Porsche, Louise outlived her brother by about a year, passing away in 1999.

Argentina by Autoar. The car would have been akin to a lengthened, roomier 356 with a 2.0-liter, rear-mounted flat-six engine. The engine was designed to sit on top of a complex transaxle featuring a version of the Porsche synchromesh system. The 372 was never built, possibly due to the significant start-up cost for production. Autoar instead began building Fiats from pre-assembled components in 1950.

Three subsequent cars were built in Argentina using Porsche power. Engines imported for industrial use were employed in the front-engine sports car built by IAME, the government-backed Aeronautical and Mechanical Industry. The Justicialista Sport was named for the ruling Party and featured an elegant fiberglass body. Only 167 were built before the Perón regime was overthrown. The Porsche engine importer, Teram, followed up with a rear-engine sports car called the Porsche-Teram Puntero. Looking like a blend of the 356 with a Volkswagen Karmann Ghia, it was also built in fiberglass and used a full range of

Porsche components. Finally, in 1960 the rather awkward and outlandish Zunder 1500 was a rear-engine fiberglass sedan also using Porsche's 1.5-liter engine and Porsche-built suspension. It was assembled by the Bongiovanni brothers who were Chevrolet dealers. Proving to be another dead end for Porsche, only about 200 cars were built.

Ignition Switch Placement

Why are Porsche ignition switches on the left? Conventional wisdom has always said it was meant to facilitate the run-and-jump 'Le Mans' start, where the driver could use his left hand to start the engine while the right hand could engage first gear. This is supported by the outboard position of the key on right-hand drive Porsche street cars. However, numerous Porsche racing cars of the 1950s and 1960s had their ignition switches inboard of the steering wheel. When Porsche shifted to right-hand drive for prototype racers (Type 907), the key was to the left with the gear lever on the right of the steering wheel.

Dan Neal's 2019 *Wall Street Journal* column quotes former Porsche Museum Director, Klaus Bischoff, saying that the outboard position of the ignition switch on the very first 356 *"saved a little bit of wire"* at a time when materials were scarce. The Gmünd-built 356 coupes then had the ignition switch in the center of the dashboard. Over the years the outboard position has become a tradition reinforced by ergonomic logic.

As early as 1951, Porsche knew they needed larger space in order to increase production. They began searching for a site to build on since the original Porsche factory as still occupied by the US military. Luckily, a site was available in the Zuffenhausen neighborhood of Stuttgart, adjacent to the Reutter site housing the 356 production line. Financing a new building for production and engineering space proved to be a challenge until the fortunate and timely relationship began with Studebaker. Design income from the American company became a major source of finance for the new building, known as Werk II. In addition to having their own factory, the product line of 1953 356s started to have a level of stability in terms of specifications, production details and options. The most significant missing link for Porsche was in the transmission department where the 356 still relied on the non-synchronized Volkswagen transaxle. In addition to its challenging operation, the Beetle 'crash box' was approaching the limit of horsepower and torque it could handle.

Luckily, Porsche had an in-house solution in the form of their own synchronizer design. A new transaxle featuring the system (which was rejected by Volkswagen in favor of the BorgWarner-based system from Opel), was tested extensively during 1952. Leopold Schmid's interlocking rings allowed for automated matching of the shaft speeds within the gearbox when changing up or down through the gears. This allowed for smooth meshing of any gear at a wide range of engine speeds. A side benefit of the new transaxle was that Porsche could select their own gear ratios specific to the 356 rather than make do with standard Beetle ratios. The manufacture of the new gearbox began Porsche's

PORSCHE POINTS

» For testing the new synchromesh gearbox, Porsche used the first Stuttgart-built 356, *Windhund*, as well as *Ferdinand*, the car that had been given to Dr. Porsche for his 75th birthday. Ferdinand eventually covered nearly 400,000 kilometers testing things like rack and pinion steering, radial tires and the first installation of a four-cam Carrera engine in a 356.

» In 1954, Heinz Nordhoff offered Ferry Porsche a job at Volkswagen, to oversee engineering. The offer stipulated that Porsche would have to close its design office and stop building cars. Luckily for all of today's Porsche owners, Ferry Porsche chose to stay independent.

» Hans Miersch was an East German shoemaker who wanted a Porsche. With design help from the young Reimann brothers and coachbuilder Arno Lindner, Miersch created an East German quasi-356 in 1954. The car used a Kübelwagen chassis and only had a 30 horsepower engine at first. It had a hard time powering the oversized, heavyweight four-seater. Porsche helped by providing some parts (which had to be smuggled into East Germany). Engineers Falk and Knut Reimann eventually built a similar car for themselves. Lindner built a small series of 12 to 14 replicas of the Miersch 356. The original 'Miersch' eventually received engine upgrades for more adequate power. Hans and his car survived the cold war and the car is now owned by Porsche collector Michael Dünninger.

relationship with *Getriebe und Zahnrad Fabrik GmbH* commonly known as Getrag. Setting up the quality control testing process for the new gearbox was entrusted to young, newly hired engineer, Helmuth Bott. Porsche synchromesh was not only a great improvement for the 356, but licensing the technology to other automakers proved to be highly lucrative for Porsche. The patents led to some direct design work as well, quietly employed by companies like Ferrari, Alfa Romeo, BMW and even Porsche's neighbor, Mercedes-Benz, for their Mille Miglia-winning 300 SLR in 1956. Auto giants like Toyota, Nissan and Chrysler also eventually licensed the Porsche transmission technology.

For the 1953 model year 356s, the new gearbox was combined with larger but lighter brakes, resulting in a much-improved car, especially with 1.5-liter engine options.

Porsche 550

The Type 550 was Porsche's first purpose-built racing car and set the basic pattern for their mid-engine competition cars over the next decade. It was preceded by a number of Volkswagen and Porsche-based 'specials'. These were racing cars with open-cockpit 'spyder' bodywork created as one-off experiments by their owners. Men like Petermax Müller and especially Frankfurt Volkswagen and Porsche dealer, Walter Glöckler, had significant successes with their specials in the late 1940s and early 1950s. The 550 was built with a lightweight steel ladder-style frame under sleek aluminum bodywork. For the all-important 24 Hours of Le Mans, a few aerodynamic 550 coupes were realized by adding removable hardtops. This included the first two 550s built in 1953. A radical new race engine (Type 547) was designed in parallel with the 550 but was not ready for the biggest competitions in the first year. The 1500 Super pushrod engine in race tune was used at first. In typically exhaustive Porsche development style, the 550s were built with an array of different two-seat body configurations, suspension designs and basic chassis layouts.

The heart of the 550 became the legendary Type 547, a four-camshaft, four-cylinder engine mainly designed by Ernst Fuhrmann. Limited to racing at first, Ferry Porsche wrote that the engine was always intended to eventually make its way into a higher-performance version of the 356. Although it retained the familiar horizontally-opposed, air-cooled cylinder configuration, it shared almost nothing with standard 356 engines. Its twin overhead camshafts were driven by a complex

LEFT: 356s posed in front of Werk II.

Ernst Fuhrmann (left) with Bruno Trostmann, testing the Type 547, four-cam engine.

set of bevel gears and shafts running from the crankshaft (rather than pushrods, belts or chains connecting to the valve train). The crankshaft itself was a unique design built by Hirth, using that company's patented interlocking bevel-style joints and roller bearings. The engine had Porsche's first twin spark plug cylinder head configuration and using twin two-barrel carburetors from Solex or Weber, the first engines made 110 horsepower from 1.5 liters.

Early results for the 550 were encouraging. In only their second race, at Le Mans in 1953, two 550s with coupe hardtops finished well. Richard von Frankenberg and Paul Frère won the 1.5-liter class, finishing 15th overall in Chassis 02. Newcomer to Porsche, Hans Herrmann, with Walter Glöckler's son, Helm, finished 16th in Chassis 01. These same two 550s were entered in the Carrera Panamericana, the great Mexican road race, in November 1953. *Carrera* is a Spanish word for 'race'. Run over five days and 1900 miles of the Pan-American Highway from the south to the north of Mexico, this was a stern test of both drivers and their machines. The 1953 contest was marred by several fatalities among competitors and spectators. The

550 coupe prepared for Le Mans, 1953.

PORSCHE POINTS ▸ In 1952, VW 60K10 owner Otto Mathé built a simple Porsche/VW-based open-wheel, single-seat racer nicknamed *Fetzenflieger*, or 'shreds flier'. It was named for small, flaming shreds from the fabric carburetor covers that the car would occasionally leave behind. Successful on dirt and especially in frozen lake ice racing, Mathé won numerous Professor Porsche Memorial Races on the frozen lake at Zell am See. The car pre-dated Porsche's own open-wheel racers.

Ernst Fuhrmann

Fuhrmann joined Porsche in Gmünd in 1947 and became a protégé of theoretician, Josef Mickl. In 1950, he earned his doctorate for a study of valve trains in high-speed combustion engines. He worked on the Cisitalia Grand Prix engine as well as the first higher-performance camshaft for the 356. During his initial period of employment with Porsche, he was best known for the Type 547 engine. The famous four-cam became known as the 'Fuhrmann Engine'. In 1956, when Klaus von Rücker was named Technical Director, Fuhrmann left Porsche to work for the Goetze engine parts company.

In 1971, he returned to Porsche as Technical Director and became CEO of the reorganized Porsche AG in 1972. This was the point at which Porsche family members removed themselves from day-to-day management of the company, so Fuhrmann was the first non-family member to lead the company. In *Excellence Was Expected*, author Karl Ludvigsen shared this revealing Fuhrmann quote on Porsche's dedication to racing: *"It seems to me that racing is particularly well suited to motivate young engineers. If you design a production car, you know in five or six years what you've done. In racing, you know in a year. Thus young engineers can gather a lot of information very quickly. At 30 years of age an engineer is already very experienced."*

Fuhrmann's tenure from 1972 to 1980 is known for the development of front-engine, liquid-cooled transaxle cars as well as turbocharging for racing cars and the 911. His perceived lack of support for the 911 was controversial both internally and with Porsche's customers. In 1980, he retired to a professorship at Vienna Technical University. Fuhrmann passed away in 1995 at age 77, not long before the last 928s were built.

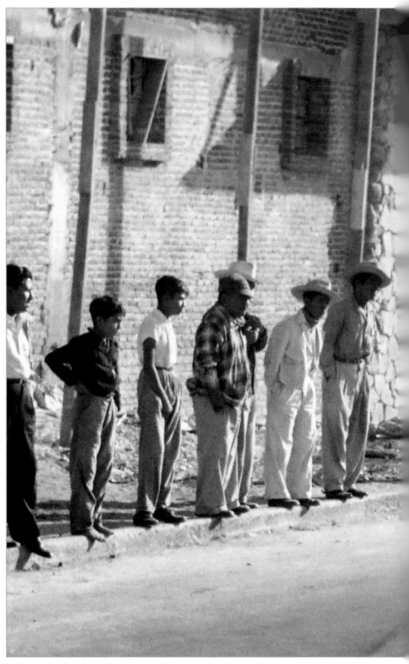

550 spyder in the 1954 Carrera Panamericana road race.

Porsche entry was organized by the Guatemalan team of Czech expatriate Jaroslav Juhan. His car failed to finish, but the second 550, driven by José Herrarte, won the 1.6-liter class as the only finisher (32nd overall).

Another famous result for the 550 came in Italy at the 1954 Mille Miglia. Using the new four-cam engine with Weber carburetors, Hans Herrmann and navigator Herbert Linge, came upon a railroad crossing with the gate down and a train approaching. However, with no ability to stop in time, Herrmann made a split-second decision to tap Linge (who was engrossed in his 'prayer book, or pace notes) on the helmet and duck under the gate! The car was low enough to make it through without a scratch and just ahead of the oncoming train. After turning north from Rome back toward Brescia, the engine was drenched in a rainstorm. However, Linge was a resourceful mechanic in addition to being a fine racing and test driver in his own right. He managed to dry out the distributors and get the engine going again. After a grueling 12 hours, Herrmann and Linge finished sixth overall and won the 1.5-liter class for Porsche.

Zora Arkus-Duntov made an appearance in the Porsche story

during this period. The Belgian-born GM engineer is known as the guiding force behind the Corvette becoming a serious sports car. He served as a driver for Porsche at Le Mans twice, in 1954 and 1955, winning the 1.1-liter class both years in 550 spyders equipped with Porsche's small-bore version of the four-cam race engine. At this stage, the 550s were tough to handle for drivers not vastly experienced with Porsche's tail-happy steering response. Based on Duntov's suggestion from his GM experience, the young Helmuth Bott began using a circular skid pad to test changes and developments to the suspension design. Continuous running on the skid pad allowed for direct comparison data as changes were made to the car. The first location for this testing was a section of runway at the Malmsheim airfield. Duntov also suggested installing a front anti-roll bar. These developments led to a *30 second* lap time improvement at the Nürburgring by July 1954, when the Porsches decimated the field in a race for 1500cc sports cars on the morning of the German Grand Prix.

The 1954 Carrera Panamericana created a legacy that lives on in Porsche lore. Hans Herrmann, who was struggling with a fever the day before the race, drove solo and by sight alone, meaning

James Dean

James Dean in his 550 spyder on September 30, 1955. (Alamy)

The American actor James Dean only starred in three Hollywood films, two of which were released after his death. He was the first posthumous winner of an Academy Award for Best Actor. His legacy as a cultural and artistic influence was perhaps enhanced by his passing at a young age and before he achieved massive fame. Like many of his acting contemporaries, Dean developed an interest in sports cars and racing. He entered his first races in a Speedster, which he then traded for a 550 spyder (nicknamed 'Little Bastard') in September 1955. On September 30, 1955, he was driving the spyder on Route 466 near Paso Robles, heading to Salinas for a race meeting. Dean's passenger was Rolf Wütherich, who was serving as a field mechanic for Porsche's US west coast representative, John von Neumann. They collided with a car that was crossing the road and making a left turn on to Highway 41. Dean could not stop in time nor avoid the oncoming Ford. Wütherich was thrown from the car, but Dean was trapped inside, suffering grievous injuries. He was pronounced dead on arrival at the hospital in Paso Robles, California.

Legends surround Dean and his 550. The wreckage of the chassis/body (550-0055) was eventually sent on a tour of the US to promote highway safety but it supposedly disappeared during transport from Florida to California in 1960. Although the histories of the transaxle (sold in 2021 on Bring a Trailer) and the engine can be traced, the whereabouts of the wrecked chassis remains an unsolved Porsche mystery.

PORSCHE POINTS

» One explanation for James Dean's 550 spyder nickname, 'Little Bastard', was that legendary stunt driver Bill Hickman had called Dean himself 'little bastard' while Dean referred to Hickman as 'big bastard'. Hickman was trailing the Porsche in Dean's Ford station wagon when Dean was killed on the way to the Salinas Road Races. Hickman said he held Dean until he could be freed from the wreckage, but that Dean died in his arms. Hickman is best known for his later work both driving and acting in the classic cop movies *Bullitt*, *The French Connection* and *The Seven Ups*.

» According to Herbert Linge, the first direct suggestion for the use of 'Carrera' as a car name came from one of the Porsche drivers who participated in the Mexican road race, Prince Paul Alfons Fürst von Metternich-Winneburg, who later became president of the FIA. Linge drove a 356 with Metternich and Manuel de Teffé in the 1952 Carrera Panamericana.

without navigation or pace assistance from a co-driver. He was slowed by two punctures on the first day. To make up for lost time, Herrmann had to drive on or beyond the limit the rest of the way. In the end, he finished third overall only 87 minutes behind two 4.9-liter Ferraris driven by Umberto Maglioli and Phil Hill with Richie Ginther. Hans won the 1.5-liter class and Jaroslav Juhan finished fourth overall in his 550 spyder only 36 seconds behind Herrmann. As in 1953, there were numerous accidents and fatalities, so 1954 was the last of the five Carreras. In honor of the 550's achievements, the 'Carrera' name would live on at Porsche as a model designation that continues to the present.

The early successes of the 550 series came from cars that were essentially prototypes from the Porsche factory, constantly being changed and updated even though they were eventually sold to private owners. After approximately 15 chassis, series-production 550 spyders became more standardized and strengthened for customer racing duty. Production of the bodies shifted from Weidenhausen (who had built the Glöckler Spyder) to Wendler (located closer to Stuttgart). The 550 went on to bring Porsche racing successes all around the world.

Porsche Speedster

Many Porsche developments were inspired or requested directly by Max Hoffman and the important export market consisting of American customers. Having a less expensive Porsche for the American market was a priority for Hoffman and the

Porsche Club of America

Founded in 1955 by Bill Sholar and a small group of enthusiastic Porsche owners, PCA has since grown to become the largest single-marque car club in the world with total membership in excess of 100,000 households. In the early years, members shared vital technical and tuning information on their cars in addition to social comradery.

The Club has become known for its signature annual event, Porsche Parade, which has been enthusiastically supported by Porsche and the Porsche family over the years. The monthly magazine, *Panorama*, was famously designed to fit in the glove box of a 356. It grew to full size in July, 2013 and continues to serve the members alongside a growing list of electronic media offerings. Porsche Club regions put on thousands of driving, technical and social events around North America each year. The unofficial motto is 'Come for the cars, stay for the people'.

PCA founder Bill Sholar. (PCA)

PORSCHE POINTS

» In 1954, the Speedster was the first Porsche to wear the crest on its hood (trunk handle). This was appropriate since Max Hoffman had prompted Porsche to create both the crest and the Speedster itself.

» Also in 1954, on March 15, the 5,000th Porsche 356 was completed.

» On August 5, 1955 the 1,000,000th Volkswagen was produced. Heinz Nordhoff was made the first Honor Citizen of the City of Wolfsburg.

356 Speedster was an effective effort by Porsche to satisfy the request. At the time, about 75% of Porsches were exported and 40% of exports went to America. Ironically, but perhaps not surprisingly, the Speedsters are generally the most valuable of 356s to modern collectors (setting aside the Gmünd-built cars and rare, special models like those with the four-cam engine). Perhaps it is the purity and elemental quality of the Speedster that appeals to 21st century owners.

In order to reduce the price, Porsche designed a roadster body simpler than the earlier America Roadster. What would normally be the rear seat area was covered by body panels, making the cockpit shorter and giving the car a hunchback look. The Speedster was stripped of unnecessary equipment and given a low, easy-to-use convertible top with no side windows (only detachable side curtains) for weather protection. Lack of weather protection was less of an issue in the rapidly growing Southern California market serviced by west coast dealer, John von Neumann. The distinctive cut-down windshield gave it a sportier look than a standard cabriolet. The dashboard was simplified but had a partly-shrouded, purposeful-looking binnacle containing the instruments. The car was given racy and supportive lightweight bucket seats. The Speedster was initially offered with two versions of the 1.5-liter engine. For

Poopers?

In 1955, Pete Lovely of Seattle Sports Cars built an English Cooper sports racer with a Porsche 1500 Super engine. Like Porsche, John Cooper was a pioneer with mid-engine racing cars. Lovely started with a streamlined Mark VIIIR that had been built for record-breaking. He won an important race at Torrey Pines, California in July 1955. Superior power-to-weight ratio gave the Porsche-engined Cooper an advantage against Porsche's own 550. At least three Porsche-powered Coopers were built including Ken Miles' R3 with a four-cam Type 547 engine, built for John von Neumann. 'Pooper' success came in the Sports Car Club of America F-Modified class as well as some unlimited races and hill climbs. East Coast racer 'Tippy' Lipe won a race at Cumberland, Maryland with his Pooper defeating Porsche spyders, including one driven by Herbert Linge.

According to author Karl Ludvigsen, Lovely said: *"We never called it the Pooper. I think that came from a Road & Track article."*

Two of the three 'Poopers' appear together in this 2007 photograph. In front is the ex-Pete Lovely car, behind is the ex-Tippy Lipe racer. (Allan Caldwell)

356 Speedster circa 1955.

better acceleration and track performance, Speedsters came with shorter 3rd and 4th gear ratios as standard. To add a bit of extra style, the front trunk handle was longer, and for the first time the Porsche crest appeared on the front hood.

Introduced in September 1954, Max Hoffman was able to price the Speedster just under $3,000, listing the tachometer and heater as 'options' even though all cars had them.

Because of its simplicity and light weight, it is no surprise that the Speedster became popular for racing, especially in the United States. In 1955, a Speedster won the Sports Car Club of America championship in the F Production class. Soon after the Speedster's introduction, all Porsches received new engines with numerous improvements and proprietary changes that took the engine further from its Volkswagen roots.

When Porsche introduced the 356A in 1955, the Speedster continued with the improvements common to 356s of that generation. In a further bit of irony, Speedsters never sold in the volumes that Hoffman expected and Porsche remained somewhat confounded by the preferences of American buyers. 4,145 Speedsters were built over six model years.

A last special version of the early 356 was the 'Continental', again suggested by Max Hoffman. These were highly-optioned, 1.5-liter 1955 coupes and cabriolets built toward the end of what are today referred to as the 'Pre-A' models. Not surprisingly, Ford objected to the use of the name made famous by the Lincoln Continental (which was planned for re-launch in 1956). Later 356 Continentals were renamed 'European'.

Porsche 356A

The 356A was introduced in 1955 for the 1956 model year. It was different enough in all the many developments to be given both the 'A' designation as well as the internal 'T1' project identifier. The second-generation 356As were developed under the T2 project. One of the main features of the 356A was an increase in engine capacity which allowed for 1.6-liter 'Normal' and 'Super' options. These 60- and 75-horsepower versions joined the existing 1.5-liter engines in the lineup. As usual, the engines were improved in many areas, focusing on smoother running and durability. A significant change was the addition of warm air ducting to the carburetor to improve performance during the engine's warm-up phase. Thanks to Porsche's skid pad work, improvements were made to springs, suspension arms, front anti-roll bar, steering and the wheel/tire dimensions. Overall handling and ride qualities improved. Numerous changes were made to the interior and dashboard, and the 356A also received a fully curved windshield in place of bent glass.

In what has become a time-honored tradition, tenured Porsche owners could grouse about the 'improvements' to the 356. The changes to the suspension made for a car that 'anyone' could drive quickly. This along with the earlier addition of a synchronized transmission and a more civilized overall package meant that Porsche aficionados might find themselves belonging to a less exclusive club.

On December 1, 1955 Porsche finally celebrated the return of the Werk I building from the US Army, and it happened to fall exactly on the 25th anniversary of Porsche's design office moving into its first office space in Stuttgart (a date that is sometimes used for the founding of the company). This allowed for design, management, customer repair and the experimental/racing department to move to Werk I, creating more space in Werk II for increased production capacity. This was badly needed given the demand for ever-improving Porsche cars. On March

---

PORSCHE POINTS

» The Type 597 Jagdwagen 'hunting car' was a modernized, four-wheel drive version of the Kübelwagen. Designed in response to a request for German military police vehicles, Porsche built prototypes at its own cost. Although a superb performer in testing, Porsche lost out on the contract to DKW (a forerunner of today's Audi), likely due to political considerations as well as the much higher cost of the Porsche design. Most of the 79 Hunters were sold for civilian use and Porsche was able to gain some compensation from the German government for their investment in the design and the prototype construction. Still, Ferry Porsche vowed never again to invest company resources in this way on a military project.

» Dr. Porsche's grandson, Ernst Piëch, married Elisabeth Nordhoff, daughter of Volkswagen leader Heinz Nordhoff, in 1959. Ernst worked for Allgaier when they produced Porsche-designed tractors in the early 1950s before moving to work at Porsche-Salzburg. 'Porsche-Salzburg' has generally been used to refer to Louise Piëch's company, Porsche Konstruktionen, which started in business with distribution rights for VW and Porsche in Austria.

	356 Pre-A 1950-1954	**356 A 1955-1959**
ENGINES	1.1, 1.3, 1.5-liter, horsepower range from 40 to 70 (DIN)	1.3, 1.5, 1.6-liter engines, 44 to 75 horsepower, plus 4-cam Carrera engine introduced
CHASSIS	Front anti-roll bar added late in 1954	Suspension developed to counter oversteer
STYLING	Original (Stuttgart) form with split or bent windshield, Speedster introduced in 1954	Conventional windshield, evolving bumpers, lights and grills, hardtop option added
OTHER	Change to synchromesh transmission in mid-1952	Improved transmission introduced late in 1958, smaller but wider wheels

16, 1956, the 10,000th Porsche was completed. 12-year-old Wolfgang Porsche drove it off the production line.

Another important development in 1955 was the introduction of the 356 *Carrera*. Mainly intended for racing, but certainly useful on the street as well, the Carrera was the marriage of the 356A with Fuhrmann's four-cam engine. The concept had an early and successful test in the 1954 Liège-Rome-Liège Rally, a grueling 3,000-mile contest. Porsche won that event with a Gmünd-built 356 coupe powered by the four-cam engine from a 550. For production sales in the 356, the engine was given a lower compression ratio and an oil tank (the engine being a dry sump design where oil is stored separately from the engine case). Carreras were also given wider tires along with dashboard switches for each ignition system and a model-specific speedometer and tachometer. The cost of the Carrera was 20 to 25% more than a standard car, but the race-bred engine could power a 356 near the 200 kph (124 mph) mark. The four-cam engine option would carry on in various forms until 1965, but it acquired more model-specific options as time went on (such as aluminum doors and deck lids). Carrera coupes and Speedsters brought Porsche a basketful of race wins and championships.

In 1957, Porsche implemented the next series of changes to the 356 with a new transaxle case and improvements to the pushrod engines. Interior and exterior appointments were upgraded as usual. In September, the 'T2' version of the 356 made its debut. The 1.3-liter engine was discontinued and the larger engines received major changes with new cylinders (iron replacing aluminum in the Normal engines), a new crankshaft design (in the Super engines) and a new oil cooling system (oil cooler bypassed when the engine is cold). Porsche changed from Solex to Zenith carburetors and made still more major changes with a new clutch and shift linkage, new steering box (another step away from Volkswagen origins), new exterior hardware and improved seats front and back.

356A on the cover of the January 1959 issue of *Christophorus*.

In 1958, the Convertible 'D' (for body builder Drauz) replaced the Speedster and was later named the 356 'Roadster'. The D was more like a 356 Cabriolet, especially with its taller windshield, but carried over some Speedster elements such as the dashboard.

Porsche 550A

The second major version of Porsche's racing spyder was particularly significant. The redesigned chassis was a multi-tube cage-like space frame (as opposed to a ladder-like structure). The tubular space frame would carry on in most of Porsche's pure racing cars from this point until 1981. In the 550A, the new

Porsche Tractors

Porsche-branded tractors are always crowd-pleasers at Porsche events. The Rennsport Reunions in 2018 and 2023 both included short races for the tractors.

The first realization of Porsche's long-held tractor ambitions came in 1950 when Allgaier began producing the two-cylinder AP 17 (for Allgaier-Porsche 17-horsepower). Hofherr Schrantz in Austria also produced tractors under license. Allgaier went on to produce 25,000 tractors before the business was sold to Mannesmann in 1956. Mannesmann set up manufacturing at the converted former Zeppelin plant near Friedrichshafen. Key Porsche manager, Albert Prinzing, moved to work on the new Porsche tractor venture. Four basic models were produced, all with diesel engines. The single-cylinder Junior produced 14 horsepower, two-cylinder Standard at 25 horsepower, three-cylinder Super at 35 horsepower and the four-cylinder Master produced 50 horsepower. All were air-cooled with cylinders in line and benefited from maximum interchangeability of parts. Production of Porsche-branded tractors ended in 1963, with Mannesmann having built 125,000 units.

Porsche Standard tractor, circa 1961.

550A coupe, shown at the 2022 Monterey Motorsport Reunion. (Dan O'Connell)

frame had three times the torsional stiffness of the previous 550 but weighed 95 pounds less. The suspension and steering were completely redesigned and tuned by testing on the skid pad. The bodywork was also updated, with new louvered hatches added to the tail for engine access. As with the first 550s, coupe hardtops were used at Le Mans in 1956, with Wolfgang von Trips and Richard von Frankenberg finishing fifth overall and winning the 1.5-liter class. Although Ernst Fuhrmann left Porsche during 1956, the Porsche team continued with improvements to the four-cam engine, pushing it up to 135 horsepower by the end of 1956. Compression ratio was raised, Weber carburetors were specified for factory entries and the distributor drive was moved

PORSCHE POINTS

- Porsche racer, author and *Christophorus* Editor, Richard von Frankenberg, was nearly killed at Avus in the Type 645 Porsche. It was nicknamed 'Mickey Mouse' for its tricky and extremely fast, even by Porsche standards, transition from understeer to snap oversteer (like a mouse fleeing from a cat). This experimental project was led by Egon Forstner, who took over as head of the calculations department from Josef Mickl. The small, tube frame, streamlined 645 never received adequate development resources, but became in many ways an inspiration for the 550A. Von Frankenberg was leading the race at Avus when the car veered to the right, climbing the steeply banked North turn. It cartwheeled off the lip of the track, throwing the driver and landing in the street below, a flaming upside-down wreck. Although unconscious, the driver was lucky to fall through a tree and land in some bushes on the earth wall outside of the North turn. Issue 23 of *Christophorus* was delayed as a result.

- Michael May was the Swiss engineer, enthusiast and racer famous for mounting a large wing on his 550 in an early attempt to create downforce. The inverted wing was positioned directly over the cockpit using angled struts. The wing even had modern-looking side 'fences'. As an engineer, May later helped Porsche with fuel injection on their open-wheel racers.

to the crankshaft from the camshafts for improved stability in the ignition system.

The 550A brought Porsche a most important race victory at the 1956 Targa Florio. Like Le Mans, the Targa was not technically part of the World Sportscar Championship in 1956, but it was still a highly prestigious event. Italian Umberto Maglioli was engaged to drive with Porsche's racing manager Huschke von Hanstein named as co-driver. Vincenzo Florio himself flagged the starters for this 40th running of the race he founded in 1906. The favorites were two 3.5-liter Ferrari 860 Monzas and a 300S Maserati. The Maserati brothers' OSCA venture also fielded a competitive 1.5-liter entry. The early part of the race was dominated by the big red Ferraris, however, Eugenio Castellotti suffered a broken drive shaft at the end of lap two. Olivier Gendebien was delayed by an accident in the other Ferrari. Piero Taruffi's poor start in the Maserati left Maglioli in the lead for Porsche. As Taruffi gave chase, a stone punctured his fuel tank, meaning he had to refuel twice each lap, at first borrowing fuel from spectators' cars! Giulio Cabianca in the OSCA gave chase, as did Hans Herrmann, having a rare drive for Ferrari, along with Gendebien.

After six laps, Maglioli elected to continue his magnificent solo drive (this may have been a requirement as there was confusion over whether von Hanstein was officially entered as a driver). After 10 laps in the hot Sicilian sun, over nearly eight hours, Umberto Maglioli won the Targa for Porsche. The OSCA was

550A at the Nürburgring in 1956, driven by Wolfgang von Trips.

550A racing in Nassau during the Bahamas Speed Week.

a close second on the road, although disqualified for using an incorrect reserve driver. This left Taruffi in second place for Maserati with Gendebien and Herrmann bringing the damaged Ferrari home in third place. It was Porsche's first overall win in a classic, World Championship-level race and the first of what would be Porsche's eleven wins in the Targa.

Porsche 718

The next version of Porsche's racing spyder, the Type 718, was also known as the RSK. 'RS' is the famous abbreviation for *renn sport*, or 'racing sport'. Unlike some later Porsches where the 'K' stands for *kurz*, meaning 'short', in this case the K denotes the redesigned front suspension. In the 718, the upper torsion bars angled downward toward the center of the car. When viewed from the front, the upper and lower torsion bars looked like a K lying on its back. The new front suspension and redesign of the forward frame gave the car a lower nose. The rear of the chassis was the same as the 550A. However, the rear suspension was redesigned using technology from neighbors, Daimler-Benz. It employed a Watt linkage with coil springs over the shock absorbers. These suspension developments moved the racing spyders further from traditional Volkswagen-Porsche layouts. The 718s are recognizable by their full-width, wraparound windscreens and air inlet grills in the tail.

The 1958 Le Mans race was a tremendous success for Porsche and the RSKs. Hans Herrmann and Jean Behra finished third overall, winning the 2.0-liter class with a 1.6-liter engine. Edgar Barth and Paul Frère finished fourth, winning the 1.5-liter class. A 550A driven by Herbert Linge and Carel de Beaufort finished

THESE TWO PAGES: 718 RSK at Le Mans, 1958, driven by Jean Behra and Hans Herrmann.

fifth. Only the 3.0-liter Ferrari and Aston Martin entries finished ahead of Porsche. 15 hours of rain, some of it heavy, helped keep the engines cool but challenged the drivers to keep their cars on the road.

For 1959, the RSK rear suspension was again redesigned, now with upper and lower wishbones. The new suspension allowed for a simplified chassis that in turn allowed for gear ratio changes without removing the entire transaxle from the car. As can happen with the fortunes of racing, 1959 was Porsche's worst Le Mans outing with no cars finishing the race (mainly due to engine problems). This was the only Le Mans race since 1951 with no Porsches finishing.

On a happier note, the 1959 Targa Florio had been a triumph for Porsche one month earlier. Using the 1.6-liter version of the

PORSCHE POINTS

» Approaching Porsche's tenth anniversary in 1958, 356-001 was tracked down in Switzerland by Richard von Frankenberg. It remains the most important car in the Porsche Museum collection.

» Roger Penske began racing Porsches in 1958 and scored more than 25 wins in Porsche cars before retiring from driving in 1964. Penske went on to partner with Porsche directly in Can-Am racing, then as an owner of dealerships, as well as racing the RS Spyders in the 2000s and now the 963 LMDh.

four-cam engine, Porsche took on the might of Ferrari directly. The first of the 14 laps was a furious battle between Jean Behra in the 3.0-liter Ferrari 250 Testa Rossa and Jo Bonnier for Porsche. Dan Gurney and Olivier Gendebien led off in the other two Ferraris. Still trying to catch Bonnier on lap four, Behra made a small mistake, slid into a ditch and rolled his car into a field. Aided by fervent spectators, the car was righted, and surprisingly, carried on. However, co-driver Tony Brooks crashed the car again and ultimately had to retire. The other fast Ferraris all hit mechanical troubles. After playing the 'hare', Bonnier's car, shared with Wolfgang von Trips, suffered suspension failure while leading on the last lap. This left the victory to Edgar Barth

PORSCHE POINTS

- At the 1959 Targa Florio, the loyal Hans Herrmann stayed with his stranded 718 RSK for 13 hours to prevent the car from being dismantled by Sicilian bandits.

- Ferry Porsche required the 718/2 to lap the Nürburgring in 9 minutes, 30 seconds (something no 1.5-liter car had done) before allowing it to race at Monaco in 1959.

- Helmuth Bott devised a form of telemetry by mounting a tape recorder and microphone in the 718/2, with the driver calling out kilometer stone locations at the Nürburgring. The engineers could then use the tape to calculate sector times within a given lap.

- Bott and team took a heavily crashed 356 tub and rebuilt the car with Mercedes front suspension and steering as a test vehicle. The car was nicknamed 'Gottlieb' for Gottlieb Daimler.

- 1959 was the only year from 1951 to the present without a single Porsche car finishing the 24 Hours of Le Mans.

- When F.A. 'Butzi' Porsche began working on designs for Porsche cars he was sometimes referred to as 'Porsche Jr.', as his father had been relative to Butzi's grandfather.

and Wolfgang Seidel for Porsche (using the 1.5-liter four-cam engine). Porsches finished first through fourth overall, handing a humiliating defeat to Ferrari on Sicilian ground. The third and fourth placed cars were 356A Carreras. Oddly, the second and fourth place Porsches were shared by the same team of three drivers – Herbert Linge, Eberhard Mahle and Paul Ernst Strähle.

Porsche 356B

In keeping with Porsche's furious pace of change and development, the 356B made its debut in the fall of 1959 as a 1960 model. Designated internally as the 'T5' (T3 and T4 were design proposals not implemented), the last 356 of the decade had a significantly restyled body. The headlights were raised, resulting in straighter front fender lines. The bumpers were raised and made much larger, to give the bodywork better protection. Front quarter windows became standard for the coupe. The taillights were raised and the tail of the body extended slightly. The rear seat area was heavily revised with the seats lowered to improve headroom and the fold down luggage shelf was now split (allowing one passenger to share space with cargo). Along with the usual round of mechanical improvements and upgrades, the 356B received new, more powerful brakes that were stronger in construction and improved for better cooling. Although it would not become available until 1960, Porsche also had a new engine announced with the debut of the 356B, the Super 90.

By 1959, Porsche's annual production of automobiles exceeded 7,000 units and the number of employees had grown to approximately 1,100 from 756 only three years earlier. The 356 had established a solid niche in the sports car market while Porsche and its customers had rolled up an impressive list of racing victories during the 1950s.

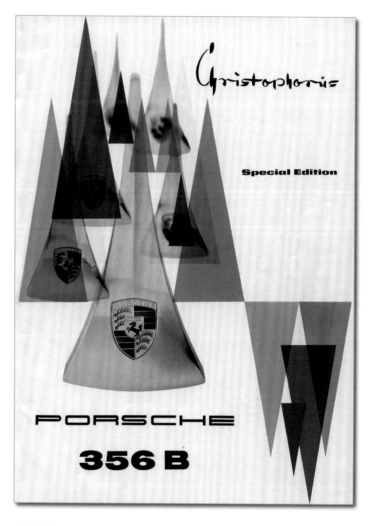

The rear end of the 356B, here seen standing behind the 356A, has also been changed. The rear end now looks longer and more extended, the bumper is raised and has overriders which are as practical as they are elegant. The chimney in the background belongs to the firm of Reutter, which makes the bodies for all coupés and cabriolets, whereas the roadster bodies are made by Drauz in Heilbronn. Reutter and Drauz were already connected with Porsche at the beginning of the 1930's

The "New Look" from Zuffenhausen

On the left the old nose of the Porsche, on the right the new one. It is clearly to be seen that the bumpers and the headlights have been raised and that special ventilation tunnels have been provided for the front wheel brakes

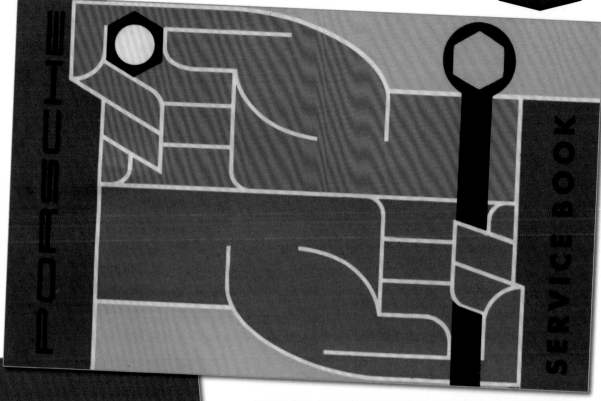

Ferry Porsche on Fashion and the Automobile

In the special edition of *Christophorus* dedicated to introducing the 356B, Ferry Porsche wrote:

"Everything connected with fashion is continually changing. In the new year last year's dress is very often completely out of fashion. One is always searching for something new, and fashion even affects automobile construction. There are firms who change the looks of their cars every year. Always new lines, new faces, new bodywork, can be found. The public wants it that way. We construct cars on a different principle. No, we aren't so conceited as to say: we never found it necessary to make any changes; what we once designed will, without alteration, still be good for decades. But when we do change something, we don't just make small alterations, following the dictates of fashion. We are, and I openly admit the fact, proud that we don't follow the current fashions in automobile construction – in spite of the fact that we have more than once influenced them: our refinements – and sometimes these are very numerous – are very often invisible ones. For we change little on the outside. The engine, the road-holding, the safety and the driving comfort: these things above all are the goal towards which we strive with regard to improvements.

In the first place we are technicians. We want to produce an automobile which stands up to the most critical technical tests. A well thought out automobile."

TOP: Colorful service book cover from a 1959 356A.

ABOVE: Mid-century modern cover for one of the early Porsche history books. Author Walter Spielberger covered the first decade of Porsche as a manufacturer.

1960s

1960	▶▶	Early 1960s Carreras
		Open-Wheelers
		Reutter and Recaro
		356 Development
1963	▶▶	Porsche 901
		Weissach
		Leopard Tank
1964	▶▶	Porsche 904
		Ferdinand Alexander 'Butzi' Porsche
1966	▶▶	'Plastic' Porsche Racing Cars
		Ferdinand Piëch
		Targa Top
		911 Developments
		911 In Competition
		Erich Strenger
		Volkswagen of America
1969	▶▶	Porsche 917

Now it's a racing car… – *Brian Redman*

By 1960, Porsche was a splendid example of the *Wirtschaftswunder* – the rapid post-World War II reconstruction and economic recovery. West Germany was one of the strongest of the recovering economies in the European 'economic miracle'. Robust economic growth continued during the 1960s and Porsche was a full participant. Although aging in basic concept, the ever-improving 356 continued to be a popular seller. In the early 1960s, Porsche expanded its competition efforts with a foray into open-wheel racing while still maintaining a presence in sports cars. Later in the decade, the racing focus would shift back to the highest levels of sports car and endurance racing. Meanwhile, the most important project of the decade was the design and development of a replacement for the 356, resulting in Porsche's defining model, the 911.

Early 1960s Carreras

An important offshoot from the 356 during this period was the competition-focused Abarth-Carrera GTL. In order to remain competitive in GT racing, Porsche decided to seek assistance from Italy in creating a customized 356, lighter in weight, with less drag and employing the four-cam engine. Coordination and building of the new cars was entrusted to an old family friend, Carlo Abarth, who had built a successful tuning and sports car business in Turin. The body design was sketched by Franco Scaglione, who started with Bertone and then became a noted freelance designer with many famous cars to his credit. The shape was much lower and more compact than a standard 356B, resulting in a lower drag coefficient. The roof was so low that the seating position had to be redesigned. Bodies were made of aluminum, contributing to the weight savings. The Abarth-Carrera also became a test vehicle for Porsche's early experiments with disc brakes.

In all, twenty Carrera GTLs were made by Abarth for Porsche. They were quite successful racing for the factory as well as private owners. Right out of the gate, the first Abarth-Carrera sold won its class at the 1960 Targa Florio, driven by Paul Ernst Strähle and Herbert Linge. At the 1960 Le Mans race, the leaky GTLs soaked their drivers in the rain which tended to collect and slosh around on the floor. However, Linge and Hans Walter won the 1.6-liter class, finishing tenth overall as the highest-placed Porsche. Updated cars won the same class at Le Mans in 1961 and 1962. As late as 1963, the Porsche factory was still entering Abarth-Carreras in some races.

356B 1600 GS Carrera GTL Abarth at the Nürburgring in July 1960.

As with the 356, Porsche continued development of their basic spyder-style sports racing car in the early years of the decade. The RS60 was updated to comply with a variety of new rules from the FIA for sports car racing. Perhaps the most significant change was the four-inch longer wheelbase. That length, combined with rear suspension improvements from the previous year, made for a fine handling car enjoyed by works drivers and privateers. A disadvantage for the RS60 was the taller windshield requirement, which impacted top speed. Some of the cars ran with raised rear decks to smooth the air flow from the higher windshield. Ongoing development of the stalwart four-cam engine helped to overcome this problem, as did Porsche's exploration of various displacement sizes with the basic engine design.

Like the Abarth-Carrera, the RS60 also had major successes from the beginning. In March of 1960, RS60s finished first and second at the Sebring 12 Hour. When the larger engine cars faltered, the factory-entered car driven by Olivier Gendebien and Jo Bonnier took Porsche's first of numerous wins at the old World War II airfield circuit. A privately-entered RS60 from Florida dealer Brumos took second place driven by Bob Holbert and Roy Schechter. The winning ways continued in Sicily at the Targa Florio where Porsche again overcame the 3.0-liter Ferraris. This time Bonnier and Hans Herrmann won the race about six minutes ahead of Phil Hill and Wolfgang von Trips' Ferrari. The durable Herrmann also drove the third place RS60 with Gendebien. Edgar Barth and a newcomer to Porsche, Graham Hill, made it three Porsches in the top five. Amazingly, Porsche was in some danger of winning the Manufacturer's World Championship for 1960 over Ferrari. The season point tally was actually tied, but Ferrari was awarded the title based on a better combination of finishing places.

718 RS61 at the Nürburgring 1000 KM in 1961. The drivers were Dan Gurney and Jo Bonnier.

PORSCHE POINTS

» Contrary to common misconception, the Abarth-Carrera bodies were not built by Zagato. Although Carlo Abarth was not anxious to dispel this myth, the bodies were mainly made by another of his local suppliers, Rocco Motto.

» In 1961, Bob Holbert and Roger Penske finished sixth at Sebring in a Brumos-entered RS61, winning the Index of Performance.

Jo Bonnier and Carlo Abate drove this 718 GTR coupe to victory at the 1963 Targa Florio.

718 F2 shown at the 1985 Porsche Parade. (Caldwell Collection)

The RS61 was very similar to the RS60 and brought success at Le Mans, winning the 2.0-liter class in 1961, Masten Gregory and Bob Holbert driving. These later 718s were popular and successful with private entrants like the Pennsylvanian Holbert who won the SCCA E Sports Racing class in 1961 and 1962. Joe Buzzetta won the same class in 1963.

Open-Wheelers

As early as 1957, Porsche was able to enter their 718 sports cars in Formula Two races. F2 was restarted during this period as a junior series to Formula One Grand Prix racing. Porsche eventually built a center-seat (*Mittellenker*) version of the 718 RSK for F2 racing. At that point cars with fenders were still allowed. Porsche's interest in 'formula' racing increased when the FIA announced that for 1961, F1 cars would be limited to 1.5-liter engines. The first Porsche racer with open wheels was confusingly named the 718/2 and started racing in 1959, using the familiar four-cam Carrera engine. These cars were also the test bed for some of Porsche's earliest experiments with fuel injection and a six-speed gearbox. In 1960, the 718/2 led to Porsche's first experiment with a kind of outsourcing, loaning a car to the Rob Walker team for Stirling Moss to drive.

Results for the 718/2s were generally solid. The first win for a Porsche-built open-wheeler came at Aintree, near Liverpool, in April 1960. Stirling Moss was followed by Jo Bonnier and Graham Hill, sweeping the podium. The German Grand Prix of 1960 was run for F2 cars as a preview of 1961 rules. On the Nürburgring Südschleife, Bonnier won for Porsche with Wolfgang von Trips second, Graham Hill and Hans Herrmann were fourth and fifth. Moss won at Zeltweg and Bonnier won a race on Ferrari's home ground at Modena.

For 1961, Porsche entered F1 Grand Prix races with either the 718/2 or the updated, longer wheelbase 787. Kugelfischer fuel injection made its debut and, not surprisingly, caused its share of problems early on. The 787 was also the first Porsche to run with a horizontal, rather than vertical, cooling fan. However, the troublesome 787 was set aside in mid-season and Porsche concentrated its efforts on the 718/2 chassis. Dan Gurney racked up three excellent second place finishes at Reims, Monza and Watkins Glen. Gurney tied with Stirling Moss, racing a Lotus-Climax, for third place in the F1 Drivers' Championship. In standard Porsche practice, three of the five 718/2s were sold to private owners and continued racing into 1964, when gentleman racer Carel de Beaufort was killed in his at the Nürburgring during practice for the German Grand Prix.

For 1962, Porsche rolled out the Type 804 Formula One car with the company's first eight-cylinder engine. Design work began as early as 1959 on a new 1.5-liter engine in anticipation of the coming change to F1 rules. The plan was to build an engine that could also be expanded to 2 liters for sports car racing. For this Type 753 engine to make 200-plus horsepower from 1.5 liters, it would need to produce high revolutions per minute but also be very compact in overall size. The eight-cylinder layout was thought to be the best choice for these requirements, but the engine retained horizontally-opposed cylinders and air cooling in traditional Porsche fashion. Like the Fuhrmann four-cam, camshafts retained shaft drive, although heavily redesigned.

Shaft drive was also used for the cooling fan, made of epoxied plastic for the first time, still horizontal and perched atop the engine. Later development of the valve angle and combustion chamber shape was an early assignment for a young Ferdinand Piëch, Ferry Porsche's nephew.

The chassis and body of the 804 were designed to be much lower and narrower than the previous open-wheel cars. The hope was to minimize drag given the small size of the engine. Main competition came from eight-cylinder British cars; BRM and those using the Coventry-Climax engine (Lotus, Cooper and Lola). Porsche sent two cars to the opening Grand Prix of 1962 at the seaside Zandvoort circuit in Holland. Given concerns about horsepower, Ferry Porsche specified that the cars would be withdrawn if they were not competitive. The drivers for the season were Dan Gurney and Jo Bonnier. Gurney struggled to get on the third row of the grid, but ran as high as third place in the race before the gear linkage failed.

Porsche sent only a single entry for the prestigious race at Monaco, where Gurney tied for third fastest in qualifying. In the race, he crashed when hit from behind. The situation was encouraging but needing further development, Porsche skipped the race at Spa. An important change was to adjust the seating position for the tall men, Gurney and Bonnier. Gurney complained that he looked like a giraffe driving the car. During Ferry Porsche-mandated endurance testing at the Nürburgring, Gurney set a lap record at 8:44.4. This was cause for optimism when the team headed for the French Grand Prix on July 8, 1962. The race was held at Rouen, a twisty 4-mile circuit. Gurney qualified 1.7 seconds slower than Jim Clark on pole position. Dan was running sixth early in the race, but then a series of crashes and mechanical failures put him in the lead and he was able to cruise home, winning on reliability rather than outright speed. This was Porsche's only Championship-level win in a Formula One race.

A week later, in front of a massive crowd at the Solitude circuit near Stuttgart, Gurney passed the great Jim Clark in his Lotus to win a non-Championship Grand Prix. Jo Bonnier finished second for Porsche on home ground. For the 1962 F1 season, Gurney placed fifth in the Driver's standings, Porsche fifth in Manufacturer points. Porsche's open-wheel adventure came to an end after the season with Ferry Porsche questioning the expense (more than $1.5 million per year) when there was little transferability of developments to Porsche's production cars. Porsche engineering resources were also stretched with development work on a replacement for the 356, plus work on

Dan Gurney in the Formula One 804, 1962 French Grand Prix.

the Volkswagen Type 3 (the first really new car from VW since World War II). There was added financial stress caused by the plan to acquire Reutter's body building operation.

Further intrigue for Porsche came during this period with Klaus von Rücker's decision to leave for BMW. Although Leopold Schmid may have expected to become Technical Director, that job went to outsider Hans Tomala, hired from Steyr-Daimler-Puch. Schmid then retired and became an independent designer.

356 Development

Although Porsche was heavily involved with designing a replacement for the 356, improvements to the original Porsche continued during the early 1960s. In 1961, Porsche released the 'Carrera 2 GS' version of the four-cam 356. This was along the lines of earlier 356 Carreras but now with a full 2.0-liter

356s on the Weissach skidpad.

engine. These cars were very expensive and are now quite rare. The street-going car produced 130 horsepower and top speed exceeded 120 mph. The 'GT' versions were intended mainly for competition with lightweight aluminum body panels, stripped-down interiors and tuned for even greater horsepower.

In March of 1960, the Super 90 became generally available in the still-new 356B. Continuing Porsche's aggressive pace of development, the newest engine had bigger inlet valves, new valve gear, a reshaped combustion chamber, the carburetor from the Carrera and an increased compression ratio. In addition to numerous other parts-strengthening measures, Porsche and Mahle improved the cylinder lining material. The engine also had improved oil and air circulation, as well as the exhaust system. It was all designed to cope with and produce something close to the advertised 90 horsepower.

The last iteration of the 356 body came in the 1962 model year with the 'T6'. The windshield was wider and slightly taller. The rear window was also enlarged. The 'frunk' (front trunk) was completely redesigned to improve luggage space, with the fuel tank no longer a box-like structure under the hood. As a result of the fuel tank forming the floor of the trunk the 356 got an external fuel filler for the first time. The front hood was reshaped to be wider and straighter at the front edge with

Reutter and Recaro

Wilhelm Reutter, a saddle maker, founded *Stuttgarter Karosserie und Radfabrik* for 'Stuttgart Bodywork and Wheel Manufacturer' in 1906, to build bodies for horse-drawn carriages as well as the newfangled horseless kind. As noted in Chapter Three, Reutter assisted Porsche with the earliest Volkswagen prototypes. During World War II, the factory was badly damaged by Allied bombing and the family lost key members as casualties of war. Building buses and postal vehicles sustained Reutter until Porsche came along in 1949. Reutter's allocation of factory space to Porsche in 1950 was critical to launching Porsche as a manufacturer. After building the majority of 356 bodies, the Reutter family decided to sell their company to Porsche in 1963 (as investment in tooling for the forthcoming 911 became an issue). Acquisition of Reutter's 1000 employees nearly doubled Porsche's total of 1372 at the time.

Reutter retained their seat manufacturing business, renaming it 'Recaro' ('re' from Reutter, 'caro' from carrosserie). Porsche remained an important client until 1997 and Recaro went on to great success in selling aftermarket sport and racing seats.

PORSCHE POINTS

» For the 1961 and 1962 model years, Porsche offered a 'notchback' 356, known as the Hardtop, with the body built by Karmann. These cars resembled 356 Cabriolets with their hardtop in place.

» The last of over 76,000 356s was a Cabriolet built in September 1965. However, in May 1966, Porsche built a final ten 356Cs by special request for the Royal Dutch Police.

	356 B 1960-1963	356 C 1964-1965
ENGINES	1.6-liter Normal, Super and Super 90, 60 to 90 horsepower	C or SC (Super or Super 90), 75 to 95 horsepower
CHASSIS	Further improvements to rear suspension	Disc brakes
STYLING	Major change to styling (T5) with higher headlights, straighter fenders, higher and larger bumpers, T6 body for 1962	T6 continued with its larger glass areas, larger engine lid, larger front trunk area
OTHER	1962 T6 dash changes	New wheels

rounded corners. The engine cover was changed to twin grills. As with earlier 356 updates, the T6 received numerous other improvements, including interior ventilation, a dash-mounted clock, better and safer seat adjustments, improved windshield wipers, an optional anti-theft transmission lock and the usual round of improvements to the engine and transaxle.

For the 1964 model year, Porsche released the 356C. It continued with the T6 body but brought the last round of major improvements, most notably, disc brakes. Porsche worked with Teves/Ate for the new brakes, developed specifically for the 356 and incorporating some previously patented Porsche ideas. A small drum inside the rear discs was used for the emergency/parking brake. The new brakes required a new full-disc wheel design, but Porsche retained the oval vents, making the wheels appear similar to those on earlier 356s. Interior comfort and switchgear improvements included new seats and new armrests with integrated door-closing handles. The suspension was tweaked to improve ride quality and reduce oversteer.

For the 356C, Porsche simplified the engine offerings which were limited to 1.6-liter 1600 'C' and 'SC' models. Porsche's rising star, Hans Mezger, worked on getting these engines to achieve their implied 'Super 75' and 'Super 90' horsepower figures. The key change was better balancing the sizes of the inlet and exhaust valves. Numerous other detail changes improved the breathing of the engines and Mezger was not afraid to employ new camshaft and crankshaft designs in engines for a 'lame duck' model. The results were a fitting and strong-selling conclusion to the 356 journey. Reaction from the automotive press was also highly favorable. Through the end of 1965 and during the transition to 911/912 production, Porsche sold over 16,000 356Cs. Given the years of development and overall success of the 356, it is easy to see why the Porsche faithful might have been, at first, skeptical about a completely new sports car.

Porsche 901

It is a small irony that Porsche's defining automobile, the 911, started out as the Type 901. When preparing to replace the venerable 356, Porsche skipped project numbers above 822 and designated the new car '901'. The reason was a matter of practical coordination with Volkswagen for parts and servicing. It turned out that three-digit numbers beginning with 9 were the only available group in the VW parts computer system. All would have been well except for an objection from Peugeot based on trademark rights in France to three-digit auto names with a zero in the middle. Rather than battle with the French, Ferry Porsche elected the expeditious solution of changing from 901 to 911 for marketing purposes. Early 911 part numbers retained the 901 prefix. About 80 cars were built as 901s before the model name change.

The final version of Karl Ludvigsen's *Excellence Was Expected*, released in 2019, substantially updated the historical understanding of how the 911 as we know it was designed. For a full account of 911 styling development, see Book 1 of *Excellence Was Expected*. According to further research with Porsche's Archive team, Ludvigsen found that the roots of the 901 started with planning for a second model line. Work on Type 695 began as early as 1954 and called for something like a true four-seat car with improved luggage space (a frustrated Ferry Porsche echoed

PORSCHE POINTS

» The very first 911 (901 prototype) came alive on November 9, 1962. This white car was nicknamed *Sturmvogel*, 'storm bird', after the Messerschmitt ME 262 jet fighter of World War II. Helmuth Bott was likely the first to drive it.

» Alois Ruf Jr. won the Porsche 75th Anniversary class at Pebble Beach in 2023 with his 901 prototype. The chassis is the sixth-ever 911 built and the oldest known to still exist.

Ferdinand Piëch and Ferry Porsche with a 901 test engine in 1963.

a common 356 complaint that not even golf clubs could fit in the Porsche). Work on the project proceeded slowly given the brisk sales of the 356. Both Erwin Komenda and the independent designer Albrecht von Goertz (of BMW 507 fame) eventually created designs that made it to the styling model stage. Goertz drew a handsome fastback four-seater somewhat like the AMC Marlin or Dodge Chargers of the mid-1960s (at least when viewed from behind). However, Ferry Porsche and Louise Piëch decided it just didn't look like a Porsche. By 1959, Ferry Porsche's son Butzi was beginning to take a significant role in styling. Like Komenda, Butzi concluded that some form of notchback, rather than fastback, roofline would be correct for a four-seater, even if the 'notch' was very slight.

Between 1959 and 1962, a series of designs were completed by Butzi Porsche and Erwin Komenda. The Type 754 design for a larger Porsche features the recognizable look of a Porsche 911 from the windshield forward and remains in the Porsche Museum collection. However, the 911's styling resulted from a proposal to update the 356 as a pure sports car, possibly a two-seater. This new body, with a clear fastback roofline, could have been sold alongside the larger 754. As Ludvigsen points out, Butzi was ably assisted by designer Gerhard Schröder and modeler Heinrich Klie in the styling work.

At the beginning of 1962, Porsche's executive team made the momentous decision that the pure sports car update would be built with a 2-plus-2 seating configuration, a 2,200 mm wheelbase and replace the aging 356. In March 1962, the revised project was designated the Type 901. The idea for a larger Porsche was dropped at approximately the same time. As Karl Ludvigsen points out, Ferry Porsche took the view that it was best to concentrate on maintaining Porsche's established niche in the German auto industry. Volkswagen made the basic transportation cars, Mercedes the expensive luxury cars, Opel and Ford the middle class, and Porsche made the sports car. This unspoken order had grown up after World War II and was still fairly clear in the early 1960s.

Along the way, Porsche had been working on a six-cylinder engine to power the larger car. In 1959, project 745 was the first attempt on a path that would lead to the 901. This design used camshafts placed very close to and below the crankshaft. It used long push/pull rods to open and close the valves. The engine also featured twin cooling fans, one for each bank of cylinders. Prototypes were built at 2.0 and 2.2-liter capacities, with the larger version reaching the target at 130 horsepower. A major concern became the performance limits for a pushrod engine in racing (where much higher rpm would be needed). At the end of 1961, Porsche started over on the flat-six with Type 821, an overhead cam design that eliminated pushrods in favor of chain drive to the camshafts. As with all Porsche engines, it would remain air-cooled, but the 821 was a wet sump, like the 356, carrying the oil supply in a pan below the engine. The designers also retained the very stout basic 745 concept for the engine case and crankshaft although it was eventually strengthened further with seven main bearings. The 821 first ran in January 1963, but only produced a disappointing 110 horsepower.

During 1963, and after the cancellation of Porsche's F1 program, the six-cylinder was heavily revised with the involvement of Hans Mezger. In April 1963, Ferdinand Piëch joined the team which became focused on developing the flat-six for use in competition (in the upcoming 904) as well as the 901. With production of both cars scheduled for 1964, the pressure was on. The engine team made numerous improvements, including revision of the tensioning system for the chain drive to the camshafts. The engine was converted to a dry sump system to prevent concerns about oil starvation in high-speed cornering (following race engine practice). Mezger made further cylinder head changes and the team managed improved carburation with a new Solex 'overflow' design (using only a single float per cylinder bank). Unlike the 754, the 821/901 engine employed the prominent single cooling fan at the rear of the engine, so familiar to all who have spent time around any air-cooled 911. The team gradually developed the new engine to produce 130 DIN (German standard) net horsepower.

As noted above, the new five-speed transmission design for the 904 was shared with the 901. The 901 inherited the strut-type front suspension from the Type 754 with a lower control arm (wishbone) and longitudinal torsion bars with a separate anti-roll bar. The non-adjustable upper mounting for the front strut was a questionable engineering decision that led to conflict between Ferdinand Piëch and Hans Tomala, hastening the latter's

RIGHT: 911/912 assembly line in 1967.

departure from Porsche. 901 steering was ZF rack and pinion with universal joints in the steering column. Collapsibility added to safety in the event of a frontal collision. The rear suspension was also strut-based, with an inner trailing arm and outer radius arm connected to the torsion bar. Drive was accomplished with half-shafts and universal joints connecting the transaxle to the rear wheels. The rear suspension received an anti-roll bar in 1966 with the introduction of the 911S. The 901 was also engineered to eliminate the need for periodic suspension lubrication, an important industry trend at the time.

The 901/911 received a handsome, but not overly luxurious, interior with the famous five dial dashboard, tachometer front and center. Compared to the 356, the glass area was much larger and amenities such as interior ventilation and luggage space were improved. Alongside the 901/911, Porsche developed

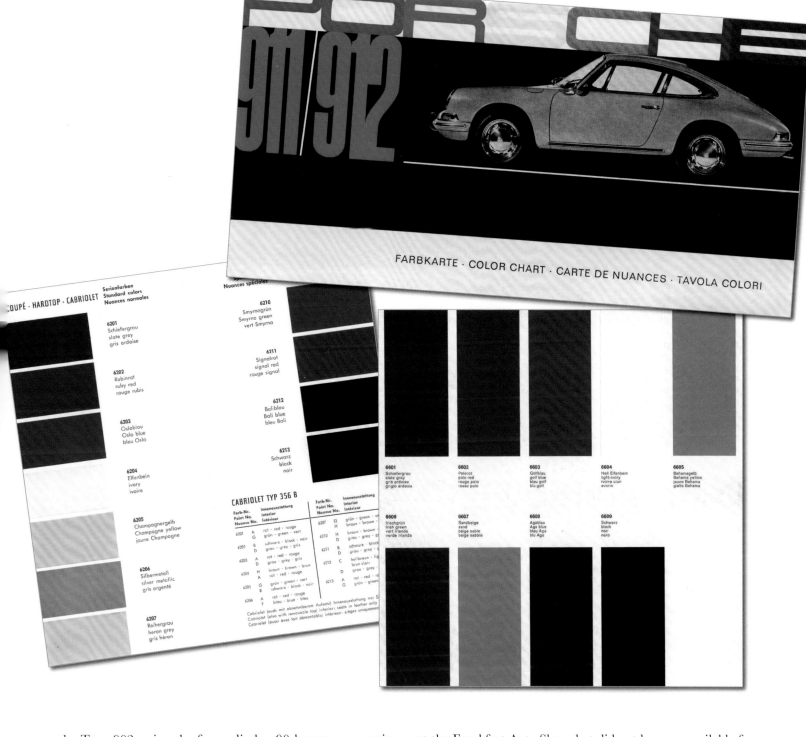

the Type 902, using the four-cylinder, 90-horsepower engine from the 356 to create a lower-priced alternative to the 911. Debuting in April 1965 and marketed as the '912', these cars were very popular upon introduction, perhaps due in part to the familiarity and reliability of the engine. The 912 outsold the 911 for the first three years of production by a two-to-one margin. Early 912s had a three-gauge dashboard but eventually were converted to the 911-style five-gauge.

The 901 was introduced to the public on September 12, 1963 at the Frankfurt Auto Show, but did not become available for sale until late in 1964. Production was delayed by construction of the 904s and ongoing development using the 901 prototypes and pre-production cars. Initial reaction to the 901 was not overwhelmingly positive, but this did not worry Ferry Porsche who believed strongly that a car receiving an immediate 'splash' of positive sentiment might not have an enduring strength of design and style. After 60 years, he has been proven correct time and again.

Weissach

In 1959, Ferry Porsche assigned his cousin Ghislaine Kaes and head of finance, Hans Kern, to begin searching for property that could be developed into a testing site. Kern had been with Porsche since 1933 (as the 13th employee). The only sites they found were much larger than Ferry's target of 15 acres. In December 1960, with a recommendation from Porsche's long-serving mechanic, test driver and Weissach native, Herbert Linge, Porsche purchased the largest parcel of all. It was 93 acres located near the villages of Weissach, Mönsheim and Flacht, west of Stuttgart. In spite of Ferry's concerns about the cost, the purchase turned out to be a wise investment in land value alone.

Construction began in 1961 with a skid pad finished in 1962. The test track, initially designed by Helmuth Bott, was completed in stages between 1967 (a 'Mountain' track including off-road/destruction sections) and 1971 (shorter, faster Can-Am circuit). The Development Center also opened in 1971. With major expansions during the 1980s (including a wind tunnel) and in 2014, the facility has since grown to a complete engineering, design and development center occupying 111 acres and employing more than 6,000 people.

Aerial view of the original Weissach skidpad.

Porsche 904

The Type 904 is one of Porsche's most beautiful and unique cars. It was drawn by Ferdinand Alexander Porsche, known by his nickname, Butzi. The Carrera GTS, as it was officially known, was Butzi's favorite design. It was largely completed on his own as head of Porsche's new styling department. The project began late in 1962 with the goal of producing a competitive racing car for the 2-liter GT class. Porsche had decided to abandon open-wheel racing, as noted above, and refocus on sports cars. Competition was on the way from Porsche's old friend, Carlo Abarth, as well as Alfa Romeo, in the 2.0-liter class. Some inspiration for the 904 came from the special 718 coupes raced at Le Mans in 1961. Edgar Barth and Hans Herrmann had finished in the top ten although second in class behind a 718 spyder. A coupe 718 with an eight-cylinder engine also won the Targa Florio in 1963. Basic drawing and modeling work in Butzi's studio for the 904 was completed in less than six months.

The 904 was naturally conceived as a mid-engine car since its main purpose was racing, although it would potentially be usable as a street car. The long and low profile with its wedge-shaped nose was meant to minimize frontal area and therefore reduce drag. The distinctive flying buttresses flowing back into the tail were influential on sports car coupe designs for years to come. The form was not only gorgeous and modern but it

Leopard Tank

Porsche Type 714 was a tank design begun in 1957 as part of a competition initiated by the West German defense department. It shared the 'Leopard' tank nickname with Porsche's early World War II design. By 1963, the Porsche team had won the competition for this swift, light-armor battle tank. Powered by a 37-liter, multi-fuel MTU V10 capable of 830 horsepower, it had a 105 mm British-built cannon. Production began in 1965 and the Leopard was purchased by numerous NATO countries. Over 6,000 were eventually produced. Prototypes were built and tested at Weissach. When the program ended in 1981, the tank shops were taken over by the race team.

also proved itself in the wind tunnel with a very low coefficient of drag.

The 904 was unique in its construction and chassis design. Rather than Porsche's typical multi-tube space frame, it had large, ladder-style rails on each side and two main cross members. Although Porsche had explored the idea of using fiberglass as early as 1953, the 904 would be the first Porsche actually built using glass reinforced plastic for the body. The molded body and interior bulkheads were bonded to the chassis. The resulting structure was light but as rigid as a steel body 356. Because the seats were molded into the bulkhead, the 904 had adjustable pedal locations and a telescopic steering column. The body/chassis units were made by the aircraft company, Heinkel.

For power, Porsche planned to use the six-cylinder engine being prepared for the upcoming 356 replacement. However, the reliable 2.0-liter version of the Carrera four-cam engine ended up being used for the majority of the 904s since the new engine was not ready for racing in 1964. As with the pushrod fours for the 356C, Hans Mezger worked his magic on the four-cam, creating an improved 180-horsepower race version for the 904. The driveline for the 904 featured an extensively redesigned Porsche transaxle with a new version of the traditional ZF differential and a new driveshaft design. The transaxle was the same basic unit as intended for the 356 replacement. For suspension, the 904 had race-proven unequal length wishbones, with coil spring shocks and anti-roll bars, disc brakes, and ZF rack and pinion steering. Standard size wheels were five inches wide, six inches at the rear for racing, with racing tires.

PORSCHE POINTS

» For crash testing, the 904 chassis/body shell was suspended from a crane and dropped on its nose!

» In 1964, the Elva-Porsche debuted. Like Cooper, Elva was a small British manufacturer of super lightweight sprint racers and they were represented in the US by Carl Haas (later famous as the Lola importer and a successful IndyCar team owner). Unlike the 'Poopers', the Elva-Porsche concept was officially endorsed and assisted by the Porsche factory. Porsche supplied special four-cam engines with a space-saving horizontal cooling fan. Elva-Porsches finished second and third in the 1964 US Road Racing Championship, a forerunner to the Can-Am, and also competed in the European Hill Climb Championship. 'Elva' is a contraction of the French *elle va* for 'she goes'.

Porsche planned to build 100 904s for homologation as a GT car. Customer sales were intended to help finance the project and the price was considered attractive at just under $7,500. The 904 was introduced to the press in late November of 1963 and Porsche began taking orders for the 1964 season. Something over 110 were ultimately built, including the prototypes and cars that eventually carried six- and even eight-cylinder engines.

The 904 enjoyed an excellent run of success during the 1964 and 1965 seasons. It won its class at Daytona and Sebring right out

Ferdinand Alexander 'Butzi' Porsche

As a namesake of Dr. Porsche, it seems appropriate that Butzi would play a significant role at Porsche, even though he never replaced his own father as CEO of the family business. Born in 1935, he apprenticed at Bosch before joining Porsche in 1958, where he trained under the prolific Franz Xaver Reimspiess and importantly, Erwin Komenda as head of body engineering. His role in the evolving Type 754 and, of course, shaping the 911, is his most important legacy along with his dedication to the philosophy of form following function. He also played a significant role in the styling of the 804, 904, 911 Targa and 914.

When the Porsche family withdrew from day-to-day involvement in the company, in 1972, Butzi formed his own company, Porsche Design, which worked on products from watches to clothing, pens, sunglasses, cookware and even ski lift gondola cars. Later in his career, Butzi contributed to the design of Porsche's first SUV, the Cayenne. He served as Chairman of the Supervisory Board of Porsche AG from 1990 to 1993. In 2003, Butzi's Porsche Design merged into the main company. Butzi passed away in 2012 after retiring in 2005.

Butzi Porsche with the 901.

904s outside of Werk II in 1964.

of the gate in 1964. In Sicily for the 1964 Targa Florio, Ferrari boycotted the race in a homologation dispute, so the main competition for the standard 904s came from eight-cylinder Porsches (a 718 spyder and a 904/8). When the faster Porsches and the privateer 3.0-liter Ferrari GTOs faltered, the standard 904s came home first and second. Colin Davis and Antonio Pucci drove the winning car. 904s won the 2.0-liter class at Le Mans in both 1964 and 1965. A six-cylinder 904 also won the 2-liter Prototype class in 1965, finishing a fine fourth overall behind three 3.0-liter Ferraris.

The 904 proved its versatility in the 1965 Monte Carlo Rally, finishing second in mostly snowy conditions, driven by Eugen Böhringer and Rolf Wütherich. Only the great Timo Mäkinen in a Mini Cooper S prevented a Porsche win. In the 1965 Targa Florio, a *Bergspyder* or hill climb special 904, powered by an eight-cylinder engine, finished second to a Ferrari prototype. A six-cylinder 904 finished third. 904s also won multiple SCCA National titles in the US. Although the 1965 hill climb season was a disappointment for the 904-based spyders, a new tradition was born. Lacking time for proper silver paintwork, the Porsche entries began appearing in German racing white. Author Karl Ludvigsen quotes Hans Mezger in *Excellence Was Expected*: *"The bodies were produced in the workshop and they were fairly crude, so this was just an easy color to paint the car without having to send it away for a proper paint shop to do it."* From this point forward, most factory-entered Porsche race cars would appear in white, or with a sponsor-specific livery.

It is worth making note here of the great Porsche tradition of never throwing anything away. An idea whose time isn't yet right, or a technical solution for a given project, might be put to use again elsewhere or at a later time. A good example was Porsche using larger capacity versions of the eight-cylinder Type 771 engine (conceived for Formula One) in 718 and 904-based sports cars. The 771 would power Porsche racers as late as 1968.

'Plastic' Porsche Racing Cars

Porsche offered no new race car for 1965, as Ferdinand Piëch took over as head of the experimental department/R&D, which included racing. Piëch decided not to build any more 904s as the car was considered too heavy, plus the FIA announced a rule change for 1966. The production number for homologation sports cars was reduced to only 50 cars. As mentioned above, Hans Tomala left Porsche in September 1965, giving overall control of Porsche engineering to Ferdinand Piëch and his able lieutenant, Dr. Helmuth Bott. Piëch's decision was to return Porsche to tubular space frame chassis for racing cars. The first project completed under Piëch's leadership was the Type 906, officially named Carrera 6. This seminal project set the basic parameters for Porsche racing cars over the following 15 years.

Unlike earlier space frame Porsche racers, the 906 made use of very lightweight fiberglass-reinforced plastic resin body panels. The 906 continued the 904 theme of lower, sleeker aerodynamic proportions. Suspension was also similar to the 904, with upper and lower wishbones and coil springs over the shock absorbers (the layout becoming standard for most racing cars by the mid-1960s). The 906 began the trend toward wider wheels for Porsche racing cars, initially at seven inches in front, nine inches in the rear. The 906 retained the 904's wheelbase but had a wider track. The 2,300 mm wheelbase would remain a Porsche standard for several years. Gullwing doors were employed to accommodate the shape of the roof over the cockpit and to make door closing easier for the drivers during a pit stop. The 906 inherited the basic six-cylinder, 901-based racing engine and transaxle as used in the 904. This engine used higher compression, twin spark plugs, bigger valves with a more aggressive camshaft profile and triple-throat Weber carburetors to produce well over 200 horsepower (at 8000 rpm). This compared to 130 for the standard 911 engine. The engine was made lighter through the use of special metals like titanium and magnesium in place of steel and aluminum.

As always, the 906 received a massive number of upgrades and improvements during its run, including Porsche's first use of ventilated brake discs. 906s running as Prototypes received fuel-injected engines (sometimes referred to as the 906 'E' for

The 906 was very popular in advertising. See (below) the open gullwing doors.

906 driven by Gerhard Mitter at Zeltweg in 1966.

> The six- and eight-cylinder 904s were built with serial numbers that had a '906' prefix. During a confusing period for Porsche Type numbers, the Carrera 6 (906) would have a completely different chassis design compared to the 904.

> In 1965, Piëch and team designed and built the 'Ollon-Villars' spyder (906-010) in less than a month to compete at the Ollon-Villars hill climb. Teams worked literally around the clock. This tube frame, eight-cylinder car, with suspension corners and wheels purchased from Lotus, set the pattern for the Carrera 6 and other ultra-lightweight spyders like the 909 and 908/03. It didn't win at Ollon-Villars, but Huschke von Hanstein did use it to set FIA records for the standing-start quarter-mile, 500 meters, and 2-liter standing kilometer. The first two were newly established FIA categories and over-50 rival, Carlo Abarth, then broke the records a month later in one of his own cars. 906-010 was converted to a 906-style coupe for the 1966 European Hill Climb Championship and helped Gerhard Mitter win the title for Porsche. Its chassis number may have led to the out-of-sequence Type number, '910'.

906 at the Targa Florio in 1970.

Einspritzung). Porsche had experimented off and on with fuel injection as early as 1951 and used it in Formula One. In 1966, working with Bosch, Porsche built a really effective fuel-injected engine for sports car racing. Using a Bosch-made injection pump, the injected flat-six had an advantage on throttle response and fuel mileage.

Like its predecessors, the 906 was a formidable competitor in the 2-liter classes at races around the world. It won on its first time out at the 24 Hours of Daytona in 1966, driven by Herbert Linge and Hans Herrmann. Herrmann had returned to the Porsche team after several years with Abarth. A 906 also brought Porsche's sixth overall win at the Sicilian classic. In the Targa Florio, the 906 driven by Herbert Müller and Willy Mairesse overcame unusually wet weather and received an assist when the favorite, a Ferrari 330 P3 driven by Nino Vaccarella and Lorenzo Bandini, crashed out of the lead with only two laps remaining. Importantly, the 906 had beaten its natural rival from Maranello, the 206 SP 'Dino' Ferrari.

For Le Mans in 1966, Porsche produced their first in a series of long-tail racers. The exaggerated length of the tail was meant to further reduce drag on the three-mile Mulsanne Straight. This had been designed and tested by ace Porsche body man and former Reutter engineer, Eugen Kolb, using the wind tunnel at the Stuttgart Technical Institute. 906 long-tails won the 2-liter Prototype class, finishing fourth through sixth at Le Mans, behind three 7-liter Fords (as depicted in the film, *Ford vs. Ferrari*). A standard 906 won the Sport class. Despite having only a 2.0-liter engine, the long-tail 906s reached speeds near 170 mph on the Mulsanne.

1967

The Porsche 910 was a development of the 906, with a new body including 'targa'-style removable roof and forward-hinging doors. It was designed to be still lighter but stiffer than the 906 and with improved handling. The 906 suspension was discarded in a favor of new design to accommodate wheels that were smaller in diameter, at only 13 inches, but wider. The wheels were cast magnesium with center-locking hubs. Smaller wheels allowed for a lower front fender profile and the entire tail/engine cover was hinged for access. The rear body had integrated ducting for brake and transmission cooling. 910s were raced with both six- and eight-cylinder engines depending on the situation and were Porsche's primary weapon in the important races during 1967.

Not surprisingly, the 910s frequently won their class at major races including Daytona and Sebring in 1967. They were also successful in hill climbs and brought Porsche yet another triumph

910 on its way to winning at the Nürburgring in 1967.

in the Targa Florio. This time, Porsche invaded Sicily with a six-car entry, three with six and three with eight-cylinder engines. The Ferrari and Alfa Romeo opposition either broke down or crashed in this high-speed battle, leaving victory to the 910/8 driven by Rolf Stommelen and Paul Hawkins. 910s also finished second and third. Perhaps more important, Porsche finally took an overall victory at their 'home' race, the Nürburgring 1000 KM. Not only did Udo Schütz and the American, Joe Buzzetta, win the race, but 910s finished in first through fourth places. Although the Ferrari factory skipped the race to prep for Le Mans, the field included Gulf-Mirages, Ford GT40s, Lolas and a Chaparral (all with engines at 5 or 7 liters).

The Type 907 was significant in the shift to right-hand seating for the driver, notably with the gear lever remaining at the driver's right hand. Since most sports car races are run on clockwise circuits, the right-hand position is better for driver changes in the pits as well as for negotiating the preponderance of right-hand bends. In the 907, the driver position was quite close to center of the car but left room for the mandatory 'passenger' seat. The 907 was also the first Porsche to use Italian-made Momo steering wheels (which later became a popular custom touch on Porsche street cars). The chassis was very similar to the 910 with improvements mainly to the front suspension. The main distinction of the 907 was its smooth, aerodynamic shape, recalling 'streamliners' as far back as the 1930s, including the Porsche-designed VW 60K10. The 907 had a narrow greenhouse with a long, swept-back windscreen and an extremely long tail, resulting in 25% less drag compared to the 910. Porsche now had a car capable of speeds above 180 mph but discovered the first hints of aerodynamic instability that would become a greater problem in the next few years. As with the 910, the 907 could race with either six- or eight-cylinder engines.

The 907 was squarely aimed at the 24 Hours of Le Mans in 1967. In that race, Ford won again with AJ Foyt and ex-Porsche driver, Dan Gurney, driving the long-tail Mark IV Ford GT. The 907 won the 2-liter Prototype class, driven by the Swiss Jo Siffert with Hans Herrmann. Siffert joined Porsche for the full sports car season in 1967 and quickly became one of the team's lead drivers. Siffert and Herrmann were fifth overall behind two 7-liter Fords and two 4.0-liter Ferrari P4s. Finishing sixth was a 910 driven by Rolf Stommelen and Jochen Neerpasch. Vic Elford, in his first Le Mans race, won the Sport class in a 906 driving with Ben Pon, finishing seventh overall. Ferrari edged Porsche by only a single point in the final 1967 Manufacturer's World Championship standings, although Porsche won the 2-liter Championship as usual.

1968

Long-tail 907s brought Porsche a triumph at Daytona in 1968. It was Porsche's first overall victory in a major 24-hour race, although a 911 had won the Spa 24 Hour in 1967. Porsche entered four cars at Daytona and the surviving trio swept the podium places. The winning car was driven by Vic Elford and Jochen Neerpasch, but team manager von Hanstein gave time in the leading car to Siffert, Herrmann and Stommelen, so in a way it was a preview of modern Daytona endurance races where cars are sometimes driven by four or five drivers. The cars ran as Prototypes with the aging but still effective Type 771, 2.2 liter, eight-cylinder engines, strengthened to last 24 hours for the first time in their service life. Top speed was now better than 200 mph.

> In 1966, Californian Al Stein built an Indianapolis 500 racer using two early Porsche 901 engines, one in front and one behind the driver. Although slow on the straights, the cornering speeds predicted the rise of four-wheel drive at Indianapolis in the late 1960s. No official qualifying attempt was made for the 1966 race but it was, unofficially, the first Porsche-powered car at the Indianapolis Speedway.

> In 1967, Gerhard Mitter won the European Mountain Championship for Porsche using open-cockpit 'spyder' versions of the 910. The final iteration was Porsche's first experiment with aluminum for the tubular space frame. Reduced bodywork for the final events gave the car its 'Miniskirt' nickname. In the exotic metals department, Porsche tried insanely expensive brake discs made of Beryllium (plated with chrome on the surface, since Beryllium is poisonous). For 1967, the hill climb championship was run under Group 7 rules (similar to the Can-Am), which is to say, there were almost no rules except the cars had to have two seats and covered wheels.

> Ludovico Scarfiotti became the first and only driver to be killed in a factory-entered Porsche. The accident happened during practice for the Rossfeld Hill Climb in 1968. He was an ex-Ferrari factory driver and twice European Mountain Champion. A small irony in Scarfiotti's death driving for Porsche was that he was a grandson of one of Fiat's founders.

Formation finish for the 907s at Daytona in 1968.

The third place 907 was Porsche's first circuit racer with an aluminum tube frame (rather than steel). The 907 also marked Porsche's first use of a 'space-saver' spare tire (to comply with the rules) and early experiments with a six-speed gearbox and driver cool suit concept. The 907 was also run in short-tail form with rear bodywork very similar to the earlier 910. The short-tail 907 brought Porsche an overall win at the prestigious, demanding 12 Hours of Sebring. Jo Siffert and Hans Herrmann won by one lap over Elford and Neerpasch in a race of significant attrition.

The 907 finished its brief but illustrious career as the factory's frontline racer by taking Porsche's eighth win in the Targa Florio. It was an epic drive by Vic Elford, partnered with Targa expert Umberto Maglioli. Porsche entered four short-tail 907 coupes against the Alfa Romeo factory's four T33/2s. Elford had to stop during the first lap to tighten a loose rear wheel nut which was then changed halfway around at the mountain service location. The new nut also came loose, sending Vic off the road and he had to install the space-saver spare to finish the first lap. Back at the pits, Vic rejoined the chase with a new rear wheel and nut having lost 18 minutes to the leading cars. Deciding to go all-out for a lap record, Vic found himself catching the leaders by the end of his fourth lap. Maglioli did an excellent job during the middle stint, but Elford convinced Helmuth Bott and Peter Falk that if he could drive the final three (rather than two) laps, they could still win. Maglioli was miffed about being pulled after only three laps, but Vic then went on a tear. He passed the 2.0-liter Alfa of Ignazio Giunti on the last lap and won at a record average speed for the Targa, even with the lengthy delays on the first lap.

In recognition of Elford's heroics, Porsche featured a photograph of him (not the car) on the factory 'victory' poster for the event.

Porsche began work on the Type 908 during 1967, in anticipation of new rules for sports car racing in 1968. Prototypes would be limited to 3.0-liter engines with homologated 'Sports' cars allowed 5.0-liter engines, but with a minimum production number of 50 cars (allowing the Ford GT40 and Lola T70 to compete). Ferry Porsche insisted on a cost-effective solution if Porsche were to compete with a 3.0-liter car. The solution was found by essentially adding two cylinders to the 901 engine layout. Having built an experimental twin-cam version of the 911 engine, Porsche elected to build an eight-cylinder, twin-cam engine which worked out to just over 2.9 liters capacity. To handle the additional power and torque of the larger engine, Porsche also had to design a new transaxle (confusingly named the Type 916, which had also been the project number for the twin-cam engine). The new gearbox was used from 1969 forward. As shown above, Porsche had been trying to win races overall with engines no larger than 2.2 liters, often ranging against cars with 4, 5 and even 7 liters. The 908, with well over 300 horsepower, would in theory put Porsche on a much more equal footing, especially at Le Mans.

The 908 chassis and body were heavily and logically based

908K with all four wheels off the ground at the Nürburgring in 1968.

on the 907, adjusted to make room for the larger engine but maintaining the same 2,300 mm wheelbase. The 908 eventually reverted to 15-inch diameter wheels and was slightly wider and lower than the 907. Short-tail 908 coupes looked very similar to the 907, but the long-tail 908s had a new design with a rear wing and movable flaps (suspension-actuated tabs at either end of the rear wing). Porsche had experimented with movable wing elements as early as 1966 on the 906. An important change came from 908 Chassis 012 on. Porsche determined that aluminum was strong enough to replace steel for the space frame tubes. This led to a significant and necessary reduction in weight since weight minimums were dropped for Group 6 Prototypes in the 1969 rules.

During the 1968 season, the 908 had its share of teething problems, including bad vibration in the engines which was eventually solved by redesigning the crankshaft. As speeds increased, aerodynamic instability reared its ugly head, frightening the drivers such that they even conspired to tank the lap times at Monza (to avoid having to drive the long-tails at Spa). The 908 did win on home ground, at the Nürburgring. Jo Siffert and Vic Elford teamed up to win using a short-tail 908. After problems in qualifying, Siffert made a rocket start from 27th on the grid and was up to second place on lap two.

The 1968 Le Mans race was moved from its traditional mid-June

> **PORSCHE POINTS**
> - In the late 1960s, the original Type 918 was a design for a mid-engine, 908-based eight-cylinder 'supercar'. The design, championed by Ferdinand Piëch, shared little with the 914. It never made it to the prototype stage.
> - Also in the late 1960s, the Type 915 was a design project for a four-seat version of the 911. This distant, early forerunner to the Panamera was never built but the project did produce the transaxle design used in the 911 from 1972 through 1986.

date to September due to civil unrest in France and elsewhere. It turned out to be a disappointment for Porsche and the 908s. Long-tail 908s were the fastest cars but suffered gearbox failure and alternator problems. The lone surviving 908 finished third behind a long-tail 907 in second. The winning car was a reliable John Wyer-Gulf GT40 driven by Pedro Rodriguez and Lucien Bianchi (substituting for the injured Jacky Ickx and Brian Redman). The 1968 Manufacturer's Championship results were also a heartbreak for Porsche. They had won five races, as had Ford in the form of Gulf GT40s. However, Porsche's win at the 500 KM Zeltweg race only counted for half points, giving Ford the victory in the season series.

1969

For 1969, Porsche developed the nimble 908 spyder since the FIA Group 6 changes eliminated the 'luggage' space, spare tire and minimum windscreen height rules. The 908/02 spyders appeared in two forms. At first, the more curvaceous body resulted from essentially chopping the roof off a 908 coupe. Later in the 1969 season, Porsche developed the flat-looking *flunder*, 'Flounder' body that was similar to Ferrari's open-cockpit 312P. The 908/02s all cracked their chassis' first time out at Sebring, but then came good for the rest of the season. The new driving pair of Jo Siffert with Brian Redman (recruited from the Gulf team), won at Brands Hatch, the Nürburgring and Watkins Glen with the new spyders. They also used long-tail 908s to win at Monza and the super-fast Spa-Francorchamps circuit in

Porsche 909

The 909 Bergspyder was a tiny race car built for the 1968 European Mountain Championship. This ultra-light hill climb special weighed less than a thousand pounds. To save weight, it was designed with no fuel pump. Instead, the fuel cell was pressurized to provide fuel to the injection system just long enough to make it up the course. This arrangement was used for one event only and in its second and final event, a fuel pump was used. The 909 didn't win, but it was the inspiration for the very successful 908/03 (see Chapter Seven). Gerhard Mitter took the Porsche factory's last Mountain Championship in 1968, winning seven of eight events mainly using the 'miniskirt' version of the 910 spyder, powered by the last of the venerable Type 771 flat-eight engines.

Belgium. Porsche took six 908/02s to Sicily and won the Targa yet again, victory going to the all-German crew of Gerhard Mitter and Udo Schütz. Porsche had already clinched its first World Manufacturer/World Sportscar Championship before Le Mans.

But the big one got away again at the 1969 24 Hours of Le Mans. Porsche sent an entry of 908s (long-tails, plus a special 908/02 spyder with an extended tail for Siffert and Redman) and 917s

Jo Siffert and Brian Redman won at Spa in 1969 with the long-tail version of the 908.

to France. The 917s were too new and the 908s all fell out except for the one driven by Hans Herrmann and Gérard Larrousse. For the last three hours of the race, the 908 fought a sensational battle against the Gulf GT40 of Jacky Ickx and Jackie Oliver. The cars were in sight of each other, often nose-to-tail the entire time. For the last stint, it was Porsche veteran Herrmann against the young star from Belgium, Ickx. Herrmann was handicapped by having to nurse the front brakes with his warning light for pad wear on the whole time and an engine that had lost a bit of performance. Ickx and Herrmann played a cat-and-mouse game of passing and re-passing over the final few laps, neither man wanting to lead on the straight. In the end, Ickx led by just a few yards, bringing the aged Gulf-GT40 over the line first. It was the closest contested finish ever at Le Mans, but bitterly disappointing for Porsche.

Ferdinand Piëch

Second of three sons of Anton and Louise Piëch, Ferdinand was destined to become a giant of the auto industry, not unlike his famous grandfather. During and after the war, young Ferdinand was known to haunt the tiny drawing office as the engineers worked in Gmünd. After his father's death, Ferdinand was sent to boarding school and later received a Master's degree from the Federal Institute of Technology in Zurich.

Piëch joined Porsche at a critical time in 1963 and worked on the development of the 911 engine.

He became head of R&D and led the run of legendary racing cars from the 906 to the 917, bringing Porsche its first win at Le Mans. He left Porsche in 1972 when the family members withdrew from day-to-day management.

After working as an independent designer, Piëch joined Audi in 1972 where he championed four-wheel drive, turbocharging, the inline five-cylinder engine and a hugely successful rally program. He eventually went on to become CEO and Chairman of Volkswagen. He acquired Bentley, Bugatti and Lamborghini among other brands for Volkswagen after saving the company from near-bankruptcy. He also remained a key shareholder in Porsche and helped maintain Porsche's independence during the difficult period of the early 1990s.

Piëch outmaneuvered Porsche's 2009 attempt to takeover VW, resulting in Porsche becoming a VW brand. His surprise 2015 resignation from the VW Supervisory Board Chairmanship came during another power struggle and shortly before the diesel scandal erupted. Piëch was known for his boundless ambition, energy and autocratic leadership style. He was also known for his vision which sometimes made it hard for colleagues and subordinates to keep up. Some speculated his drive was born of resentment for having been born a Piëch rather than a Porsche. The connection between the Piëch-VW culture and the massive diesel emissions scandal will likely be a topic of speculation and analysis for decades to come. Ferdinand Piëch died in 2019 at age 82.

LEFT: Ferdinand Piëch.

Targa Top

In the September 2017 issue of *Excellence* magazine, a story on Porsche sales manager Harald Wagner discussed the origin of the name given to Porsche's removable roof panel 911. Wagner worked for Porsche from 1954 to 1988. When the 911 came along, it was a coupe and even a sunroof was not initially available. Wagner was worried that 30% of 356 sales had been cabriolets. Given general concerns about future safety regulations, Porsche elected to engineer a removable roof 911, rather than a true cabriolet. Since the new car was neither a coupe nor a convertible, it needed a name. Wagner and his team looked for the name of a race track that hadn't already been used by another manufacturer. They came upon the Targa Florio but worried that people would refer to the car as a 'Flori'. Wagner suggested dropping 'Florio' and just calling it the 'Targa'.

As the Excellence article pointed out, the team didn't quite realize how good the name was. First, at that point, Porsche had already won the Targa Florio five times. These were prestigious, World Sportscar Championship-level wins for Porsche. Second, the word *targa* means plate, plaque or shield in Italian, so it was symbolic of the new 911's protective roll hoop. With the roof off, the 911 Targa even resembled the open or semi-open-cockpit cars that most manufacturers raced in Sicily. 'Targa' has since become something of a generic term, as in targa top, for cars with a removable roof panel.

1967 911 Targa.

911 Developments

The first major change to the 911 arrived in 1966 for the 1967 model year. The 911 'S' was a substantially upgraded version. It offered 30 horsepower above the base model with higher compression, bigger valves and a new camshaft profile allowing for higher engine speeds. The 911S had a higher-performance carburetor from Weber (all 911s were converted to Weber carburetors earlier in 1966). 1967 was also the first model year for the classic Fuchs forged alloy wheel option and in the later 1960s, wheel width grew from 4.5 to 5.5 and then 6 inches. The 911S also received Porsche's first ventilated brake discs for a street car.

The Targa (removable roof with roll hoop) option also debuted in the 1967 911s. Early Targas were equipped with soft plastic, zip-out rear windows. Fixed glass rear windows became an option in 1968 and the soft window was phased out during 1969. Factory sunroofs remained an option but the Targa was much more popular, at times reaching 40% of 911 sales.

1960s

PORSCHE POINTS

» Porsche built one 901 Cabriolet prototype, in June 1964. Although judged too expensive to build with adequate rigidity, the car then became the prototype for the 911 Targa. It was rescued as a non-runner (no drivetrain, suspension, wheels) by friend of the factory, Manfred Freisinger, who installed a period-correct drivetrain. As one of two 901 prototypes known to survive, it sold for over $600,000 at an RM auction in 2017.

» Daimler-Benz had acquired a controlling interest in the post-World War II remnants of Auto Union in 1958. In the mid-1960s, Volkswagen acquired majority ownership and combined Auto Union and its brands with NSU (acquired in 1969). VW gradually unified products under the Audi brand leading to the formation of Porsche-Audi for US marketing in the late 1960s.

» 911 serial numbers followed 356 practice with simple sequential numbers, starting at 300001, from 1964 through the 1967 model year. For 1968, Porsche changed to a coded 8-digit VIN with the last four digits being the sequential chassis number. 1969s have a nine digit number, then through the 1970s and 1980, Porsche used 10 digits, but often changed the coding from year to year. In 1981, Porsche adopted the 'World Production Code' (WPO for Porsche), 17-digit coded VIN.

» In 1966, students at the University of Aachen designed a sporty, low-slung, aluminum body 911 concept with a forward-hinging canopy roof and doors. It brought a touch of 1960s exoticism (or kit-car kitsch) to the 911. The 911 'HLS' was then realized by coachbuilder Hans-Leo Senden. Porsche's involvement with the project remains unclear, but the car did sit (rotting) with Porsche for nearly 50 years before being acquired for restoration by Manfred Hering. The status of the *Klappdach* 'folding roof' 911 is unknown as of this writing. The University of Aachen project also produced a similar prototype with a Targa-style roof that was also restored by Hering.

PORSCHE DECADES | 1960s

In the 1968 model year, Porsche introduced the 911 'T' (for Touring) as the first reduced-price 911. Performance remained quite good even with a detuned, 110-horsepower engine that could run well on lower-grade fuel. The 911S was dropped from the US lineup temporarily, due to emissions requirements, but returned later in 1968 for the 1969 model year. In the US, the 911 'L' (Luxe) was a standard 911 with all the 'S' trim and interior upgrades. Also in 1968, Porsche's first two-pedal option appeared. Called the 'Sportomatic', this transmission allowed shifting without a clutch pedal. The driver's initial movement of the gear lever electrically disengaged the clutch allowing for a change of gears. This was a distant forerunner to modern semi-automatic transmissions. The system was complex and expensive to repair in its day and surviving examples are fairly rare. The 'Sporto' was popular at first, but became a special order option in 1970, lasting until 1980.

For the 1969 model year, Porsche introduced a new range of 911s, the T, 'E' for *Einspritzung* or 'fuel injection' along with the top performance 'S'. The E and S shared a new mechanical fuel injection system. The T or E could be optioned with 'S' trim and amenities. 911 fenders were slightly flared to allow for the 6-inch wheel option. Perhaps the most important change for 1969 was the longer wheelbase, extended by 55 mm or about 2.5 inches. This helped to reduce the tendency for the 911 to oversteer and reduced front-to-back pitching.

The 912 was discontinued in 1969 to make way for the 914 (see Chapter Seven). In September 1969, for the 1970 model year, Porsche introduced larger 2.2-liter engines for the 911.

Fuchs Wheel

The Otto Fuchs company had worked with Porsche on tank wheels in the 1940s. With the introduction of the 911, Porsche sought to develop a light alloy wheel. Fuchs proposed a forged alloy five-spoke wheel which was introduced for the 1967 model year. It was the first forged aluminum wheel available for a production car. Forging versus casting allowed for greater strength in the metal, combined with reduced weight. Porsche's Heinrich Klie was responsible for the now-classic five-spoke styling. Finishing the wheels by polishing, anodizing or painting (especially on the spoke surfaces) resulted in many style options over the next 25 years.

Janis Joplin

Rock singer Janis Joplin had one of the more famous Porsches, a 356C Cabriolet painted in a psychedelic 'History of the Universe' mural design appropriate to the late-1960s flower-power era. The design was created by her friend and roadie, Dave Richards. After Joplin died in 1970, the car was eventually inherited by her brother and sister. They restored the well-used car and returned it to its original Dolphin Gray color, but in 1994 the Richards artwork was recreated. This was done by the Denver Center Theatre Company art department for lobby display at performances of *Love, Janis*, based on sister Laura Joplin's book. The car was then displayed for many years at the Rock and Roll Hall of Fame. It was sold at auction in 2015 for $1.76 million.

Janis Joplin's 356 at the Amelia Island Concours, 2019. (Jay Gillotti)

911 In Competition

In the later 1960s, the 911 worked its way into Porsche's very active racing program. Privateers began campaigning 911s as well. In 1966, Jerry Titus was able to win the SCCA D-Production Championship in the US driving a Vasek Polak-entered 911. At Le Mans, a privately-entered 911 finished 14th and won the 2-liter GT class. Also in 1966, Günter Klass won the Group 1 European Rally Championship driving 911s. Vic Elford, who desired to move away from Ford in England, convinced Huschke von Hanstein to provide a 911 for the 1966 Tour de Corse, an all-pavement rally in Corsica where Vic finished third overall. This event brought Vic into the Porsche fold and he went on to win the Group 3 European Rally Championship for 1967 while Porsche also won Group 1. 911s began to rack up numerous national-level rally championships.

The 911 'R' (Racing) was an experimental program to create an ultra-light 911 that might have been homologated as a GT car (a sort of forerunner to modern the modern GT3 concept). In the end, Porsche elected to build only 20 beyond the four prototypes, rather than the 500 needed for homologation. The 911R would have cost twice as much as a 911S and Porsche judged the demand too low. Still, the 911R was a superb performer with its lightweight chassis 'tub', aluminum and fiberglass body panels, Plexiglas windows, stripped interior and an engine very similar to the 906. It remains one of the most desirable and valuable of all 911s.

The winning 911R, with the victory laurels from the 1967 Marathon de la Route, poses in front of its prototype cousins.

PORSCHE POINTS

- 1968 911s are easy to identify by their side reflectors just above the bumpers ahead of the front and behind the rear wheel arches. Also in 1968, windshield wipers were changed to being parked on the driver's side so that the driver's view would be swept immediately when switched on.

- Porsche designed the VW EA266 as a possible Beetle successor. Looking much like a forerunner to the Golf/Rabbit, the EA266 was a mid-engine hatchback with its water-cooled, inline four-cylinder engine lying on its side under the rear seat. VW was prompted by competition from the Opel Kadett and ultimately chose to go with the now-familiar front-wheel drive, transverse front-engine design of the 1970s Golf and its siblings.

- In 1966, Porsche experimented with a twin-cam version of the 911 six-cylinder engine. Although not a success, the engines were tried in the 911R at a few events.

- As Vic Elford pushed rally success for Porsche and the 911 in 1967, he could customize the transaxle for every event. With all the various gear options and three different final drive ratios, Vic had some 43 (!) different gear ratios to work with.

PORSCHE DECADES | 1960s

Trailing Throttle Oversteer

'Oversteer' is the term for when the rear wheels slide outward in a turn and a car begins to spin. Reducing power (trailing throttle) in a corner can be particularly dangerous, especially in a rear-engine car. The 911 had a reputation for oversteering although experts like Vic Elford pointed out that it was really an understeering car when driven correctly. 'Understeer' is the term for when the front wheels slide straight ahead while turned (pushing or plowing). Over the years, Porsche worked to mitigate oversteer in the 911 but the real solution is to enter turns at a proper speed, not lift (or worse, brake) in panic while cornering.

Although the 911R never achieved great success in racing, it did bring Porsche some glory in a record-setting exercise. A Swiss team had planned on using a 906 to set several records at the Monza oval. However, when the 906's suspension succumbed to the pounding of the uneven Monza banking, Porsche sent two 911Rs (one for spare parts), driven on the road from Stuttgart to Italy as emergency substitutes. Over a six-day period at Monza, the Swiss drivers, Jo Siffert with Rico Steinemann, Dieter Spoerry and Charles Vögele, set numerous speed records for a 2.0-liter car including 15,000 and 20,000 kilometers, 10,000 miles, and 72 and 90 hours. The average speed for 20,000 kilometers was just over 130 mph.

In August 1967, Porsche had taken a team of Sportomatic 911s to the Nürburgring for the Marathon de la Route, an 84-hour race that succeeded the old Liège-Rome-Liège Rally. With a rally-spec 175-horsepower engine, the surviving 911R won the event having driven over 6100 miles at an average speed of 73 mph. The race used the entirety of the North and South loops of the course. Vic Elford, learning every blade of grass at the 'Ring, drove the winning car with Hans Herrmann and Jochen Neerpasch. The 911R won the Marathon again in 1968, driven by the reliable Herbert Linge with Willi Kauhsen and Dieter Glemser. Gérard Larrousse won the 1969 Tour de France Auto and the Tour de Corse with the 911R.

The greatest of the early 911 competition successes came in 1968 at the Monte Carlo Rally. After nearly winning in 1967, Vic Elford and navigator David Stone used a rally-prepped 911T with a 180-horsepower engine to win the event overall (yet another first for Porsche during this era). A 911 also finished second, driven by Pauli Toivonen, as Porsche overcame the established competition from BMC Mini Coopers, Lancia and Alpine-Renault. The 1969 911 victory in the 'Monte' brought Swedish driver Björn Waldegård tears of joy.

1969 Tour de Corse victory poster. Success in top-level rallies proved the 911's mettle as a competition car.

Volkswagen of America

During the 1960s, Porsche transitioned away from Max Hoffman and began managing their own dealer network in the US. However, Porsche's cooperation with Volkswagen (including work on the upcoming 914 project) led to a new partnership in 1969. VWoA (initially led by future VW Chairman, Carl Hahn) introduced the Audi brand to the US market and most Porsches were then sold through Porsche-Audi dealerships. Some VW dealers also began to carry the Audi brand. This arrangement carried on until the mid-1980s.

Erich Strenger

Christophorus magazine's first art director also became famous as Porsche's poster artist.

Recruited by Richard von Frankenberg, Strenger started in 1951. Porsche wanted to produce posters celebrating important race victories. Intended mainly to decorate showrooms and workshops, they were an important part of Porsche's 'win on Sunday, sell on Monday' advertising strategy. Strenger brought a bold modern art creativity and graphic sensibility to the posters which have now become prized collector's items. Strenger also worked on sales brochures and factory manuals. After 37 years with Porsche, he retired in 1988 and passed away in 1993.

Bold graphics of the 1969 Spa victory poster are emblematic of Strenger's style.

Porsche 917

One of the most important of all Porsche cars, the 917, was born of a seemingly minor rule change by the FIA. In 1968, the FIA was concerned that endurance sports car racing based on 3-liter, Group 6 Prototypes might not produce sufficient entries. It is not certain if some manufacturers (like Porsche) lobbied for a change to Group 4 homologation rules. In any case, the minimum production number for Group 4, 5-liter Sports cars was reduced from 50 to 25. Ferdinand Piëch and his team quickly seized the opportunity even though they were heavily invested in the new 3.0-liter 908 project. In June of 1968, Piëch and a small team of his engineers moved forward in secret on the Type 917 with its Type 912, 12-cylinder engine.

This audacious move was based in part on the sheer number of racing cars that Porsche was already building. Building 25 917s was not a tremendous reach considering the numbers of 906/910s, 907s and 908s that had been constructed. Efficiency was achieved by adapting the existing 908 (aluminum chassis, general shape and layout) to a much larger engine, initially at a 4.5-liter capacity. The engine, Porsche's largest to date, used many dimensions from the existing flat-eight, but with four additional cylinders. Hans Mezger, concerned about vibration and lubrication issues with such a large engine, designed the signature central power take-off. A system of gears would take power from the center of the crankshaft to a separate output shaft running back to the transaxle. The maze of gears also ran the valve train and most of the engine's accessories including the cooling fan perched atop the engine. Unusually for Porsche, the Type 912 engine was not a boxer, but rather a 180-degree V12 (where opposing pistons move in the same direction). The project number (912, indicating 12 cylinders) was a clever bit of subterfuge allowing for potential confusion with the 912 street car which had started as project 902. Such was Porsche's confidence that parts were ordered for the engines without any prototype being built. It worked out well with the first engine making more horsepower than the engine's 528-pound weight. After tuning and refinement, the 917 would race with approximately 580 horsepower during 1969.

In less than nine months, the 917 was designed and construction started. The first car stunned the racing world when unveiled on March 12, 1969 at the Geneva Salon. After wrangling with the sporting commission of the FIA, Porsche was required to complete all 25 cars for homologation inspection, which took place on April 21, 1969 with the cars famously lined up in front of Werk I like a squadron of fighter planes. The 917s were all presented with long tails featuring wings and movable, suspension-actuated flaps designed to level the car in the corners. The 917 was offered with an interchangeable short-tail option. By coincidence, the movable flaps were outlawed by the FIA after Le Mans in 1969 as a result of problems with suspension-mounted wings in Formula One.

The 917 had made its track debut at the Le Mans test weekend but all was not well. The car was tremendously fast but dangerously unstable. Although some suspected chassis flex or suspension issues, the low drag strategy was reaching the outer limits of aerodynamic understanding and the car was producing lift at the rear wheels. Porsche persevered with a tentative first race outing at Spa (long-tail) and then the Nürburgring (short-tail). At the 24 Hours of Le Mans, the race the 917 was clearly meant to win, Porsche's expectations were low with such a new and tricky handling car. Sadly, the first privateer 917 crashed on the first lap, killing its owner/driver (the Englishman, John Woolfe). Rolf Stommelen played the 'hare' displaying the 917's tremendous speed until the clutch failed. Vic Elford had requested the other 917 entry with a cunning plan to drive cautiously with fellow Englishman, Richard Attwood. Even with careful driving, they used the speed advantage to build a four-lap lead by Sunday morning. Unfortunately, the transmission case cracked allowing oil to disable the clutch. As noted above, Porsche nearly won the race anyway, with Hans Herrmann narrowly beaten by Jacky Ickx in the Gulf GT40.

Even before Le Mans, Ferry Porsche's growing concern about the cost of racing, particularly in the form of engineering time and focus, had planted the seeds for a major change. For the 1970/71 seasons, Porsche decided to outsource the factory racing entries to none other than John Wyer and his Gulf-sponsored JW Automotive Engineering team. As author Karl Ludvigsen documented in the last edition of *Excellence Was Expected*, the financial considerations were mitigated by Volkswagen's substantial subsidy of Porsche's racing budget (with the provision that Porsche's racers continue to use air-cooled engines). Although it may have been suspected at the time, this sponsorship was only confirmed decades later in books by Ferdinand Piëch and Volkswagen Chairman Carl Hahn (an eventual successor to Heinz Nordhoff who died in 1968).

The still spooky-handling 917 managed its first win at the season-ending 1000 KM race at Zeltweg, Austria in August 1969. Jo Siffert and Kurt Ahrens were the winning drivers but much work was left to be done. In October, at the same track, Porsche convened a test session, the first to which the JWAE personnel were invited in preparation for the 1970 season. Gulf team engineer, John Horsman, famously noticed the splatter of dead gnats on the front of the car but almost no bugs on the tail. In a superb feat of 'shade-tree' engineering, Horsman and team reshaped the tail of the 917 using sheet aluminum, rivets and duct tape. They were supported by Porsche engineers, Peter Falk and Helmut Flegl, who were under orders from Piëch not to increase drag but who also instinctively knew that the car needed more downforce on the rear end. By the end of the test, using the cobbled together 'Horsman' tail and Porsche's own redesign of the nose, plus Firestone tires, the 917 was five seconds per lap faster. Falk and Flegl had the unenviable task of explaining to Ferdinand Piëch the 'bad' news that drag would have to increase in favor of downforce but the car was much faster! Equally important it was stable and controllable for the drivers.

Porsche worked quickly to refine a new tail for the 917 *kurz*, for 'short' tail. Among many benefits, the upward sweeping ramps of the new tail allowed for a lighter, more open configuration at the rear. This not only improved cooling for the transaxle but also allowed all of the exhaust to be routed to the rear of the car (rather than having a partial exhaust exit ahead of the rear wheels). This made the car less ear-shattering for the drivers and eliminated problems with fuel vaporization. In a secret test at Daytona in November 1969, the 917 was five seconds under the existing lap record.

After his first seven laps at Zeltweg with the new tail, Brian Redman had pronounced *'now it's a race car'*. Little did anyone know how prophetic that statement would be…

> **PORSCHE POINTS**
>
> ≫ In 1968 and 1969, Porsche was beaten at Le Mans by the Gulf team Ford GT40, Chassis 1075. In 1984 and 1985, the Joest-Porsche team repeated this feat, winning Le Mans two years in a row with the same car (956B Chassis 117). Joest and Porsche repeated the feat again in the 1990s with the WSC 95 car derived from the ex-TWR-Jaguar XJR-14. The Bentley 'Old Number 1' was the first car to win consecutive Le Mans races (1929-1930) and it has recently been determined that Ferrari's 275P Chassis 0816 was the winner in 1963 and 1964.

Twenty-five 917s lined up for homologation inspection, April 21, 1969.

1970s

1969 » Porsche 914
911 Developments
1970 » Porsche 917
Le Mans
Porsche 908/03
Hans Mezger
Family Business
1972 » Porsche 917 Can-Am Turbos
Safari Rally to Safari Builds
IMSA
1973 » Porsche 911 RS/RSR Carreras
Impact-bumper 911
Tony Lapine
1974 » Porsche 911 Turbo
1976 » Porsche 934/935
Type 2304/05 Wiesel AWC
Manfred Jantke
Porsche 924
Porsche 936
1977 » Porsche 928
Porsche 911 SC (Super Carrera)

I'm telling you, we're going to do it!
— *Ernst Fuhrmann, as quoted by Norbert Singer*

Porsche began the 1970s with a new entry-level car, the 914. It also had larger engines for its 911 line, and the reshaped 917K was ready to race. The 917 started the decade with a bang at Daytona in January 1970. The Gulf team entry driven by Pedro Rodriguez and Leo Kinnunen won by *45 laps*. Jo Siffert and Brian Redman finished second for the new Gulf-Porsche team even with a long pit stop to replace the clutch. Porsche also moved into the new decade with a new factory building, adjacent to Werk I, capable of increasing production beyond the 1970 rate of nearly 17,000 cars. By the end of the decade, Porsche would offer two lines of water-cooled, front-engine sports cars along with the 911.

Porsche 914

The *Vierzehner*, 'Fourteener' was Porsche's first real alternative and entry-level model. Unlike the 912, which it replaced, the 914 was a completely new chassis/body design, conceived by Ferry Porsche and Volkswagen's Heinz Nordhoff in 1965. Porsche's goal was to extend the brand at a price below the 911. VW was tempted by a more visible partnership with Porsche for a true sports car to bring customers into dealer showrooms. The concept and basic design were certainly sound, as modified and highly developed 914s prove even to this day. However, the 914 was compromised by several factors, not all within Porsche's control.

First, the joint marketing project, where 914s were branded 'VW-Porsches' everywhere but the US, caused some identity confusion. Was it a VW or a Porsche? Although the 914 had less Volkswagen content than the early 356s (for example), it was too easy to deride the 914 as the 'Volks-Porsche'. Second, the gentleman's agreements between Ferry Porsche and Heinz Nordhoff changed when Nordhoff began to have serious heart trouble, passing away in April 1968. His successor, Kurt Lotz,

Porsche 914/6 in Phoenix Red.

had not been given time to learn all the intended details of a project that was rather small in the grand scheme of Volkswagen's business. This led to a major problem with the cost accounting for 914 bodies built by Karmann for Porsche to sell as the six-cylinder 914/6. A further economic problem came with the revaluation of the Deutschmark. With the US intended as a major market for the new car, pricing and profitability became huge challenges as the value of the mark increased relative to the US dollar. Finally, the early 914s were not as well developed as journalists and customers might have expected from Porsche.

The styling of the new 914 came in for its share of criticism, although it has aged rather well as an example of functional, period-modern industrial engineering. Primarily the work of Heinrich Klie with Butzi Porsche, the 914 was unlike earlier Porsches, with its sharper lines and relative lack of Komenda-style curves. However, it did feature innovations like pop-up headlights and Porsche-style cues like the front fenders and 904-inspired flying buttress B pillars.

The 914 was always intended generally as a mid-engine sports car capable of good performance either with a VW four, or Porsche six-cylinder engine. It could have replaced the VW two-seater, the Karmann Ghia, although that model continued almost through the 914's production life. Since Porsche had long worked with VW-owned Karmann, it was logical to assign production of the 914 to Karmann's factory in Osnabrück. Conceived with a removable 'targa' top, the 914 was designed with a very strong chassis using a roll hoop for greater rigidity and crash protection. Front suspension and steering were very similar to the 911 and could use VW and Porsche components. The rear suspension was Porsche's first coil spring-over-shock/strut arrangement for a street car. Combined with a trailing arm ahead of the rear wheels, the suspension allowed for the installation of either four- or six-cylinder engine ahead of the transaxle. With almost no weight ahead of the front or behind the rear wheels (low polar moment of inertia), the 914 was responsive and great in the turns (born to autocross).

The 914/4 used the 1.7-liter, electronic fuel injection engine from the relatively new VW Type 4. The Bosch-VW electronic injection system helped to pass ever more demanding emissions requirements in America. Mid-engine design allowed for excellent storage space in both front and rear trunks. A key feature of the 914 was the fiberglass removable top designed to fit easily in the rear trunk, still leaving some room for storage. The 914 had a roomy but low interior with a good driving position and excellent visibility for a mid-engine car. On the downside, early 914s had a fixed passenger seat (seat back integrated with the firewall) and a vague gear shift action that was universally criticized.

For cars to be sold directly by Porsche, completed body/chassis/interior units were shipped to Porsche from Karmann. Porsche then installed the suspensions and drivetrain, using the 2-liter, six-cylinder engine from the 1969 model 911T (for the 1970 model year, the T received the new 2.2-liter Porsche six). The 914/6 could be equipped with Fuchs alloy wheels for a more 'Porsche' look although many were sold with steel wheels and VW-like hubcaps. The 914/6 certainly outperformed the 914/4 but fell victim to its proximity in price to the 911T. For relatively few extra dollars, customers could purchase a 2.2-liter, 2-plus-2

> **PORSCHE POINTS**
>
> » The 914 joint venture between VW and Porsche was named *Vertriebsgesellschaft*, 'sales or distribution company'. A large sales and parts distribution warehouse was built at Ludwigsburg, north of Stuttgart for the VG. This arrangement, not to mention the US Porsche-Audi distribution agreement and the 914 itself contributed to persistent rumors that VW would acquire Porsche.
>
> » A trio of Signal Orange 914/6 GTs dominated the 84-hour Marathon de la Route in 1970, finishing first, second and third.
>
> » Four-cylinder 914s have their ignition switch on the right. 914/6s have the key in the traditional left-hand Porsche location.
>
> » The 1970 Le Mans GT-winning 914/6 was entered by Sonauto, owned by Auguste Veuillet, driver of the first Porsche at Le Mans in 1951.
>
> » After making the rounds on the show circuit, Giugiaro's 914-based 'Tapiro' was sold to a Spanish industrialist. A group of left-wing terrorists planted a bomb under the car and it was heavily damaged by fire. Re-acquired by Ital Design, the remains of the Tapiro were turned into a lawn sculpture at their HQ building.
>
> » The 914 served as a platform for one of GM's many mid-engine Corvette concepts. In the case of the 914, it was powered by Wankel rotary engine. Rivaling the 911 for model longevity, the Corvette finally changed to a mid-engine design in 2019.

Porsche promotional photograph of a 914/6 continues the theme of spring flowers.

seating 911. Porsche only sold 2,657 914/6s in the 1970 model year and 432 for 1971. The model was phased out and only available by special order in the final 1972 model year. With less than 3,500 built, the 914/6 almost immediately caught the attention of those in-the-know for its future collectability.

Porsche used the 914 platform for some intriguing prototypes. The '916' was a hardtop 914 with 914 GT fender flares and suspension, powered by a 2.4-liter 911S engine. Only 11 were built and they were scooped up by Porsche family members or close friends of the factory. The experimental department also built two 914/8s with 908 race engines to explore the 914's potential as a supercar. One was presented to Ferry Porsche for his 60th birthday. Independent designers also used the 914 as a platform for intriguing concept cars, including Giorgetto Giugiaro's gorgeous, wedge-shaped Tapiro with its Mangusta-like gullwing doors and engine covers. Another was the Murene by French designer Jacques Cooper for the Hueliez company. Porsche's old friend Albrecht von Goertz and the Italian coachbuilder, Frua, also offered their interpretations on the 914 platform.

For the 1972 model year, the 914 received a much-needed, adjustable passenger seat. The 1.7 engine was improved to run better on lower cost, regular gasoline. An improved 914 arrived for the 1973 model year, with a 2-liter engine producing a target 100 horsepower (less when tuned for US emissions). This version of the VW Type 4 engine was engineered specifically for the 914 by Porsche and not used in any other models. A new transmission with a shorter, improved gear linkage was another key development. VW-Porsche sold more than 27,000 914s in 1973, the closest ever tally to the original target of 30,000/year. The 1974 base model received a mildly improved 1.8-liter engine. Also during 1974, 1,000 appearance-package, Limited Edition 914 2.0s were sold. Later known as the Bumblebee and Creamsicle for their color combinations, these cars received front and rear anti-roll bars along with Mahle cast alloy wheels. For the 1975 model year, large rubber impact-bumper covers appeared on the 914s, integrated with the shape of the nose. For the final, 1976 model year 914s, the base engine was dropped.

Over 115,000 four-cylinder 914s were sold, more than 80,000 in the US. It was an important learning experience for Porsche if not very profitable. Porsche did not return to the concept of a mid-engine street car again until the mid-1990s.

911 Developments

The early to mid-1970s would be a period of rapid developments and changes to the 911, much of which was needed to keep up with evolving emissions and safety requirements in the all-important US market. The 911 entered the 1970s with a larger bore, 2.2-liter engine and the 911S was the first Porsche street car with gross horsepower reaching the 200 mark. The 2.2-liter engines were the first to change their Type numbers from 901 to 911. Also new for 1970 was a wider selection of bright and interesting colors, thanks to a new paint shop facility.

1971 was a relatively poor year for Porsche, in terms of sales. This was due to several economic factors including inflation impacting basic production costs and exchange rates. Inflation and the relative weakness of the dollar made Porsches ever more expensive in the US where a 911S cost about $9,000 (more than three times the price of a VW Beetle). However, things improved for the 1972 model year. The 911s received a longer stroke 2.4-liter engine and a new transmission. Carburetors were discontinued in the US as the 911T received mechanical fuel injection. Larger capacity, lower compression engines were meant to maintain performance levels with the use of lower octane, unleaded fuels (needed to meet increasingly stringent emission standards). However, with overall power and torque gradually increasing, the Type 915 gearbox was introduced. It was a sturdy design based on the unit used in the 908 spyder racing cars and stayed in use through the 1986 model year. Among many improvements, it put the first four gears in the familiar 'H' pattern, with 5th gear up and to the far right. Also for 1972, the oil tank was moved ahead of the right rear wheel and given an outside filler. This led to potential disaster if gasoline station attendants mistakenly tried to add fuel at the oil location. In 1973, the oil tank returned to its old location and the outside filler disappeared. 2.4-liter 911s are the last of the 'long hood' body style and well preserved or restored examples are considered highly desirable by collectors (especially the mid-year 1973 911T models with Bosch's new CIS, for continuous fuel injection, system).

The early 1970s were a period of continuing success for the 911 as a competition car. In 1970, a 911 won the Monte Carlo Rally for the third straight year, Björn Waldegård driving the winning car as he had in 1969. With the increase in size of the base engine to 2.2 liters, GT racing 911s were now in the 2.5-liter class. As a result, Porsche and its customers began experimenting with numerous versions of larger 911-based engines. Cars in this category brought GT class wins at Le Mans in 1971 and 1972. John Fitzpatrick won the European GT category in 1972 driving for the Kremer brothers while in the US, Hurley Haywood won the IMSA GTU class driving for Brumos. This success was a prelude to the legendary exploits of 911-based racers from 1973 into the early 1980s.

The expansion of private customers representing Porsche in all manner of race series prompted the founding of the Porsche Cup in 1970. Cash prizes were awarded based on a weighted point system accounting for various categories of racing around the world. The first winner was Gijs van Lennep based on his successes driving the Martini Racing team's 908 and 917 during 1970.

PORSCHE POINTS

» In September 1970, Porsche Club of America, Pacific NW Region member Dave Nurse bought the 150,000th Porsche produced, a 911S. Nurse was the facility manager of Boeing's Seattle and Renton, WA factories.

» The 911 'ST' was a series of competition cars with engines up to 2.5 liters. Custom built with lightweight shells for GT racing and rallying in 1970/71/72, less than 25 were sold. STs were forerunners to the 911 RS and became the inspiration for legions of 911 outlaw and tribute builds. However, the 'ST' designation was not used in period.

» In July 1971, NASA launched Apollo 15 with the first Lunar Rover. Boeing and NASA studied Dr. Porsche's wheel hub motor concept during design.

Porsche 917

The 917K, with its revised nose and tail, sped into the 1970 season with a crushing victory in the 24 Hours of Daytona. As noted in Chapter Six, the official factory entries were run by John Wyer's Gulf-sponsored JWAE team. A back-door, factory-supported team appeared in the form of the Piëch family's Porsche-Salzburg operation. 917s were also sold to true privateers.

The Gulf team dominated the season, winning six of eight Championship races entered with their blue and orange 917s. Ferrari competed with its own 5-liter sports racer, the 512S, but only managed one victory, a narrow win at Sebring led by Mario Andretti. With support from two wins by the 908/03s in the Targa Florio and Nürburgring 1000 KM, Porsche easily won their second World Manufacturer's Championship.

The all-important 24 Hours of Le Mans set up as a titanic battle with no less than 11 Ferrari 512s lining up against seven 917s. Porsche entered two special long-tail 917s under the Salzburg and Martini teams. Vic Elford put the Salzburg 917 LH with a 4.9-liter engine on the pole, becoming the first to lap Le Mans at greater than 150 mph average speed. The improved long-tail was much more stable than it had been in 1969 and speed on

Le Mans

In 1971, Steve McQueen's documentary-style racing film, *Le Mans*, was released. The film combined race footage from the 1970 24 Hours of Le Mans with staged scenes completed between June and November of 1970. *Le Mans* is controversial for its lack of traditional plotting and the star's erratic and egocentric behavior during filming. It is also treasured for its captivating look at a peak era of motor racing with the glorious 5-liter, 12-cylinder Porsches and Ferraris battling at high speed. Filming was supported by Porsche, fresh from their first overall win in the big race. As with many of Porsche's racing activities in period, the advertising value of the film and the association with one of Hollywood's biggest stars are beyond calculation.

Start of the 1970 24 Hours of Le Mans. Vic Elford (#25) and Jo Siffert (#20) lead into the Dunlop Curve in their Porsche 917s.

Winner of the 1970 24 Hours of Le Mans, 917 Chassis 023, driven by Hans Herrmann and Richard Attwood.

the Mulsanne straight was greater than 230 mph. In the race, heavy rain, crashes and mechanical failures winnowed the field. Porsche veteran Hans Herrmann along with Richard Attwood inherited the lead and had a nerve-wracking time holding it for 14 hours in their 4.5-liter, Salzburg-entered 917K. They had posted the slowest qualifying time of any 917 in the field but proved the old adage, to finish first, first you must finish. In the end, Porsche achieved their long-sought first overall victory at Le Mans. The long-tail 4.5-liter Martini entry, famously painted in the 'Hippie' swirls of green over blue finished in second place. Porsche won every class and all the prizes available in the race. Ferdinand Piëch's expensive gamble on the 917 paid off.

Even before the 1970 season, the FIA had decided to eliminate the 5-liter 'Sports' cars from endurance racing competition. 1971 was the last year for the 917 in this form of racing, but Porsche continued to develop and improve the cars. JWAE won five of nine Championship races with their 917s, including the fastest sports car endurance race ever run. At Spa-Francorchamps, the cars driven by Pedro Rodriguez/Jackie Oliver and Jo Siffert/Derek Bell averaged an incredible 154 mph over the 1000 KM distance on the public road circuit. Vic Elford and Gérard Larrousse won at Sebring for the Martini team (having taken

The 917/20 at Le Mans in 1971, driven by Reinhold Joest and Willi Kauhsen.

For the famous 1971 917/20 'Pink Pig' livery these are the translations for the cuts of meat:

Hals (Neck)	Rückenspeck (Back Bacon)
Haxen (Feet)	Rüssel (Snout)
Hirn (Brain)	Schinken (Ham)
Kotelett (Chop)	Schulter (Shoulder)
Lende (Loin)	Schwanz (Tail)
Ohrlappen (Earlobe)	Wamme (Belly)
Rippe (Rib)	

over from Porsche-Salzburg). Porsche again put on a crushing display at Le Mans. A trio of 917 LHs were doing better than 240 mph on the Mulsanne but couldn't last the distance. Porsche's second Le Mans victory went to Helmut Marko and Gijs van Lennep in a Martini-entered 917K. Unbeknownst to the drivers, the car had an experimental chassis made with magnesium tubes in place of aluminum. With warm, dry weather throughout, the winning car set a Le Mans distance record that lasted until 2010. Porsche again won the World Championship although 3.0-liter Alfa Romeo spyders won three of the eleven races. With the Championship limited to 3.0-liter Prototypes for 1972, Porsche withdrew from the series to concentrate on racing the 917 in the North American Can-Am as well as further developing the 911 for competition.

> **PORSCHE POINT** » At Daytona in 1970, Brian Redman drove a stint in the Rodriguez/Kinnunen Gulf 917 that won the race (while his own car was in the pits for clutch replacement). After the repair, Redman was passed on track by his own car, driven by Jo Siffert.

Porsche 908/03

Inspired by the 909, the final iteration of the 908 was conceived and built to contest just two races in the 1970/71 seasons. The 908/03 was similar to the 'Flounder' 908/02 spyder, but even more compact. The key development was moving the gearbox ahead of the diffcrential. This brought all the weight within the wheelbase but pushed the driver even further forward (feet in front of the front wheels). It made for a highly maneuverable car with the 908's reliable 350 horsepower and total weight at just 1,200 pounds.

The car was intended only to race at the twisty Targa Florio and Nürburgring 1000 KM races (where the 917 would be ill-suited). 908/03s won Porsche's home race both years and won the Targa in 1970. The cars went to Sicily wearing suit of cards identifiers in tribute to Dr. Porsche's 1922 Saschas. Privately-owned 908/03s went on to race all through the 1970s and into the early 1980s, eventually using turbocharged six-cylinder engines.

Porsche team 908/03s at the 1970 Targa Florio. Left to right, Jo Siffert, John Wyer, Ferdinand Piëch and Vic Elford.

Hans Mezger

The youngest son of an innkeeper in Ottmarsheim saw his first auto race at Hockenheim in 1946, won by Hans Stuck. The man most identified with Porsche engines graduated from the Stuttgart Technical Institute in 1956 to 28 job offers (but not from Porsche). Inspired by the 356 and wanting to work on sports cars, he applied to Porsche and was offered a job working on diesel engines but ended up in the calculation department to be closer to the sports cars. Early in his career, he developed improvements to the Fuhrmann four-cam, 356 pushrod four, and the eight-cylinder F1 engine. He was also heavily involved in the production six-cylinder 901/911 engine project.

For much of his career, Mezger also had overall responsibility for race car design and construction. His masterpiece was the flat-12 engine for the 917. The turbocharged version of the 917 not only provided crushing levels of performance, but also led to Porsche's use of turbos in street cars. A further Mezger triumph was the TAG F1 engine designed for the McLaren team. Between 1984 and 1986, this engine powered three Driver's and two Constructor's World Championships. Mezger remained the guiding light for Porsche engine development, especially in the 911, and his basic architecture survived into the 2000s with the 911 GT3 and Turbo.

Mezger's quote in an SAE presentation summarizes a key aspect of Porsche philosophy: *"We know what we can take over from earlier designs and we also know what we do not want to take over. One thing is evident; the more experience, the smaller the risk in a new development, and the quicker – and thus also the cheaper – the development process. Porsche likes development work, and its participation in racing is one among several reasons for this."*

Hans Mezger retired from Porsche in 1994 and passed away in June of 2020 at age 90.

Hans Mezger in 1984.

Company designations in German

- **KG** *Kommanditgesellschaft*, meaning limited partnership. Porsche was a limited partnership early in its history, so readers will see references to Porsche KG in books or articles on the history before 1972.

- **AG** *Aktiengesellschaft*, meaning stock company in German. Porsche converted to a stock company in 1972 as part of Ferry Porsche's plan to remove all of the Porsche family members from day-to-day management of the company. The full, formal name is listed as Dr. Ing. h.c. F. Porsche AG, which stands for *Doktor Ingenieur honoris causa Ferdinand Porsche*. This refers to Ferdinand Porsche's honorary degree(s) as a Doctor of Engineering.

- **GmbH** *Gesellschaft mit Beschrankter Haftung*, meaning limited liability company and is used mainly for privately-held companies. The first formal iteration of the Porsche company, in 1931, was designated as a GmbH.

- **SE** Porsche SE refers to the current version of the Porsche family holding company that owns approximately 53% of the voting shares of Volkswagen (which, in turn, partially owns the Porsche brand and manufacturing/design company, Porsche AG). SE stands for *Societas Europaea*, or European Company in Latin. The full name is Porsche Automobil Holding SE.

Porsche 917 Can-Am Turbos

The North American Can-Am series featured sprint races run to FIA Group 7 rules, which placed very few limits on the cars. There was no minimum weight and no limit on engine size. Porsche saw it as a development and promotional opportunity in their largest export market. The 917 engines could grow to 5.4 liter capacity, but that was nowhere near the 8-liter, Chevrolet-based monsters that dominated in Can-Am racing. Porsche turned to a new and important solution – exhaust-driven turbocharging. The concept had been around for decades, particularly in aircraft, and successfully applied in oval-track racing. However, to apply it in sports car racing, where engine speeds are highly variable, was a tremendous challenge.

Porsche employed one turbocharger for each cylinder bank of the 917's flat-twelve, initially using the 4.5-liter version, but eventually moving to the 5-liter for the 917/10 used during the 1972 season. The engine was capable of delivering about 1,000 horsepower, depending on boost. However, fuel metering through the mechanical injection system proved to be a significant challenge when adapting the proven 917 engine to forced induction. In the pre-electronic era, 'mapping' of the fuel/air mixture had to be achieved by mechanical means, plus there had to be a reliable system for maintaining air flow in the induction system at low rpm while relieving excess pressure at high rpm. Turbo 'lag' (where the increase in power is delayed slightly while the turbochargers build up boost pressure) required the drivers to adapt their driving style, including the use of left-foot braking while keeping the throttle open to maintain boost level.

Porsche selected the Roger Penske organization to conduct the official factory race operations in the 1972 and 1973 Can-Am series. 917/10 spyders were sold to other teams (initially with 5.4-liter, non-turbo engines). Penske's driver/engineer, Mark Donohue, proved invaluable in assisting with the development of the 917/10. In addition to engine development, the aerodynamics of the body shape and the suspension geometry were substantially revised in collaboration between Porsche and Penske. After finishing second in the first race at Mosport

Family Business

By 1970, tensions and rivalries within the Porsche and Piëch families became a significant concern for the future of the company. With eight of Dr. Porsche's grandchildren as shareholders, it was natural that there might be differences of opinion, ambition and talents for those taking an active role in managing the family business. Between the autumn of 1970 and the spring of 1972 a gradual series of changes largely resolved the issues.

The family members all agreed to withdraw from day-to-day roles in running the company. The new rule was that all had to agree to any one member returning to an active role (which was unlikely). Ernst Fuhrmann, father of the four-cam engine, returned to Porsche eventually becoming head of the management board (CEO). Heinz Branitzki succeeded Hans Kern as head of finance and commercial departments. Helmuth Bott became head of R&D. The families formed Porsche GmbH as a holding company for shares in Porsche KG (the limited partnership) which converted to Porsche AG (stock company) by the end of 1972. Ferry Porsche was chairman of the supervisory board from that point forward.

Author Karl Ludvigsen quotes Ferdinand Piëch regarding the improved family relations in *Excellence Was Expected*: *"We're happy to meet with each other at all the relevant birthdays and various festivities. We still have our ups and downs, but not so angrily."*

PORSCHE POINTS

» Before committing to turbocharging for its 917 Can-Am cars, Porsche experimented with a 16-cylinder version of the Type 912, flat-12. Possible capacities ranged from 6 to 7.2 liters. A surviving 16-cylinder engine resides in 917 Chassis 027 at the Porsche Museum.

» Penske racing secretly provided a McLaren M8D (Denny Hulme's 1970 car) to Porsche in November 1971, for evaluation at Weissach during the development of the 917/10.

» Valentin Schäffer, Porsche's ace mechanic-turned-engineer, was a key collaborator with Hans Mezger from the beginning of Porsche's work on turbocharged engines, starting with the 917/10. Schäffer adapted the IndyCar, AiResearch pressure-relief valve to limit maximum boost in the first 917 turbos. As early as 1969, Schäffer had experimented with a turbo engine in a 911.

Mark Donohue in the Penske Racing 917/30. *"The perfect racing car."*

Park in Canada, Donohue was injured in a testing crash at Road Atlanta. George Follmer was drafted by Penske to drive the remaining 1972 Can-Am races. In the early races, the battle between the Chevy-powered McLarens and the 917/10 was fairly even. However, as the season wore on, the Porsche advantage grew. In the end, Penske and Porsche won six of nine races and Follmer, with five victories, won the championship.

In 1973, Porsche rolled out the awesome 917/30 for Penske and Donohue. With a longer wheelbase, revised aerodynamics (including a longer tail based on Le Mans experience) and an even more powerful engine, the 917/30 proved nearly unbeatable. Depending on turbo boost level, about 1,100 horsepower was available for race conditions. Maximum horsepower on the dynamometer was approximately 1,500! A minor accident in the first race and a fuel cap leak in the second prevented Donohue from winning (those races were won by 917/10s). Starting with the third race, at Watkins Glen, Donohue and the 917/30 romped to victory in the final six races of the season, securing Porsche's second Can-Am Championship.

For 1974, the SCCA adjusted the Can-Am rules for fuel consumption, making it unlikely for the 917/30 to be competitive. Porsche withdrew from the series although Penske did enter

> ≫ In 1972 and 1973, as a weight-saving measure, Mark Donohue had no reverse gear in his Can-Am 917s. When questioned by Porsche on what would happen if the car went off track and ended up facing a barrier, Donohue said he was only paid for going forward.
>
> ≫ Mark Donohue pronounced the 917/30 *"the perfect racing car"* and speculated that if he drove it *"as hard as I could for the whole race, I'd end up five laps ahead of everyone."* Laguna Seca 1973 was the only Can-Am where he lapped the entire field at least once.
>
> ≫ In 1974, Porsche built a road-going 917 for Count Gregorio Rossi of Martini & Rossi. Chassis 030 had been used extensively for testing an early form of ABS. Legend has it the car was granted Alabama registration on condition it never be driven there.

their remaining car for Brian Redman to drive at Mid-Ohio in a one-off entry where Brian finished second overall to Jackie Oliver in a Shadow. Turbocharging quickly made its way into Porsche street cars and continues to this day as an important technology for increasing horsepower.

917/30 Closed Course Record

Following in Porsche's long tradition of involvement with speed records, Mark Donohue's Racemark company modified the Penske 917/30, Chassis 003, with assistance from Porsche, to set a new closed course speed record. In August 1975, with Donohue driving at the Talladega Super Speedway in Alabama, the record was set at 221.12 mph even though the turbocharged 917 engine was not originally designed for continuous top-speed running. Sadly, Donohue was killed the following week in practice for the F1 Austrian Grand Prix. The record stood until 1979 when broken by a Mercedes C111-IV *Rekordwagen* at the Nardò test track in Italy.

Safari Rally to Safari Builds

With all its success in competition during the 1970s, the 911 never quite won the actual East Africa Safari Rally. Founded to celebrate the coronation of Queen Elizabeth, this was one of the toughest off-road tests in the world, racing over 3,000 miles of unpaved roads and trails. A Porsche was first entered in 1968 and the factory made a significant effort in 1971. 911s finished in second place in 1972 and 1974. A full factory effort with Martini-sponsored 911 SCs resulted in second and fourth places in 1978. A 911 finally won the 2019 'Classic' version of the Safari.

Recent years have seen significant growth in the popularity of 'Safari' builds, customized 911s built for off-road and overland driving in tribute to the original Safari Rally. Porsche's own 911 Dakar was announced in 2022.

Porsche 911 RS/RSR Carreras

With Ernst Fuhrmann taking over day-to-day management of Porsche in March 1972, and with the end of the 917 program for endurance racing, focus shifted to the 911. The production cars had to be updated to maintain performance against increasing emissions and safety regulations. On the racing side, Porsche desired to create a high-performance 911 for homologation as a Group 4 Special GT car. With 911 production planned to last at least until 1980, Porsche also needed the advertising benefits of a successful racing program.

IMSA

The International Motor Sport Association was founded in 1969 by ex-SCCA Executive Director, John Bishop in conjunction with Bill France of NASCAR. IMSA gained in popularity during the 1970s as an alternative to SCCA professional racing series', mainly the Trans-Am. The 1980s were a golden age for IMSA as the leading sports car racing series in America. Porsche 935s, Porsche-powered March GTPs and the Porsche 962 were highly successful but the series attracted participation over time from Ford, GM, BMW, Jaguar, Nissan and Toyota. After a split with the American Le Mans Series in the 2000s, this form of racing reunified in 2013 and IMSA lives on as the governing body for the WeatherTech SportsCar Championship.

The first step in a long line of mid-1970s developments for racing the 911, was the 2.7-liter 911 RS Carrera, announced in September 1972. The revival of the 'Carrera' name was significant at this point and not without some risk. Homologation would require building, and selling, 500 of these special lightweight, higher horsepower 911s planned for racing in 1973. In charge of the project was the fast-rising young engineer, Norbert Singer. In addition to Porsche's typical facility with weight reduction, the 911 Carrera introduced innovations such as larger wheels at the rear (necessitating more shapely rear fender flares) and the *Bürzel*, or 'duck tail' spoiler. The spoiler had the effect of reducing aerodynamic lift as well as overall drag and improving engine cooling for the standard 911 shape. For the engine, Porsche borrowed bore and stroke dimensions from the 917/10 along with 'Nikasil', the Mahle nickel-silicon alloy process for cylinder lining technology from its big race car. Standard horsepower was 210 in a car weighing only 1,985 pounds.

Porsche's styling department emblazoned the new 911 Carrera with distinctive side graphics boldly stating the car's name (as a body-color relief from the side stripes). For road use, the RS could be purchased in its stripped form or as a 'Touring' version with some interior comforts (and weight) added back. For competition use, some RSs were taken from the assembly line and finished in the Werk I customer sport department as Carrera RSRs (for *Rennsport Rennwagen*). The RSR received a full competition-spec engine capable of 300 horsepower from 2.8 liters. RSRs also received numerous other performance modifications including still-larger rear wheels, 917-based brakes, racing seat with harness, interior roll bar, fire suppression system and more. The popularity of the 911 Carreras surprised Porsche and in spite of some difficulty getting the cars certified for road use, eventual production reached 1,580, including the RSRs (which were clearly documented as 'not for public highway' use). These legendary 911 Carreras remain among the most desirable of all 911s among collectors.

In 1973, the new 911 racers achieved towering successes, immediately cementing their legend. At the 24 Hours of

Daytona, the RS had not yet received official homologation approval, so the RSRs had to run as Prototypes against pure racing cars from Matra and Gulf-Mirage. Porsche sent RSRs to both Peter Gregg's Brumos team and Penske Racing with Mark Donohue and George Follmer driving. The pure racers from Matra and Gulf were much faster but could not last the distance. It set up for a close battle between fierce rivals running the new RSRs. The Penske crew built up a substantial lead by 5 AM when a piston failed due to manufacturing error. This left the Brumos team of Peter Gregg and Hurley Haywood with a massive lead on the field. Despite concerns about the possible piston problem, the Brumos team carried on and won by 22 laps over a Ferrari 365 'Daytona'. It was the first overall win in an FIA World Championship-level race for a 911-based car.

But the RSRs were just getting started. Brumos' dynamic duo won again at Sebring in March, teaming with Dr. Dave Helmick in his RSR but without any competition from pure FIA sports prototype racers (Sebring was not part of the Championship series that year). Michael Keyser and Milt Minter were second in another RSR. A third triumph for Porsche in major 1973 races came in May. In Sicily, the Targa Florio was run for the final time as an FIA-sanctioned, World Championship sports car race. Home favorites Ferrari and Alfa Romeo sent their Group 6 Prototypes to claim the honors for Italy. Porsche entered two Martini & Rossi-sponsored RSRs with newly-designed, full 3.0-liter (2993cc) engines along with a 2.8-liter that had been used as the practice car. The RSRs ran as prototypes in Sicily due to several non-homologated developments.

Despite Brian Redman's admonitions to 'go carefully', co-driver Jacky Ickx (attempting the Targa for the first time) crashed out in their Ferrari 312PB on the third lap when avoiding a rock in the middle of the road. The other Ferrari, driven by Arturo Merzario and local hero Nino Vaccarella came to grief on the first lap when Merzario had a flat rear tire. He struggled back to the pits on the rim but the damaged drive shaft let go on lap two. Alfa Romeo had countered Ferrari in 1973 with a flat 12-cylinder of its own, installed in a tubular chassis with sleek new bodywork. The Targa was the first appearance for these new cars, driven by Rolf Stommelen/Andrea de Adamich and Clay Regazzoni/Carlo Facetti. Regazzoni left the road during practice and somersaulted down a hillside, leaving the car too badly damaged to race. The surviving Alfa 33TT12 led the race until lap four when de Adamich was forced off the road by an errant Lancia Fulvia, damaging the suspension.

From that point, Gijs van Lennep and Herbert Müller took control with their RSR. They were fortunate that the Dunlop tire rep, Dieter Glotzbach, became nervous about the hot weather and called for an unscheduled right rear tire change at the first fuel stop. It turned out the tire had an audible puncture! The Porsche pit team also had to deal with a false report of problems

LEFT: 911 RSR on the way to victory at the 1973 Targa Florio.

with the leading car from an Italian commentator saying: *Otto kapute!* The emergency pit location up in the mountains then reported that the car went by their post as normal. Müller and van Lennep carried on to score Porsche's 11th overall victory at the Targa. They beat the second place Lancia Stratos by about six minutes. Leo Kinnunen and Claude Haldi finished third in the other 3.0-liter RSR while the 2.8-liter practice car took sixth overall. During a week of practice, with five to six laps a day, Van Lennep memorized all 720 corners of the 44-mile circuit to avoid mistakes such as "hitting a house (!) or a spectator."

Not surprisingly, the RSR dominated in its class at Le Mans through 1976. During the mid-1970s, 911 Carreras racked up fistfuls of victories in SCCA Trans-Am, IMSA GT and numerous European national race and rally championships. As late as 1977, Hurley Haywood, driving with amateur enthusiasts Dave Helmick and John Graves (Ecurie Escargot) won the third victory for an RSR at the Daytona 24 Hour.

An interesting offshoot of the RSR program was the IROC version, purchased by Penske for the made-for-television, International Race of Champions series. These cars were taken from the 1974 series of 3.0-liter RS 911s, but given some RSR features, including full race-spec engines. They also featured a new rear spoiler design. The flatter, tray-style spoiler improved on the aerodynamic features of the *bürzel* and would become a familiar feature on 911s for years to come.

The ultimate development in the RS/RSR program was the 1974 turbocharged version built for Le Mans. Because of the FIA formula for forced induction engines, a turbo RSR could not have an engine larger than 2.142 liters. This line of development appealed to Porsche based on proposed rule changes coming in 1976 for Group 5 'silhouette' racers and (as it turned out) for Group 6 Prototypes. The single turbo engine inherited technology and inspiration from numerous earlier Porsche racing and street car engines. Installation of the turbo required significant modification to the rear part of the chassis. The car was also aggressively lightened, down to just 1,818 pounds, motivated by an engine that could produce over 450 horsepower. The RSR Turbo did run into some trouble with its very large rear wing and tail structure. Ernst Fuhrmann was concerned its appearance would detract from the advertising connection to the road-going 911. However, he also insisted that the engine have an intercooler (a radiator-type device that cools the hot, compressed air coming from the turbocharger before it enters the engine). The tail was redesigned to make space for the intercooler and the wing was painted flat black to disguise its appearance in photographs.

> **PORSCHE POINTS**
>
> ›› An early project for Butzi Porsche's independent design company was a watch meant to be given as recognition to key employees and important Porsche customers. Watches became one of the most important Porsche Design products.
>
> ›› A 911 RS prototype was the first street-going Porsche to use an anti-lock braking system (ABS). Chassis 0781 was assigned to Dr. Fuhrmann as part of the general RS testing program and also received special treatments from the styling department under Tony Lapine.
>
> ›› The famous ducktail spoiler on the Carrera RS was the product of development work by engineer Hermann Burst along with Tilman Brodbeck and stylist Rolf Weiner, working under Harm Lagaaij. The team wanted to create the benefits of a cut off, 'Kamm' tail while maintaining the basic shape of the standard 911.
>
> ›› The 911 RS 'Carrera' side script famously changed from positive lettering over a wide stripe to negative relief lettering with a thinner stripe for the production cars. The layout was an early project for future head of styling Harm Lagaaij.

> **PORSCHE POINTS**
>
> ›› Racing improves the breed in all sorts of ways. In December 1972, Porsche tested the 917/30 at the Paul Ricard circuit in the south of France. While testing the 917, Mark Donohue also pitched in on 911 RSR development with Gijs van Lennep, Herbert Müller, Norbert Singer and 917 engineer Helmut Flegl.
>
> ›› At Daytona in 1973, Hurley Haywood survived a crash with an unfortunate seagull that destroyed the windshield. A new windshield had to be sourced from a 911 street car in the paddock.
>
> ›› The RSR 911 was the first to rely on coil springs for the suspension although the torsion bars had to remain in place for rules compliance. This was an early example of Norbert Singer's creativity in interpreting the racing rules.

The car was challenging to drive with its extra-large rear tires, significant rear weight bias, and tremendous horsepower (although it came on with turbo lag). Using the third iteration of the new turbo-six, with a horizontal rather than vertical cooling fan, Gijs van Lennep and Herbert Müller finished second overall at Le Mans in 1974. They might have won the race when the sole surviving Matra prototype had gearbox trouble on Sunday morning. However, the French car had a Porsche-designed gearbox. Porsche could not reasonably refuse their client's request to assist with a pit lane rebuild of the gearbox and that car, driven by Henri Pescarolo and Gérard Larrousse, won the race. The RSR Turbo led briefly before it lost fifth gear and the Matra returned to the track. The RSR Turbo was the first turbocharged car ever to race at Le Mans.

Porsche only raced the RSR Turbos six times during 1974, then set them aside, knowing all-new cars would need to be prepared for the 'silhouette' formula, which was delayed until 1976. Porsche did provide a small number of the 2.1-liter turbocharged engines to customers still racing the 908/03. Various versions of these 908s continued racing with some success all the way until 1981.

PORSCHE POINTS

>> The Carrera RSR of 1973 was the first 911 to produce 300 horsepower from the flat-six engine. Such was the success of the Carreras that Porsche finished third in the Manufacturer's World Championship behind pure racing cars from Matra and Ferrari, and ahead of the Gulf-Mirage team.

>> For the RSR Turbo, Dr. Fuhrmann proposed to Norbert Singer that the 911 monocoque (or 'tub') be dipped in acid to thin the metal and save weight (as had been done for some American racing cars). The idea was explored but then rejected when the subcontractor required building a larger vat to accommodate 911 and the cost was estimated at 50 to 80,000 marks.

>> In 1975, Peter Gregg and the Brumos team developed a controversial 'long-tail' version of their IMSA 911 RSR which foreshadowed Porsche's eventual direction with the 935.

>> At Daytona in 1977, Hurley Haywood wore his co-drivers' helmets for some of the night driving. The switcheroo was tolerated by IMSA in the interest of safety since Hurley's teammates were less experienced drivers.

Impact-bumper 911

Of all small-volume sports car manufacturers, Porsche designed the most elegant solution to the US impact-bumper regulation (required to withstand a 5 mph impact without damage). For the 1974 model year, the 911's front hood was shortened to make room for a nicely integrated bumper (with hidden pistons allowing the bumper to 'give' and then return to position). Accordion-style rubber bellows integrated the bumper edges with the front fenders. The 911s also received a larger, hefty rear bumper with bulky rubber pads that was somewhat less graceful. Deeper, more substantial rocker panels connected the new look from front to back. This overall shape of the 911 would remain basically unchanged for the next 15 years. Although chrome exterior window trim was still available, the 911 moved toward a blacked-out look around its glass areas. Chrome trim was removed from the interior and along with various other interior improvements, the 911 received excellent new seats with integrated headrests.

For 1974, the 911 engine grew again, now to 2.7 liters, to keep the balance between performance and emissions. All 911 engines were fuel-injected, most making use of the Bosch K-Jetronic 'Continuous' injection system. In 1974, the 911 range included a base 911, the 911S and the 'Carrera' (somewhat equivalent to the previous T, E and S range). US Carreras were quite different from their European siblings (due to emissions requirements). A similar lineup with some variations continued through 1976. For 1975 California cars, thermal reactors (a precursor to catalytic converters for emission compliance) caused numerous heat-related problems in the engines. Heat stress could cause general problems with the 2.7-liter including cylinder head studs pulling out of place and excessive oil leaks. Most of the existing cars with these engines have long since been rebuilt and re-engineered to overcome the original problems.

Porsche sales declined drastically in 1974, hit hard by the OPEC oil embargo and subsequent fuel crisis. Reliability and quality problems were not helpful either during this period. A brighter spot was the 1975 Silver Anniversary 911S, celebrating Porsche's 25th year as manufacturer. A significant development in 1976 was full dip, hot-galvanization zinc-coating of the steel components of 911 bodies to prevent rust. This technology proved highly effective and over the following decade became common throughout the auto industry. In 1977, the 911 received improved valve guides and new timing chain tensioner/guides to address ongoing reliability issues. Some felt that the 911 was past its prime, the basic concept outdated and due for a replacement but the top-of-the line 911 Carrera was upgraded to a 3.0-liter, fuel-injected engine (essentially a

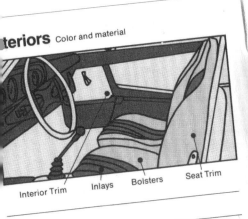

Interiors — Color and material

Interior Trim | Inlays | Bolsters | Seat Trim

914 MODELS

Color	Interior Trim	Seats Trim	Inlays
Black Brown Beige	Leatherette	Leatherette	Leatherette Basketweave
Black Brown Beige	Leatherette	Leatherette	Corduroy Fabric

911 & CARRERA MODELS

Blue Black Copper Red Brown Beige	Leatherette	Leatherette	Leatherette Basketweave
Blue Black Copper Red Brown Beige	Leatherette	Leatherette	Shetland Fabric**
Blue Black Copper Red Brown Beige	Leatherette	Leatherette	Tweed Fabric**
Blue Black Copper Red Brown Beige	Leatherette	Genuine Leather	Genuine Leather Perforated
Blue Black Copper Red Brown Beige	Genuine Leather	Genuine Leather	Genuine Leather Perforated

NOTE Genuine Leather optional at extra cost. All interiors are available with any exterior color. Specifications subject to change without notice.

**Bolsters and Head Restraint Trim in Twill Fabric.

911 & Carrera Models (Coupe & Targa)

Light Yellow, Lime Green, Orange, Chocolate Brown, Grand Prix White, India Red, Peru Red, Desert Beige, Mexico Blue, *Aubergine, *Copper Brown Metallic, *Gazelle Metallic, *Salmon Metallic

914 Models

Light Ivory, Bahia Red, Olympic Blue, Saturn Yellow, Phoenix Red, Ravenna Green, Zambezi Green, Sunflower Yellow, Signal Orange, *Black, *Gold Metallic, *Marathon Blue Metallic, *Alaska Blue Metallic, *Silver Metallic

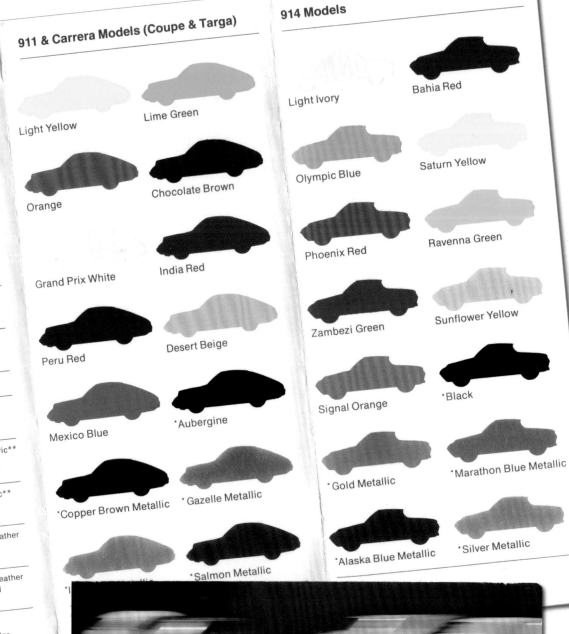

Porsche brochure pages from the mid-1970s. 911 Turbo at right.

1970s

Turbo Carrera — 911 S Targa — 911 S Coupe — 912 E

In a world of compromise, PORSCHE doesn't.

In 1948, the first Porsche was built to meet the standards of an individual who would not accept compromise in any sense of the word. That individual was Ferdinand Porsche. Today, Porsche cars are still being built for people who will not accept any form of compromise in quality, workmanship or performance.

Yet in spite of its high standards, or perhaps because of them, Porsche stands out among today's practical cars as a distinctively sensible automobile.

It is extremely compact on the outside, yet luxuriously comfortable and pleasing on the inside.

It offers high, seemingly limitless, performance at amazingly low fuel consumption. It provides maximum safety for its occupants, yet requires a minimum of maintenance. In short, a Porsche is as sensible as it is rewarding.

The rewards of owning a Porsche come to you each day you drive it. Every time you approach your Porsche, you feel a surge of pride in knowning what the car is, and what it can do. (Ask any Porsche owner.) You know, for example, that you are driving a car that can win fiercely competitive races, if you so choose. On the open road, it will accelerate and track as an extension of yourself. Yet it is well-mannered for city driving or for leisurely meandering, as your wishes dictate. In short, your Porsche is ready to do what you want it to do. Any time. This is the product of a dictum of no compromise.

You may choose from three models: the 911S, the Turbo Carrera and the 912E.

The PORSCHE 911S is the PORSCHE.

We have been making a 911 since 1964. Although today's 911S looks very much like the original, it has been vastly improved. It is built with more sound-deadening material and the engine operates at 1/3 less rpm's than older models, to produce the same speed. The result is, of course, a quieter, smoother ride.

The current 911S is also easier to drive. Its higher-torque engine, combined with higher gear ratios, gives you more performance with less gear shifting. So you get all the responsiveness you need, without paying the price of driver fatigue.

If you want the 911S in a convertible, you want the Targa. It is a unique design that provides all the exhilaration of an open car without sacrificing the safety and comfort of a coupe. It has a built-in roll bar, a fixed rear window, and snug fitting folding top.

The Turbo Carrera. The ultimate PORSCHE.

The Turbo Carrera is a very special 911 model that has a turbocharged engine. Turbocharging uses the engine's exhaust gases, which would otherwise go up in smoke, as a source of energy to pre-compress the engine's intake air. This additional energy provides an enormous performance boost on demand. Yet this car is completely docile in city traffic and leisurely country driving.

Standard equipment includes such luxury features as air conditioning, a stereo radio with power antenna, power windows, a headlight washing system, a two-speed rear window with wiper, fog lights and front and rear seat belts.

In all respects, the Turbo Carrera is a true personal car.

The 912E. A popular model

Porsche enthusiasts will recall that 1965–69, as a four cylinder 4[...] many 912's on the road, in ex[...] loyal following. And so we have [...] quite different from its predece[...]

The engine is a design that to[...] Also, the 912E has the modern [...] So if you'd like a thoroughly up-t[...] Now that you've read about our [...] never considered yourself a "spo[...] quality, precision and responsiven[...] Porsche firsthand.

And nothing's easier. Your Porsch[...] experience.

If you're like us, it's an introduction [...]

Technical Data. 1976

1976 Porsche	912 E	911 S	Turbo Carrera
Engine			
Number of cylinders	4	6	6
Displacement	1971 cc	2687 cc	2993 cc
Compression ratio	7.6 : 1	8.5 : 1	6.5 : 1
Horsepower-SAE net	86	157	228
Max. torque/rpm (ft. lbs-SAE net)	98/4000	166/4000	245/4000
Lubrication	forced circulation	dry sump with separate oil tank, thermostatically controlled oil cooling, full flow oil filter	
Fuel supply	electric pump	electric pump	2 electric pumps
Mixture	electronic fuel injection	CIS injection	CIS injection, turbo-charged
Electrical			
Voltage	12	12	12
Battery rating	44 amp/hr.	66 amp/hr.	66 amp/hr.
Alternator	980 watt	980 watt	980 watt
Ignition	Capacitive Discharge (CD)	Capacitive Discharge (CD)	Capacitive Discharge (CD)
Transmission			
Standard	5-speed	4-speed	4-speed
Optional	–	3-speed Sportomatic	–
Chassis			
Suspension	4-wheel, independent	4-wheel, independent	4-wheel, independent
Springing	torsion bars	torsion bars	torsion bars
Brakes	4-wheel disc brakes, internally vented	4-wheel disc brakes, internally vented	4-wheel disc brakes, internally vented
Wheels	5½ x 15 steel	6 x 15 pressure cast alloy	Front-7 x 15 forged alloy Rear-8 x 15 forged alloy
Tires	165 HR 15	185/70 VR 15	Front-185/70 VR 15 Rear-215/60 VR 15
Performance			
Top speed (approx.)	110.6 mph	134 mph	152 mph
Acceleration (0–62 mph)	13.5 sec.	7.8 sec.	5.8 sec.
Miles per gal. U.S.	approx. 25 mpg	approx. 24 mpg	approx. 22 mpg

In addition, the 911S model is available in two body styles; as a Coupe or as a Targa convertible with removable roof panel.

Your Porsche dealer will be glad to discuss with you the standard equipment furnished on each Porsche model as well as the options which may be selected to help personalize your Porsche. Some of the vehicles and equipment shown or mentioned in this catalog may not be available in some areas. We reserve the right to make specification and equipment changes without notice.

33-71-660 10 Printed in Germany. SVA, Ludwigsburg.

	911 Early 1964-1973	**911 Mid-Year 1974-1977**
ENGINES	2.0, 2.2 and 2.4 liter carburated or mechanical fuel injection, horsepower range 125 to 190	2.7 and 3.0-liter, emission control issues, CIS fuel injection, horsepower range 150 to 175
CHASSIS	Original short wheelbase lengthened for 1969 model year	Galvanized for rust protection starting in 1976 model year
STYLING	Pure, original, long hood; Fuchs wheels and Targa roof introduced in 1967	Impact-bumpers can be polarizing, Carrera fender flare option
OTHER	Sportomatic offered from 1968 model year, introduction of RS, RSR in 1973	Redesigned seats, introduction of 911 Turbo with 245 horsepower

Anatole 'Tony' Lapine

Born in Latvia, Tony Lapine apprenticed at Daimler-Benz before his family moved to America in 1951. He worked in America for GM (including on some Corvette projects) and then in Germany for Opel where he became head of the Opel Research Center. Ferry Porsche was impressed with Lapine's advanced 'skunkworks' projects at Opel and hired him to lead Porsche's styling department in 1969. His team, including ex-Opel employees Dick Söderberg and Wolfgang Möbius, became known for pure style exercises like the liveries for Le Mans racers but they also had to take Porsche into the future by styling the 924 and 928. The impact-bumper 911 was one of the first major projects to emerge from the styling studio under Lapine and remains a classic. The last major project under Lapine was the Type 964 version of the 911, another critical project for Porsche that involved major restyling of the bumpers.

Lapine was known for his irrepressible, sometimes irreverent style and quotes such as: "If the engineers come in and they like it, you have to start again." He also worried if the marketing or sales side liked a design concept since their view might not represent what would appeal to customers in the future. Although he was never a great fan of the 911, he became as dedicated as any of his colleagues to Porsche tradition and ethos. In 1975, he visited Maffersdorf in Czechoslovakia and brought back a container of soil from Dr. Porsche's birth site to celebrate the centenary.

Lapine was an accomplished driver in his 356 Carrera and an enthusiastic amateur racer. Lapine retired from Porsche in 1988 after a heart attack and passed away in 2012.

2.7 Carrera MFI

Never imported to the US, due to emissions, the 'Euro' Carreras with mechanically fuel-injected, 2.7-liter engines are considered a hidden gem among 1970s 911s. They have short-hood, impact-bumper bodies often with ducktails or later tray-style spoilers. Built in 1974 and 1975 (with a small additional series in 1976 available only through Porsche's 'special wishes' department), these cars were often finished to highly interesting specifications and style. Road performance was close to the Carrera RS since the engines are the same. These were the last mechanically-injected street cars built by Porsche.

911 Turbo, without the turbocharger). In the typical Porsche fashion of continuous improvement, the 1977 911s had better clutch pedal operation and braking, new suspension options, a much-improved climate control system and better door locks. A happy moment came on June 3, 1977 with the celebration of the 250,000[th] Porsche car built.

Porsche 911 Turbo

Often referred to by its '930' project number, the 911 Turbo was arguably the first Porsche 'super car'. In the great Porsche tradition of transferring technology from racing to its street cars, turbocharging made the transfer from the 917 program very quickly to the 911. In fact, 911 Turbo development work began in 1972 while the turbocharged 917s were still racing.

General Motors had introduced turbocharging in street cars as early as 1962, although not with great success. The German-based Swiss, Michael May, a Porsche customer and tuner, had

912E

Conceived as an interim model for the US market only, the 912 'E' bridged a gap between the 914 and the 924 as the entry-level Porsche. During the 1976 model year, these 911-based cars were sold with a fuel-injected, 2-liter, four-cylinder VW-based engine similar to that of the 914. It used the L rather than D-Jetronic Bosch injection system. The 912E was appealing in its lightweight driving character and fuel economy even though significant 911 content was removed in the interest of selling at a lower price.

Typically colorful Porsche promotional photograph for the 912E, complete with sailplanes in the background.

been experimenting with turbocharging for racing and street cars starting in the late 1960s (with the BMW 2002 and Ford Capri, as well as consulting with Toyota on their Group 7/Can-Am cars raced in Japan). Given May's proximity to Stuttgart, the Porsche engineers would have been aware of his work. Much more important was Ernst Fuhrmann's direct push to adapt turbocharging to the 911. It was a tremendous challenge to package the turbo and all its related plumbing in the already cramped 911 engine room. Since Fuhrmann had made room in the 356 for his famous four-cam engine, his engineers could take that as inspiration and now the boss wanted a turbocharged 911 for his own use. Fuhrmann was also concerned that BMW might introduce a turbocharged super car.

Although 911 turbo development was a difficult challenge during 1973, Porsche had enough confidence to show a 'concept' version at the Frankfurt and Paris auto shows in September and October (BMW introduced their 2002 Turbo model at the same time). Based on a 3.0-liter Carrera RS, with its tray-style spoiler and gorgeous fender flares, Porsche spoke of a potential for 160 mph top speeds in the most luxurious Porsche yet built. Among all the technical challenges, Porsche faced a hurricane-force headwind with the energy crisis and economic environment during 1974. BMW and Mercedes both backed off from turbo street car projects during this period. Pricing would be a concern as this new 911 would probably cost $25,000 ($125,000 in today's money). Porsche would be under pressure to build and sell at least 400 cars in 1975 in order to achieve homologation for Group 4 racing in 1976. Ultimately, the decision was to proceed with

> **PORSCHE POINT**
> BMW beat Porsche to market with a turbocharged street car, the 2002 Turbo. The car was developed under industry legend, Bob Lutz, during his tenure at BMW. Running headlong into the energy crisis, BMW soon dropped the turbo option and only 1672 cars were made.

Porsche's first 'super car' in recognition of Porsche's dedication to the sports car business.

Working in close collaboration with Hans Mezger and the racing engineers, along with experienced engine man Robert Binder, a young Herbert Ampferer was put in charge of the final configuration of the new engine. Ampferer had come to Porsche from one of Dr. Porsche's legacy companies, Steyr-Daimler-Puch. The engineering team decided on a 3.0-liter, aluminum case engine with Nikasil-lined cylinders, a relatively low compression ratio (for piston health) and a single turbocharger located at the left rear of the engine. Extensive work was done to manage the flow of air into, out of and around the new engine, meter the fuel properly, and balance performance (including turbo lag) for the street. Peak horsepower for the European version of the 911 Turbo was set at 260.

The car received a new, four-speed transmission designed to handle the greater torque and power load. The suspension was also upgraded and strengthened, with an eye toward future racing models. With its rear wing, front spoiler and flared fenders, the 911 Turbo was impressive to behold. It was also impressive to inhabit with a luxurious, mostly leather interior and air conditioning standard. Options were few, but a sunroof or limited-slip differential could be added. It was the heaviest 911 but also the fastest yet and with a price that ended up coming in above $27,000. The script on the engine cover proudly proclaimed 'Turbo Carrera'.

In the first few years of its life, the 930 received its share of typical Porsche development. For 1977, larger wheels and a redesigned

> **PORSCHE POINT**
> The first 911 Turbo was presented to Louise Piëch for her 70th birthday. It was silver and without tinted windows so that she could see colors accurately on field trips for her painting. The decorative Porsche script on the sides of the car matched the MacLachlan tartan seat inserts. The car had a prototype 2.7-liter engine, 'narrow' body with a whale tail but no 'Turbo' badge on the rear deck.

ABOVE: 1979 Porsche 911 Turbo in Moonstone at the Luftgekühlt event in 2023. (Patrick Kelley)

OPPOSITE: Promotional photograph of a 1978 3.3-liter 911 Turbo with a 911 SC in the background.

wastegate (to eliminate problems caused by condensation) were introduced. Along with the standard 911s, the Turbo also got a new, vacuum servo-assist for the braking system that reduced effort and improved pedal feel. Porsche also added a boost gauge to the tachometer to give the driver a visual reference to the turbocharger's work.

For 1978, the 'Carrera' designation was dropped and the 930s were simply identified with 'Turbo' script. Braking was improved again with stronger four-piston calipers and cross-drilled brake discs. The engine size was increased to 3.3 liters with an intercooler, following racing practice once again. The larger capacity engine improved low-end torque in the rpm range where the turbo had the least effect. 1979 was the last model year for the 930 in the US until 1986 when emission-related issues were finally resolved.

Porsche 934/935

Reacting to new FIA rules for sports car racing that were planned for 1976, Porsche worked on three different cars, coded by the last digit of their Type number. The 934 was planned for Group 4, based on the 930, so essentially a Group '4' 930. The 935 was a 930 heavily modified for Group 5 and the 936 would be a Group 6 Prototype.

The 934 (sometimes referred to as the Turbo RSR) was a GT racer based on the 911 Turbo production volume for homologation. Certain modifications were allowed and the 3.0-liter turbocharged engine resulted in a minimum weight set at 2,470 pounds by the FIA equivalency formula. The engine was upgraded with stronger pistons, special camshafts and a larger turbo, with new water-to-air intercooling (used to save space within the car's required, stock profile) and K-Jetronic fuel injection. Porsche also adopted a horizontal fan for better cooling. With greater turbo boost available, the engines were rated at 485 horsepower in standard form.

The standard suspension set-up was adapted for racing with adjustable shocks and coil springs all around along with adjustable anti-roll bars. The 934 also had upgraded brakes with cross-drilled rotors and center-lock racing wheels supplied by BBS. Fiberglass fender flares covered the wheels. The cars were built from standard production 930s with soundproofing and rear seats removed. A roll cage and racing seat were provided.

934s were only raced by Porsche customer teams. At a price around $45,000, these cars were considered an excellent value although challenging to drive given significant turbo lag. A 934 won the European GT Championship for 1976 with Toine Hezemans driving for the Georg Loos team. 934s also finished

> **PORSCHE POINTS**
>
> » The water-to-air intercooler on the 934 was the first use of water in a Porsche 911 for anything other than windshield washer or battery fluid.
>
> » During 1976 and 1977, Porsche built some 934s to near-935 spec (except for the engine) for racing in IMSA and the Trans-Am series. These cars became known as 934/5s (or 934 and a half). Some private teams did their own conversions to race upgraded 934s in Group 5.

first and second in the SCCA Trans-Am series, George Follmer just ahead of Hurley Haywood in Vasek Polak team entries. The only slight disappointment came at Le Mans where the standard 934s lost out to normally aspirated 911 RSRs in Group 4.

The 935 project was led by the crafty Norbert Singer, who became famous for his creative approach to the FIA rulebook, exploiting every word in the regulations to Porsche's advantage. In the 4.0-liter class, the turbocharged engine could be up to 2.85 liters with a minimum vehicle weight of 2,138 pounds. The engine had mechanical fuel injection with a still higher boost limit, allowing something close to 600 horsepower. Porsche's traditional torsion bars were replaced by coil spring struts all around for the suspension and larger diameter wheels to get more rubber on the road given the 17 inch limit on the width of the rear tires. The 935 also raced with a locked differential and titanium drive shafts from the 917.

The body shape of the 935 constantly evolved but always made use of a very large rear wing above a spoiler-shaped box containing the intercooler. The rear fender flares were wide and curvaceous while the nose of car evolved quickly from a standard-looking 911 fender set up to the now-famous 'slant' nose with the fenders shaved down and the lights relocated to the large front air dam. To save weight, the 935 featured a mostly fiberglass body with Plexiglas side and rear windows. The car ultimately required lead weight ballast in the front to bring it up to the minimum weight.

The 1976 season brought two new and important drivers to Porsche: Jacky Ickx and Jochen Mass. They were the lead drivers for the 935 factory program and went on to long and distinguished careers racing for Porsche. Literally from the first race, at Mugello in Italy, private teams were engaged in building their own versions of the 935. The Kremer brothers had their own build to compete against the factory, finishing second at that first race. Ickx and Mass won the first-ever race in the 935's history in their Martini-sponsored factory car. 1976 proved to be a year of tough competition in Europe, mainly from BMW with their turbocharged CSLs. Porsche also had to contend with challenges from the officials as interpretations of the silhouette rules were negotiated. In the end, Porsche won four races to BMW's three, giving the 'Makes' Championship to Stuttgart.

For 1977, Porsche built customer 935s based on the factory's 1976 car. Porsche also built a small number of their own development 935/77s with twin-turbo engines and new rear body sections. In another creative interpretation of the rules, the stock rear window was left in place but covered by a second rear window that altered the shape of the roofline and tail for better aerodynamics and airflow to the intercooler. With a combination of wins by privateer 935s and the factory cars, Porsche again won the World Championship for Makes.

Group 5 Porsche factory-entered 935 at Watkins Glen in 1976.

>> In 1976, the mayor of Stuttgart honored the four Championship-winning Porsche racing cars, two 935s and two 936s, after they were driven on the road from the Porsche factory to the town hall. This repeated the similar ceremony after Le Mans in 1970 with Arnulf Klett, however, the mayor was now Manfred Rommel, son of the WWII general.

>> In 1977, Porsche built the ultra-lightweight 935 'Baby' for the 2-liter class of the German DRM Championship. Designed and built in less than two months, the 'Baby' was radically altered for sprint racing and had to be ballasted up to the 1,620-pound minimum weight. The front and rear 911 structures were replaced by tube frames. Powered by a tiny 1.425-liter turbo version of the flat-six, 'Baby' was the first Porsche to try carbon-surface brake discs and pads (although they were abandoned due to insufficient cooling). In the prestigious, televised race at the Norisring, Ickx had to stop due to heat exhaustion on a very hot day and with almost no bulkhead material to shield heat from the engine compartment. But the 'Baby' won its second and final race, a support event at Hockenheim during the German Grand Prix weekend.

>> The familiar KKK designation for Porsche turbos stood for Kühnle, Kopp & Kausch AG, the firm that purchased the turbocharger business from Eberspächer.

935 at the Nürburgring in 1976 driven by Rolf Stommelen and Manfred Schurti.

K3 version of the 935 on the way to winning the 1979 24 Hours of Le Mans.

In 1978, Porsche built a most outlandish Group 5 car known as the 935/78 but nicknamed 'Moby Dick'. It featured a much more swooping and elongated body shape attached to the (mostly hidden) 911 tub. The whole car was lowered such that the transaxle had to be turned upside down to avoid excessive inclination of the drive shafts. A key innovation in this car was the change to *water-cooled*, four-valve cylinder heads with the heads welded (rather than bolted) in place. All previous 911 engines had managed with two valves per cylinder in part due to cooling requirements but high boost pressure had made the flat-six prone to head gasket failures. 'Moby' upped the ante to 750 horsepower and powered down the Mulsanne straight at 227 mph! The 935/78 won its first race, driven by Ickx and Mass at Silverstone, by seven laps. However, the factory chose to race it only four times in total with just the one victory before Moby was retired to the Porsche Museum.

The 935 carried on in racing long after Porsche built a final, small batch of cars at the end of 1978 (with the inverted transaxle set-up). As noted above, private racing teams and fabricators built their own versions of the 935 (including copies of Moby Dick). Perhaps most famous was the Kremer Brothers 'K3' that won the 24 Hours of Le Mans overall in 1979. 935s won the IMSA championship in 1978, 1979 and 1980 and they were still competitive in IMSA racing as late as 1985.

Manfred Jantke

As a child refugee whose father never returned from the Russian front, Manfred Jantke nearly died while walking 250 miles across Germany. He started work at the leading German car magazine, *Auto Motor und Sport* in 1959 and became Chief Editor by 1970. Jantke did some racing in the popular VW-based Formula Vee class. In 1972, he became head of PR for Porsche. Intending to stay only a few years, he ended up working at Porsche for two decades. As Porsche's racing manager he was heavily involved in every type of competition and recruited talented new drivers like Stefan Bellof. Jantke became a dedicated Porsche man and in many ways a true successor to Huschke von Hanstein. Disenchanted with leadership in the early 1990s, Jantke left Porsche and returned to journalism, race commentary and consulting.

Type 2304/05 Wiesel AWC

One of Porsche's last major military projects was the 'Weasel', a track-drive, lightweight armored weapons carrier meant for dropping with airborne troops. This maneuverable and quick mini-tank was tested by the likes of Jochen Mass and Jacky Ickx. Designed for the West German army, prototypes were built from 1975 to 1978. Lack of funding delayed production until 1985. About 500 were eventually built in two versions with the larger *Wiesel 2* entering service in 2001.

Porsche 924

Although the 924 became a replacement for the 914 as the entry-level Porsche, the project did not start out that way. Similar to the 914, the 924 had a troubled gestation, with Porsche and Volkswagen again encountering difficulty in a development process that crossed multiple leaders at VW.

The origin of the project was in the joint sales venture between Volkswagen and Porsche formed to handle the 914 (the *Vertriebsgesellschaft*, or VG, in Ludwigsburg north of Stuttgart). The original project number under VW was EA425. However, in October 1973, Volkswagen's withdrawal from the VG could have led to the 924 design becoming an Audi sports car since VW had funded the development costs. This would have deprived Porsche of a badly needed entry-level car for its dealers. At that point, the OPEC oil shock made the 924 less attractive to a company that was extremely busy developing and launching its own Rabbit/Golf and Scirocco models.

But also by this time, Ferdinand Piëch had become an Audi executive. He helped negotiate a deal between the Porsche families and Volkswagen to take over the 924 and gradually repay the development costs which were moderate for VW but huge relative to Porsche's finances. Porsche would own the design and have sole responsibility for quality control with VW acting as a manufacturing contractor. The cars would be built at the ex-NSU, now Audi factory at Neckarsulm (just north of Ludwigsburg and a short distance from Stuttgart). In spite of its difficult gestation, the 924 did prove to be a starting point for a long line of successful four-cylinder cars.

Important goals of the project were to create a vehicle with larger and more usable interior space and to make greater use of standard VW components compared to the 914 (thereby improving profitability). There was also a decision to pursue a family resemblance to another car that was in the planning stages (the car that became the 928). The 924 was certainly a radical departure for Porsche in several ways. Its engine was water-cooled and mounted in front. However, as a Volkswagen/Audi project, it would have been preceded by front-engine, water-cooled cars like the Audi 80, Passat/Dasher and Golf/Rabbit. Front-wheel drive (like its Audi and VW cousins) was rejected as inappropriate for a sports car.

The '4' in the 924 (as it was eventually named) was a reference to the Audi-sourced, 2-liter, overhead cam, inline four-cylinder engine. Similar to the 914, the engine was produced to Porsche's specification by Volkswagen. It was a versatile engine, used by Audi, VW and eventually by AMC in America. The extensive and Porsche-specific engine configuration process was led by Herbert Ampferer. In Europe, the 924 made a respectable 125 horsepower (just a bit less than the first 2-liter 911s). In the US, emission controls were a major handicap, reducing the first 924s to just 95 horses.

The transmission/final drive unit was mounted in the rear of the car for better weight distribution and handling character. The transaxle was linked to the engine by a rigid tube containing the driveshaft, with the clutch installed at the front, behind the engine. This concept had been used in the 1960s by Ferrari in its 275 GTB and 365 Daytona sports cars. It even recalled Dr. Porsche's late-1920s design for a Mercedes Grand Prix car. A similar design for the coming 928, along with the 924, would open the era of 'transaxle' cars at Porsches referring to front-engine, rear-transmission products.

For its suspension the 924 received rear trailing arms from the VW Super Beetle combined with a Porsche-designed torsion bar tube. The front suspension was McPherson strut coil spring

---

PORSCHE POINTS

» A Porsche 935 won the 1979 Le Mans race and in the second place 935, one of the drivers was a certain PL Newman, the actor making his only appearance in the 24-hour classic.

» The 'father' (or as one might call it today, the 'Executive Sponsor') for the 924 was Paul Hensler. Hensler's career spanned Porsche tractors, 911 engine development and race engineering during the 917 period.

units from the Beetle with Rabbit/Scirocco lower wishbones. Anti-roll bars were optional. Rack and pinion steering also came from the 924's Rabbit/Scirocco cousins. While equipped with front disc brakes, rear drum brakes opened Porsche to some criticism on the 924. It is a tribute to Porsche's engineering prowess that mostly stock Volkswagen components could be assembled into a chassis that was immediately recognized for its handling capability.

Styling for the 924 was largely credited to newcomer Harm Lagaaij along with Dick Söderberg. The styling team maintained a Porsche family resemblance and significant difference from the hard-edged Giorgetto Giugiaro-drawn lines of the Rabbit/Scirocco. The aerodynamic shape featured a plunging nose with pop-up headlights, large frameless rear 'bubble' window for storage access, and integrated body color impact-bumpers. A removable roof panel for sunroof-like driving was an option. As with the 911s at this time, the bodies were fully galvanized for rust protection. Heating, ventilation and air conditioning along with many other interior components from the VW parts bin were arrayed in a functional, modern layout with Porsche bucket seats.

Production of the 924 started in late 1975 and US cars arrived later in 1976 as 1977 models. The initial reaction could be summarized as 'great handling car in need of further refinement and more power'. This aligned to some extent with Porsche's view of the 924 as a 'brand extension' targeted for a new client group with a planned evolution over 10 to 15 years (as with the 356 and 911). Those new customers, including those who could not afford a 911, benefited from rapid improvements. In classic Porsche fashion, 924 developments came fast and furious including, in early 1977, an automatic transmission option. A catalytic converter upgrade for the US helped bring horsepower up to 110. Initial sales met expectations and the planned production of approximately 100 cars per day. Also in 1977, Porsche made 3,000 'Championship' edition 924s in white with Martini stripes and special interior trim to celebrate the double sports car racing World Championships of 1976. Among other special edition 924s, the bright red 1979 'Sebring' edition included a vivid tartan interior pattern.

In 1978, for the 1979 model year, Porsche introduced the 924 Turbo (Type 931/932). The new engine design was so significantly changed that final assembly was moved to Porsche

Handsome and sleek profile of the 924 in this Porsche promotional photograph.

The 924 Turbo is shown at right with a standard 924 on the left.

at Zuffenhausen. With a new cylinder head and good for 170 horsepower in Europe, the 924 was edging into 911 territory. All but the earliest 924 Turbos received upgraded five-bolt wheel hubs and bearings (from the 911), disc brakes on all four wheels and standard anti-roll bars. The cars were identifiable by the NACA duct in the hood (to cool the turbocharger itself) and rubber rear lip spoiler. A flaw in the plan resulted from adapting the KKK turbo designed for use in trucks to a passenger car (used for short trips with more frequent heat cycling). Although performance won praise from the press, warrantee claims became a problem. For 1979, a five-speed transmission became a desirable option for the 924. The new gearbox was placed ahead of (rather than behind) the differential and gave the 924 a more flexible and fun performance envelope.

The 924s would begin to phase out in 1982 when production of the VW/Audi engine ended.

Porsche 936

As noted above, the 936 was built for the 1976 season as a Group 6 Prototype racer. The car was a surprise to the racing world, not unlike its close cousin, the 917. In fact, only a small team of less than 20 Porsche employees knew that such a car was being designed. All of Porsche's top engineers pushed back on Ernst Fuhrmann's desire to build a Group 6 car (which was somewhat contradictory to Fuhrmann's own desire to keep a close connection between Porsche's racing and street cars). However, Fuhrmann saw opportunity with Porsche's 2.1-liter turbocharged engine from the 911 RSR Turbo in combination with existing components and development work on the most recent 'plastic' Porsche racing cars. According to Norbert Singer's account in *24:16*, Fuhrmann said: *"I'm telling you, we're going to do it!"* Such was Fuhrmann's commitment to the 936 that he loaned the 917 veteran, Helmut Flegl, from the critical 928 project to lead the 936 team.

Designed and built in just five months from September 1975 to January 1976, it is a tribute to Porsche that a 'parts bin' special, made mostly with existing parts from other cars, could achieve the level of success reached by the 936. Of course, raiding the parts bin saved both time and money for the 936 project. The aluminum tube frame was familiar from Porsche racers dating back to the 906. The new chassis design was akin to the 917/10 but with a slightly longer 2,400 mm wheelbase and the rear section modified to accept the smaller engine. One major change was rigid engine mounting, making the engine a stressed member of the chassis and increasing overall rigidity of the chassis by 28%. The body shape also benefited from all of Porsche's work on open-cockpit racers in the preceding years. Suspension, brakes and steering came directly from the 917

Can-Am cars, along with a transaxle from the non-turbo 917 which was proven to handle more horsepower than projected for the 936. Minimum weight for Group 6 cars was 1540 pounds so the extra heft of some parts was not an issue. The engine was similar to the 2.1-liter from the 1974 Turbo RSR, but tweaked to produce 520 horsepower.

For the somewhat confusing FIA schedule for 1976, only Group 5 cars were eligible for the manufacturer's ('Makes') championship. In fact, the Group 5 races were run in a completely separate series from Group 6, with Prototypes having their own schedule and competing for the World Sportscar Championship. At this point, the 24 Hours of Le Mans was an outlier, not part of either Championship, but still clearly the most important race and open to the various FIA categories.

The 936 was painted stealth black for its first test at Paul Ricard in the south of France, hoping to keep its existence secret. A few onlookers were told it was just an old race car being checked for a client while a sharp-eyed journalist was promised an exclusive story in exchange for keeping the 936 secret. The Martini livery was placed over the flat black for the first race at the Nürburgring. Principal competition came from the relatively new 2-liter Renault prototypes with turbocharged V6 engines and Porsche's own customers racing with aging 908/03s. At the Nürburgring, Rolf Stommelen qualified the 'black widow' 936 in second spot, behind the fastest Renault. In this 300 KM sprint race, the cars were driven single-handed. The Renaults took each other out in a crash and Stommelen led easily until slowed by a jammed throttle cable. Stommelen finished third in class and fifth overall but Reinhold Joest saved the day for Porsche, winning the race in his turbocharged 908/03. For the remaining races, the 936s appeared in photograph-friendly white Martini livery. Porsche went on to win the remaining six 1976 events, securing the World Sportscar Championship, although opposition was not exactly deep on the grids.

As ever, Le Mans was the important race and for 1976 Porsche entered both existing 936s. Chassis 002 appeared with a large airbox perched behind the cockpit (similar to those in fashion for Formula One cars at the time). This intake provided cooling air for the engine as well as the turbo intercooler. Chassis 001 ran in its standard 1976 configuration without the airbox. The field was relatively thin with opposition from only a single Renault, two Cosworth-powered Mirages (of the type that won in 1975), Alain de Cadenet's Cosworth-powered Lola T380 and Porsche's own 935. 936-001 failed to finish but Chassis 002 ran fast and steady except for a 34-minute stop to repair a cracked exhaust. The 936 won by 11 laps over the best of the Mirage

The 936 won on its first attempt at Le Mans, in 1976. Gijs van Lennep at the wheel.

1976 Le Mans-winning 936 (left) with a Manufacturer's Championship-winning 935.

M8s. Winning drivers Jacky Ickx and Gijs van Lennep had to contend not only with hot weather, but extreme heat from the oil cooler that conducted to the pedals and burned their feet. The Porsche crew tried various improvisations to improve the situation but in the end, the drivers had to suffer for their art. The 936 became the first turbocharged car to win Le Mans and van Lennep chose the occasion to retire from racing.

For 1977, Porsche only entered the 936s at Le Mans. Although they were faster on the straight, now achieving some 215 mph down the Mulsanne, competition from Renault was much more formidable. Renault brought four cars after an extensive testing program and the Arizona-based GTC Mirages now had Renault turbo engines.

The French cars took an early lead in the race despite Didier Pironi's car catching fire and retiring on lap one. Jacky Ickx kept his 936 close to the Renaults for the first few hours of the race. The real drama started around 8 PM on Saturday night when Ickx' co-driver and fellow three-time Le Mans winner, Henri Pescarolo, suffered engine failure due to a broken connecting rod. The blown engine left Ickx without a car and the Renaults running 1-2-3. Thinking fast, the Porsche team transferred Ickx to the other 936 to continue the race with Hurley Haywood and second-generation Porsche racer, Jürgen Barth. The second of the 936s had lost 20 minutes to replace the fuel pump then another 30 minutes for a head gasket early in the race. This left the car in 41st place. Barth and Haywood were back up to 15th place by the time Ickx joined, but miles behind the leaders.

Ickx was given the direction any true racer would love – drive flat out until the car breaks. Between 8:20 PM and 9:10 AM, Barth and Haywood only drove one shift each. Ickx spent the maximum time allowed in the cockpit during that period and drove the equivalent of *five* F1 Grand Prix races. His average lap times were within two seconds a lap of his qualifying time during the same period. The unsung Barth's double-shift was also a great drive, matching Ickx' pace. You don't expect to pass 40 cars in a race, even at Le Mans, but you also never know what might happen.

By 2 AM, Patrick Tambay's Renault had engine failure. At 3 AM, Patrick Depailler pitted his Renault for a 30-minute gearbox rebuild. This brought the 936 up to second place, six laps down to Jean-Pierre Jabouille and Derek Bell. At 9 AM, Jabouille parked with a holed piston as the hot weather continued to take its toll on the field. This put the 936 in the lead by two laps. After the lone remaining Renault of Jacques Laffite and Depailler retired, Ickx and company were now 19 laps in the lead after being down by 15!

Le Mans is a long race and even with the 936 in cruise mode, Haywood suffered a holed piston with an hour left to go. The previous four hours had seen the hottest temperatures of the week. Hurley rolled into the pit lane in a cloud of smoke. But all was not lost. The mechanics went to work blocking the affected cylinder and disconnecting the turbochargers. With ten minutes left to go, Porsche sent Barth out to do two slow laps as required by rule (the car's final lap time had to be within a certain percentage of its penultimate lap time). They taped a clock to the steering wheel so he would know the exact time. No one was sure if the car could finish, but it did. Ickx' total time in the winning car ended up being 11 in the final 18 hours of the race. This fourth Le Mans win (and third consecutive) was perhaps Ickx' greatest drive. It also turned out to be one of Porsche's finest moments. This was the first win by a three-driver crew at Le Mans and Haywood joined AJ Foyt as the only Americans to win Le Mans on their first attempt.

For 1978, the 936s again ran only at Le Mans but a significant change came with the switch to water-cooled, four-valve cylinder heads (like the 935/78). Porsche built a new car, Chassis 003, and entered 002 and 003 in the new spec to oppose Renault. Chassis 001 was also entered in 1977 spec with the two-valve engine. Renault had again undertaken an extensive development program to win Le Mans and avoid national embarrassment (they intended to concentrate on F1 from that point forward). This time the 936s raced honorably but finished second and third to the Renault A442B, driven by Didier Pironi and Jean-Pierre Jaussaud, who saved the honor of France.

Porsche 928

In the 1970/71 period, the long-range future of the 911 was in doubt because of its air-cooled, rear-engine configuration and questions about future emissions, noise and safety rules. Also in 1971, Volkswagen abruptly canceled the EA266 Beetle replacement project that Porsche was working on. The basic architecture of this mid-engine, water-cooled design might have formed the basis of a 911 replacement if Ferdinand Piëch had continued as head of R&D. The cancellation left a Porsche engineering team without a job. These resources were

ABOVE: Audi 100 test mule for the 928 chassis and drivetrain, shown at the 1985 Porsche Parade. (Allan Caldwell)

LEFT: Porsche 936 in the Esses, on the way to winning the 1977 24 Hours of Le Mans.

redeployed toward the impact-bumper 911 and the design of an entirely new, modern car. As Ernst Fuhrmann took over day-to-day management of Porsche, he enthusiastically supported the 928. Talented engineers working on the project eventually included Wolfhelm Gorissen, Wolfgang Eyb, Peter Falk and Helmut Flegl.

Like its stablemate, the 924, the 928 was conceived as a front-engine, water-cooled, transaxle layout with 2-plus-2 seating. The big difference was that the 928 would be Porsche's first completely fresh, ground-up design. It inherited nothing from other Porsches except the great skill of its engineers. For the engine, Porsche settled on a 90-degree V8 to fit under a traditional Porsche hood line. The OPEC oil shock of 1973 knocked the capacity down from 5 to 4.5 liters but the engine was

> **PORSCHE POINTS**
>
> » Porsche racing cars had to survive 1,000 kilometers of testing on the Weissach 'destruction' course, without breaking, before being approved to race. 936 Chassis 001 not only did all of the testing including survival of the destruction course, but it also raced three times in 1976, holding second at Le Mans for many hours. It then went on to win Le Mans in 1977 and finished second there in 1978. In 1979, it sat on the pole and raced for 19 hours, then in 1981 it finished 12th at its fifth Le Mans race.
>
> » Porsche used a Mercedes SL as a test mule for the 928 drivetrain. They later used three Audi 100s for 928 chassis, engine and transmission development.

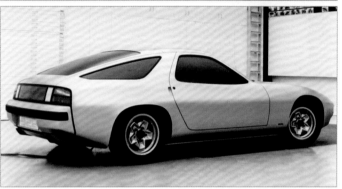

928 styling model circa 1976.

always envisioned as quiet but torquey, lower-revving and built from light alloy components. A key innovation was integrating the aluminum cylinder liners with the die-cast engine block. The engine was given a conservative two valves per cylinder and was the first Porsche engine designed from the outset with fuel injection. Maximum output of the first 928 was 240 horsepower.

As noted, the 928 had its transmission at the rear although ahead of the final drive (the 924 had its transmission behind the rear axle). This placement was similar to Ferrari practice in their front-engine sports cars since the mid-1960s. Placing the weight of the transaxle within the wheelbase was optimal for handling as well as rear storage space (making room for the spare tire). It also allowed a quiet, direct drive top gear ratio in the all-synchromesh box. A limited-slip differential was optional. The clutch was located at the forward end of the drive shaft similar to the 924. A Mercedes-sourced automatic transmission was also optional.

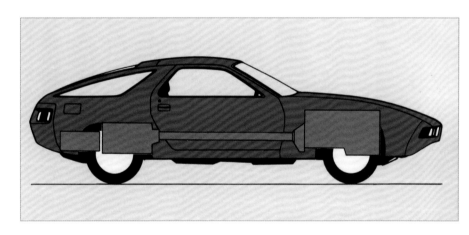

LEFT AND BELOW: 928 brochure illustration showing the location of the transaxle at the rear wheels (left) and the rear suspension, Weissach axle assembly shown below.

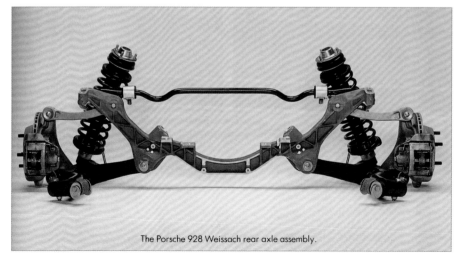

The Porsche 928 Weissach rear axle assembly.

928 winter testing in February 1976.

For the first time in a Porsche street car, there were no torsion bars. A reduction in manufacturing cost for coil springs made them a better choice. Coil springs in combination with strut-type shock absorbers at all four wheels also allowed for more precise suspension tuning. Similar to Porsche racing cars dating to the early 1960s, the 928 front suspension was designed around upper and lower wishbones. Steering was power-assisted ZF rack and pinion. One of the most famous innovations was the 928's rear suspension, which came to be known as the 'Weissach' axle (an acronym for the German *Winkel Einstellende, Selbststabilisierende Ausgleichs-Charakteristik* for 'angle-adjusting, self-stabilizing compensation characteristic'). The goal was to prevent the rear tires 'toeing' outward when slowing from high speed in a turn, creating oversteer. The engineers also wanted to preserve desirable ride and noise characteristics. Configuration options and materials were tested in an Opel Admiral 'mule' with rear seat steering. The team developed a method for the lower rear wishbone to self-adjust by pivoting inward slightly under deceleration, offsetting any rear steering effect. Nearly all of Porsche's top engineers had a hand in the design and development of the rear suspension with the core solution credited to Hans-Hermann Braess. Although it slowed development of the 928, the Weissach axle concept was eventually adapted to the 911 in the Type 993.

The basic shape of the 928 was completed as early as 1973, Wolfgang Möbius credited as the primary designer. The sleek, functional shape was timeless, largely free of gimmicks and managed to look like a Porsche despite its front-engine location. The sloping Porsche hood had recessed headlights that raised up for operation. The aggressive, forward-sloping 'B' pillar in the greenhouse not only added strength to the roof but it also gave the 928 a slightly futuristic profile. Another innovation in the 928 was the use of elastic body-color paint on polyurethane bumper covers. This became an industry standard in later years. The 928 had cast aluminum alloy, five-hole 'telephone dial' wheels which became part of the car's signature look. Behind those wheels were newly-designed single-caliper brakes created to reduce weight and avoid overheating brake fluid.

The 928 was conceived to have significantly more interior room compared to the 911, including a rear hatchback storage compartment. Electrically-adjustable seats were an option and the 928 received another innovation with its adjustable steering wheel. The main binnacle moved up or down with the 911-style three-spoke wheel to keep the key instruments in sight. The dash was equipped with an extensive system of diagnostic and warning capabilities to alert the driver in the event of a malfunction. The width of the car created extra roominess for the front passengers although some felt it was too wide for easy entry and exit. The

Period-futuristic styling of the 928 with its signature telephone dial wheels.

rear seats were nestled cozily against the transaxle hump. Even in this larger car, rear passenger leg room was far from sedan-like. Many 928s had boldly styled interiors including the use of the new 'Pasha' seat insert pattern, an optical-art design meant to suggest the waving of a checkered flag.

Initial press reaction to the 928 was very favorable. Understandably, it made the 911 seem old-fashioned (which it was in many ways). 'Car for the '80s' was another typical reaction to the 928's forward-looking styling, technology and performance, although a few critics panned the look of the car. *Autocar* in England suggested the automatic transmission 928 was a *"grown-up super car."* However, even in the early days of the 928, there were significant questions about whether it could replace the 911 in the hearts and minds of the Porsche faithful. In fact, more than 50% of 928 buyers were new to the Porsche brand. That was both good news and bad depending on the point of view. Interestingly, the 928 was similar in many dimensions to a Corvette, but weighed about 300 pounds less and had about 30 more horsepower. Admittedly, the C3 Corvette was an aging design by 1977 and the Porsche was a much more expensive car.

Perhaps the more luxurious 'grand touring' character of the 928 made it less viable as a 911 replacement in the eyes of the Porsche faithful. While a 911 could be used for touring, it was considered more of a pure sports car. Ferry Porsche himself eventually came to criticize the 928 for its size and complexity. The lack of a racing program may also have hurt the 928 in the eyes of Porsche loyalists. As it turned out, the 928 and 911 would co-exist throughout the 928's 18-year production life.

PORSCHE POINT ›› The first US showing of the 928 was in July 1977, at Sea World in San Diego, as part of PCA's annual Porsche Parade event. Posed on a platform in a pool, it looked as if the car was floating on the water.

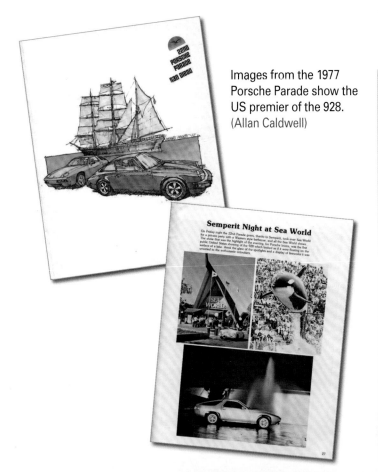

Images from the 1977 Porsche Parade show the US premier of the 928. (Allan Caldwell)

Porsche promotional photograph of the 928. Note the body-colored bumper covers front and back.

Porsche 911 SC

For the 1978 model year, Porsche introduced what might have been the 'last' 911. It's possible that these excellent cars were thought of in that way by those who engineered and built them. There was no guarantee that the 911 could meet future emissions and crash safety standards. Some authors and observers felt the 911 had been neglected, with engineering resources diverted to the 924, 928 and 911 Turbo. This was a period where the 928 was viewed as a possible replacement for the 911, which by 1978 was about 15 years old (essentially equal in lifespan to the 356). In 1977, Ernst Fuhrmann stated that customers would decide the fate of the 911, but at the same time he predicted only two or three years more before the 911 could be 'terminated'. In 1978, the stance had changed notably, as quoted by Karl Ludvigsen in *Excellence Was Expected*: "*We will build this car as long as people want to buy it. Today we produce 45 units a day. Only when the production volume falls below 25 per day will we have to end it.*"

Had they been the last 911s, the SCs would have seen the model out with a bang. Some writers felt the SC qualified as a 'drastic' improvement over their predecessor impact-bumper 911s. The cars received a new 3.0-liter engine with an aluminum alloy case. Thermal reactors were gone, replaced by catalytic converters to reduce heat-related problems in the engine room. All 911s

The 911 SC coupe and Targa versions.

Rally-prepped, Martini-sponsored 911 SCs took on the 1978 Safari Rally in Kenya. The SCs finished second and fourth overall.

were also now equipped with an auxiliary front-mounted oil cooler in addition to the usual round of numerous technical improvements. The SC received seven-inch width rear wheels and stylish fender flares from its Carrera predecessor. Porsche simplified the non-turbo 911 to a single model (in a given market) for the first time, although it could still be optioned in many different ways including coupe versus Targa.

SC flaws were relatively few with some engines doing 150,000 miles and more without a rebuild. There was a weakness with the rubber-center clutch mechanism and exploding airbox issues (any backfire of the engine could shatter the plastic intake plenum shared by all of the cylinders). Most cars have long since been retrofitted to mitigate these problems and many have also been upgraded with later, oil-fed chain tensioners from the 1980s 911. Initial press reception was very complimentary to the 911 SC and customers gladly kept Porsche's daily production well above the economically feasible minimum.

1980s

1980 ▸▸ High Performance 924s
       Peter Schutz
1981 ▸▸ 1981 24 Hours of Le Mans
       911 Developments and Cabriolet
1982 ▸▸ Porsche 956
       Porsche 944
1984 ▸▸ Porsche 962
1985 ▸▸ 944 Turbo
       924S
       TAG Turbo F1 Engine
       Helmuth Bott
       Porsche 959
       Peter Falk
       PDK Transmission
1980s ▸▸ 928 in the Eighties
       Porsche Tuners
1987 ▸▸ Porsche CART/IndyCar Program
1989 ▸▸ Porsche 989

Seeing a chart on the wall that showed 944 and 928 development stretching far into the future, but the 911 line stopping at the end of 1981, Schutz took a black marker and extended the 911 line to the end of the chart, then off the chart on to the walls of Herr Bott's office!

1980s

The 1980s brought Porsche a new CEO, an excellent entry-level car and unparalleled achievement in sports car racing. Peter Schutz became the first outsider to lead the company. The 944 became a huge sales success. The 956 and 962 dominated the world's race tracks like no other car in history. Porsche also ventured into new and advanced technology areas, including the first all-wheel drive Porsches. The venerable 911 carried on the air-cooled tradition and finally got a cabriolet top after nearly 20 years of the model's existence. Despite all the great success, the end of the decade would see Porsche facing external and internal challenges that made the 1990s, perhaps, Porsche's most difficult decade.

The 1980s also began and ended with Porsche eyeing the ultimate prize in American racing at the Indianapolis 500. Investigation of an Indy engine project began as early as 1975. By 1979, Porsche was planning to team up with 935 customer Interscope Racing and driver Danny Ongais. The engine was a 2.65-liter version of the 935 engine and would have raced in an Interscope Parnelli IndyCar. After long and complex negotiations that also coincided with the split in IndyCar racing between sanctioning bodies USAC and CART, Porsche was forced to cancel the program in 1980. For the Indy 500, USAC limited the maximum boost for the Porsche engine to a level that would have been uncompetitive without an expensive redesign. A silver lining to Porsche's first Indy episode was the engine itself. Converted back from running on methanol to gasoline, and with twin rather than single turbocharging, the 2.65-liter six-cylinder went on to power Porsche's win at Le Mans in 1981 and the 956 Group C cars.

Prior to 1980, Ernst Fuhrmann expressed to Ferry Porsche a willingness to step aside. Relations between Fuhrmann and the Porsche family had been deteriorating. He bristled at times in the face of perceived interference from Ferry Porsche and Ferdinand Piëch. He had also reached an age where he felt he could not see through another major product development cycle. He might have benefited from being more deferential to the Porsche family and, perhaps, failed to show sufficient loyalty to the 911 (a product beloved by the Porsche family and considered

Danny Ongais with the Interscope Parnelli IndyCar powered by Porsche. A rules change prevented it from racing.

1982 Porsche brochure photos show the expanding R&D complex at Weissach and the test track, circa 1970s.

the 'true' Porsche by loyal customers). The future of the 911 is, of course, much easier to see with hindsight.

In January 1981, Peter Schutz became Porsche's new chairman. He was a controversial choice. As an American citizen with no direct experience in the automobile business (and having never owned nor even driven a Porsche), his appointment was shocking at first. However, Ferry Porsche had developed a personal rapport with Schutz during the recruiting process. He thought the company might benefit from Schutz' broad industrial experience and fostering a less engineering-dominant culture (in spite of the fact that Schutz was an engineer by training). Schutz benefited from being highly deferential to the Porsche family and from enthusiastic interaction with Porsche customers. His internal relations were sometimes problematic from a management style point of view, but two significant actions during 1981 scored big points with the Porsche faithful both inside and outside of the company.

First, when asking about Porsche's plans for racing at Le Mans in 1981, Schutz learned that the focus was on the 944-based Turbo, which had little chance of winning the race overall. He authorized the revival of the 936s from Porsche's museum for a Le Mans entry. Success at Le Mans inspired the racing-focused Porsche

Peter Schutz

Born in Germany in 1931, Schutz' Jewish family escaped Nazi oppression by emigrating first to Cuba and then the United States. Schutz became an auto mechanic and studied mechanical engineering at the Illinois Institute of Technology. His career included years with Caterpillar Tractor and Cummins Engine in addition to teaching and operating a flying school. In 1978, he moved to Cologne to work for KHD, an industrial engineering company with a diesel engine business. His tenure at Porsche was marked by great sales success and dominance in sports car racing. He is also credited as a savior for the 911, a car whose future was in limbo when Schutz started at Porsche in 1981.

A black mark on Schutz' tenure was his attempt to reinvent Porsche's sales model in the US. In 1984, Porsche elected to end its contract with Volkswagen of America for US distribution. Part of the new plan involved changing the role of US dealers and establishing 'company stores' in major metro areas to handle new car sales (dealers would still have been able to funnel orders to the Porsche Centers). This plan caused such an uproar (not to mention the threat of legal action) that Porsche was forced to abandon all but the change at the distribution level. In the end, the establishment of Porsche Cars North America, with distribution points in Reno, NV and Charleston, SC proved successful for Porsche dealers, customers and enthusiasts as it provided more of a direct link to Porsche in Stuttgart. This is episode is dealt with in a forthright and educational manner in Schutz' book, *The Driving Force*.

Schutz was realistic about the impact of exchange rates on perceptions of his performance as well as Porsche's actual financial performance. When the dollar was strong versus the mark, Porsche did extremely well. As the dollar declined, Porsche's reliance on the US market put pressure on company profit. Schutz left Porsche at the end of 1987 with a mixed legacy, not unlike his predecessor, Dr. Fuhrmann. These men faced a common business challenge: managing a family business as a non-family member.

PORSCHE POINTS

» When searching for a new CEO during 1980, Porsche considered the head of Ford in Europe, Bob Lutz, who decided he was better off staying with Ford. Lutz went on to serve in executive roles at Chrysler and GM.

» In 1981, the Volkswagen Beetle became the first car to reach 20 million examples produced (the milestone car was built at the Puebla, Mexico factory).

» In the early 1980s, Porsche's engineering and design business took on one of its more notable projects by designing the cockpit for the Airbus A310. One of the main challenges was the elimination of the flight engineer position and configuring the controls for a two-person flight crew. For the first time, monitors were used in place of some analog instruments.

» The Lada Samara, a Soviet-built hatchback, used a Porsche cylinder head design. Sales in Western Europe were hampered by poor build quality.

employees, customers and fans (see sidebar). Second, Schutz' now legendary visit to Dr. Bott's office one afternoon 'saved' the 911. Seeing a chart on the wall that showed 944 and 928 development stretching far into the future, but the 911 line stopping at the end of 1981, Schutz took a black marker and extended the 911 line to the end of the chart, then off the chart on to the walls of Herr Bott's

High Performance 924s

Although it was an entry-level model, the 924 inspired racing and rallying versions in the classic Porsche tradition. As early as 1976, design work started on a streamlined, turbocharged 924 meant to challenge speed records set by the pure, prototype Mercedes C-111. Although that project was ultimately canceled (the test car survives in the Porsche Museum collection), other high-performance 924s variants were created.

The 924 Carrera GT was an intercooled, 210-horsepower 924 Turbo with fiberglass fender flares for wider wheels (inspiring the 944 body shape). Built for homologation, these cars could be driven on the road or adapted for racing and rallying. The 924 Carrera GT 'Prototype' was a full race version of the Carrera GT complete with roll cage and mechanical fuel injection. Since the Carrera GT was not yet homologated, it had to run in the GTP class. One of these cars finished sixth overall at Le Mans in 1980.

The 924 Carrera GTS was a further homologation version of the Carrera GT completed in 1981. These cars could also be used on the road but were intended for competition duty with 245 horsepower. The Carrera GTS Clubsport upped the ante to 275 horsepower with a larger intercooler.

Perhaps the ultimate 924 was the Carrera GTR, a further derivative of the GTS. This was a turn-key racing car built by Porsche to compete in the IMSA GTO and SCCA Trans-Am categories as well as Le Mans. In 1982, one of these cars finished 16th and won the IMSA GTO class at Le Mans. Tube frame versions of the GTR were eventually built and raced in IMSA and the SCCA into the mid-1980s.

924 Carrera GTS group prepared for homologation in 1981.

office! Literally, with the stroke of a pen, Schutz saved the 911 and possibly the Porsche company itself.

Peter Schutz recognized that Porsche was not really an automobile company in the conventional sense. He saw Porsches as leisure and lifestyle products, not unlike a boat or airplane (Schutz himself was an accomplished private pilot). As quoted by Karl Ludvigsen in *Excellence Was Expected*, Schutz said: *"The price of a Porsche is not viewed in comparison with a regular utility auto, because you are not selling just a car. A Porsche is a technological artwork that is probably more entertainment than transportation."*

911 Developments and Cabriolet

Peter Schutz sensed a gathering gloom over Porsche at the impending end of the 911 line.

For Schutz, one important reason to keep the 911 going was its profitability compared to the 924 and the 928. In addition, Schutz could not ignore the loyalty and dedication of Porsche's core customers to the 911.

One of the many positive outcomes for the 911 under Peter Schutz's tenure was the 911 SC Cabriolet. Launched for the 1983 model year, the first fully open, drop-top 911 was designed and developed in a the short time span of about 18 months. A prototype featuring all-wheel drive was shown at the Frankfurt Show in September 1981, predicting another significant development for future 911s. The convertible 911 made its official debut the following March at the Geneva Show. Initial response was generally favorable and 911 buyers now had a choice of coupe, Targa or cabriolet versions. The convertible soon reached about 40% of 911 production before settling at about 25% during the 3.2 Carrera period. The original top came in for some criticism, but it was improved with electric operation for

1981 24 Hours of Le Mans

For 1981, new CEO Peter Schutz asked his motorsport team if they had anything that could win Le Mans overall. At his prompting, Porsche revived their museum-piece 936s and entered Le Mans with a version of the engine developed for the Indy 500. Revised to run on gasoline and with twin turbos, the engine made a reliable 630 horsepower. It would go on to power Porsche's forthcoming Group C racers.

Derek Bell had not even sat in the car before practice and qualifying but almost immediately set his fastest laps at Le Mans since 1971 in the 917 LH. Derek teamed with Jacky Ickx for a second time (they won Le Mans in 1975 for Gulf). The Ickx/Bell team had a dream race, starting from the pole, never putting a wheel wrong and never having to lift the tail on the car. They led from the fourth hour on and won by 14 laps. The second 936, driven by Jochen Mass, Hurley Haywood and Vern Schuppan finished 12th after clutch and fuel injection trouble. Hurley still nominates the 1981 version of the 936 as his favorite all-time racing car. This was the last major victory for a Porsche with a tubular space frame chassis.

PORSCHE POINTS

» When Porsche won Le Mans in 1981, it was the third 24 Hour triumph for the 936. Each of three 936 chassis built by Porsche won the great race once. A spare chassis, 004, was used by the Joest team at Le Mans in 1980 and finished second to local hero Jean Rondeau's GTP (the only driver to win Le Mans in a car of his own construction).

» The Sebring 12 Hour in 1983 saw a surprise win for a 934 driven by Wayne Baker, Jim Mullen and Kees Nierop. All of the faster 935s and Prototypes either broke down or were delayed by the typically punishing Sebring conditions. However, the winning car had started life as a 935. In fact, chassis #930 009 0030 had finished second at Le Mans in 1979 (driven by Rolf Stommelen, Dick Barbour and Paul Newman), then won the 24 Hours of Daytona in 1981 driven by Brian Redman, Bobby Rahal and Bob Garretson. For 1983, the car was somewhat loosely back-dated to IMSA GTO 934 spec.

Derek Bell in the race-winning 936/81 at Le Mans.

the 1986 model year. The 'Cabrio' carried on with the introduction of a new 911 engine in 1984 and convertible 911s have remained a staple of Porsche's sports car lineup for the past 40 years.

For the 1984 model year, Porsche gave the standard 911 the vaunted *Carrera* nameplate. The engine was substantially revised and enlarged to 3.2 liters by lengthening the stroke with the use of the crankshaft from the 911 Turbo. The engine case itself was strengthened and fuel injection was upgraded from the CIS system to Bosch's Motronic Digital Motor Electronics system (computerized engine management was a relatively new concept at the time). Timing chain tensioners were improved (earlier 911s are often upgraded with these units), manifolds improved for better air flow and the alternator upgraded. Although the cars gained some weight, the new 911 was more powerful and more fuel efficient, thanks in part to the engine's electronic control unit.

As always with Porsche, the Carreras received detail changes for every year of production in addition to

A full cabriolet version of the 911 SC arrived for the 1983 model year.

	911 SC 1978-1983	911 3.2 Carrera 1984-1989
ENGINES	3.0-liter (3.3-liter Turbo introduced), aluminum block, CIS injection, horsepower range 180 to 265 (Turbo)	3.2-liter, Motronic engine management introduced, horsepower range 200 to 285 (Turbo)
CHASSIS	Cabriolet introduced for 1983	Brake wear sensors, larger anti-roll bars for 1986, 1988 Clubsport option, 1989 Speedster
STYLING	Impact-bumper similar to Mid-Year, with flared fenders	Impact-bumper, integrated fog lights, Turbo-look option
OTHER	All have catalytical converters, improved timing chain tensioners, improved ignition and fuel injection	Hydraulic timing chain tensioners, third brake light for 1986, G50 transmission and hydraulic clutch introduced for 1987, end of development cycle for original 911

The 3.2 Carrera was the final development of the original 901/911 concept.

some major improvements. 1984 and 1985 models saw upgrades to the standard and the sport seats. Ergonomics also improved with better heating and air conditioning. Further Turbo content was made available to non-turbo buyers in the form of the M491, 'Turbo-look' package. This combined styling and mechanical upgrades from the 911 Turbo with the standard 3,164cc engine.

For the 1987 model year, the 911 received a new transaxle and a hydraulic clutch replacing the traditional Beetle-like cable operation. Replacing the 915 transmission, the G50 unit was designed to handle the higher power output planned for future 911s. The box was also intended to improve on the notchy, sometimes vague shifting experienced in earlier 911s and reduce shifting effort. Reverse gear was synchronized for the first time.

During the final years of 3.2 Carrera production, Porsche offered several 911 variants. These included the lightly stripped-down 3.2 Clubsport and 'sports equipment' cars that offered higher performance than the standard 911. There were two special

PORSCHE POINTS

» CASIS (Computer-Aided Shift Indicator System), the yellow upshift light familiar to owners of some 1980s Porsches, was a compliance scheme that grew out of Porsche's independence from Volkswagen of America for distribution. Porsche was no longer able to benefit from positioning within VW's Corporate-Average-Fuel-Economy numbers in the United States.

» One of the rarest RS Porsches is the 1984 911 SC/RS. Only 21 were built, for Group B homologation as a rally car.

» In 1984, non-voting shares in Porsche AG were offered to the public for the first time, resulting from a desire by some members of the Porsche/Piëch family to liquidate their inherited ownership of the company.

TOP (BOTH): Porsche-powered Mooney completing its around-the-world flight, piloted by Michael Schultz and Hans Kampik. (Rob Powell)

MAGAZINE COLLAGE: Porsche-Mooney and Cessna aircraft in 1980s aviation magazines.

PORSCHE POINTS

» Peter Schutz's interest in aviation led to a 911 aircraft engine project (PFM 3200). Although several were tested in Cessnas and installed in other lesser-known general aviation models, the project was not a financial success. Mooney did build 41 M20 planes using the Porsche engine in 1988/89. Engine performance and service commitment became controversial and in 1999, owners were advised that parts and support would end in 2005. In 2004, Hurricane Charley destroyed or damaged a number of the planes that were awaiting replacement, non-Porsche engines at Mod Works in Punta Gorda, Florida. FAA Type certificate was officially surrendered in 2007.

» The 1987 911 was the first to use BorgWarner synchros in the transmission, in place of the traditional Porsche ring synchro system. Porsche already had experience with the BorgWarner design from the Audi transmission used in the 924.

editions, one to commemorate the 250,000th 911 and another to celebrate the 25th anniversary of the 911 in 1989. Perhaps most intriguing was the return of the Speedster. Based on the Cabriolet, the 911 Speedster had a lower windscreen, lower, emergency-use top, fiberglass tonneau cover and Turbo fender flares. Recalling the 356 original, Porsche produced just over 2,000 3.2 Carrera Speedsters and they remain prized by collectors.

Like the Carrera, the 911 Turbo was improved year after year. With a rise of gray market import 930s to the United States, Porsche returned the Turbo to the US market officially for the 1986 model year. Greater availability of 94 octane fuel allowed for engine tuning and catalytic converter installation that resulted in acceptable power loss compared to the non-catalyst cars built for Europe. Power loss due to emission standards for 1980 had been the reason Turbo sales were discontinued in the US. Work on chassis bracing to handle higher horsepower allowed Porsche to offer Targa and Cabriolet Turbos starting in 1987. Perhaps the ultimate 1980s style statement came with option M505, the slant nose body with rear fender air intakes and boxy rocker panels. In 1989, the 911 Turbo finally received a five-speed transaxle thanks to work intended for the next generation of 911s. 1989 was a year of transition as the Type 964 was introduced and Porsche built the last of the 911s representing 25 years of evolution since 1964.

> **PORSCHE POINT**
> ≫ The 1989 Panamericana concept car was presented to Ferry Porsche as his 80th birthday present. Although Ferry disliked the car, it was considered fun to drive by the design team and it did its job by pointing the way to the 993.

Porsche 956

For 1982, the FIA adopted an entirely new formula for top-level sports car racing. Replacing the old categories, the Group C formula was designed to be more open, especially to

BELOW: 956 wind tunnel model with, from the left, Norbert Singer (back to camera), Manfred Jantke and Peter Schutz.

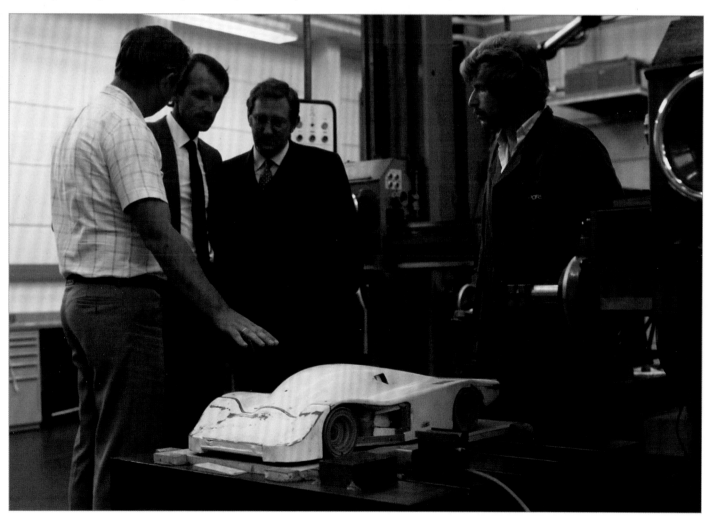

956s finish in formation, first, second and third at Le Mans in 1982.

manufacturers. Engines were unrestricted (as long as they were based on a production block) but were limited to 600 liters of fuel for a 1000 kilometer race (2,600 liters at Le Mans). Porsche took this opportunity to introduce an entirely different kind of racing car, the 956. Finally abandoning the trusty tubular chassis, Porsche produced a modern, monocoque chassis made from riveted and bonded aluminum. This form of construction was much more rigid than previous Porsche race car chassis. Porsche also made the 956 their first 'ground-effect' car, using tunnels under the car to create downforce. The tunnel shape creates a low pressure area at high speed, sucking the car down to the road. Ground-effect was an innovation started by Lotus in Formula One in 1978. Creating adequate room for the tunnel exits meant inclining the flat-six engine within the chassis.

For the engine, Porsche carried over the 2.65-liter, twin-turbo 'Indy' engine that had successfully powered the 936 to victory at Le Mans in 1981. The key to success in Group C was to make and use as much power as possible without running out of fuel. This placed an emphasis on electronic engine management. A further strategy to maximize performance in Group C was to take the corners faster (increasing lap speed without using more fuel). In this respect, using ground-effect downforce was essential, keeping the cars glued to the road.

Similar to the 917 and the 936, the 956 was developed in a relatively short time. The initial design and development work took only nine months, from June 1981 to March 1982. The car first turned a wheel at Weissach on March 27, 1982. Further testing at Paul Ricard gave the drivers their first experience with a new level of cornering power in a sports racer. The 956 proved very fast but was handicapped in its debut race, at Silverstone in May 1982. In the six-hour race, the 956s had to cover an extra 100 kilometers on the fuel allocation for a 1000 kilometer race. The winner at Silverstone was a 'grandfathered' Group 6 Lancia. However, for the 1982 24 Hours of Le Mans, all went well for Porsche's newest racer. The Porsche factory put on a crushing display with three 956s finishing in formation, first, second and third. The winning car was driven by Jacky Ickx and Derek Bell.

Porsche soon made 'customer' 956s available for sale and they proved to be excellent turn-key racing cars for private teams. During much of the 1980s, 956s and the successor 962s often filled sports car grids around the world. At Le Mans in 1983, the Porsches put on a close and exciting race with the #3 956 of Al

> In June 1983, Dr. Bott called for a test to see if the 956 was faster than the mighty 917/30 Can-Am car at Weissach. Derek Bell was the driver and the test was attended by Ferry Porsche and Peter Schutz, journalists Paul Frère and Jerry Sloniger along with engineers Peter Falk and Norbert Singer. The 917/30 came from the Museum collection and despite its massive horsepower advantage, the 956 was four seconds per lap faster. The 956 also broke the Weissach skid pad record set by a 917/10.

> At Le Mans in 1983, the only top ten finisher that was not a Porsche 956 was the ninth-place BMW-powered Sauber C7. As the advertisement said: 'Nobody's Perfect'.

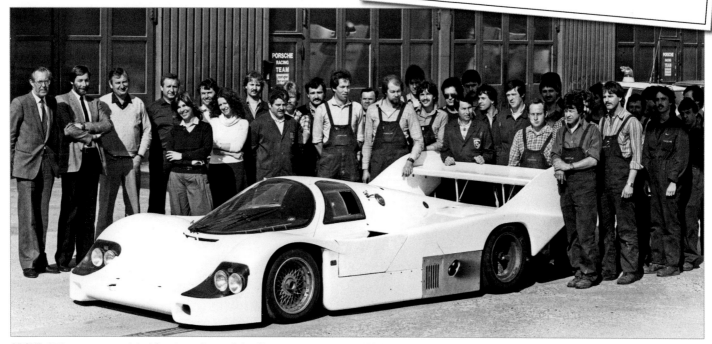

ABOVE: 956 team assembled for the rollout of the first customer car, at Weissach in January 1983.

TOP RIGHT: 1983 Porsche-Audi newspaper advertisement celebrating Porsche dominance at Le Mans.

THIS PAGE: The start of the 1985 24 Hours of Le Mans (top), and the winning 956 with just under an hour left to race (above). (Jay Gillotti)

Holbert, Hurley Haywood and Vern Schuppan barely holding off Ickx and Bell at the end. Holbert was driving for the last two stints with an obviously sick engine. Overheating of the water-cooled cylinder heads was caused by a blocked radiator resulting from repair of a door that had blown off with less than two hours remaining. Derek Bell was closing in over the last 90 minutes despite having severe brake problems. The #3 car staggered across the line only 64 seconds ahead at the finish. Nine of the top ten finishers were Porsche 956s, prompting Porsche and Audi to issue the famous 'Nobody's Perfect' newspaper advertisement in the US.

Among the most successful of 956 privateers, Reinhold Joest's team won the 1984 and 1985 24 Hours of Le Mans with the very same car, 956B Chassis 117. In 1984, the Porsche factory boycotted the race due to a disagreement about changes to the fuel consumption rules. Klaus Ludwig and Henri Pescarolo were the winning drivers. In 1985, Joest beat the factory's new Group C 962s thanks to Joest team development of their own programming for the engine management system, improved underbody aerodynamics and some fuel-saving teamwork with the English Richard Lloyd Racing team. What appeared to be close racing during the first half of the race turned out to be cooperative 'drafting' between the rival 956s to save fuel. Ludwig anchored the driving team with Paulo Barilla and 'John Winter' (Louis Krages). The RLR 956 had a chassis built by the team rather than Porsche.

The 956 was certainly one of Porsche's most successful racing cars, contributing to World Endurance Championship titles from 1982 through 1986. In 1983, Stefan Bellof set an absolute record at the Nürburgring during qualifying for the final WEC event held on the Nordschleife. His time in a factory-entered 956, at 6:11.13, stood for some 35 years. He also set the race lap record at 6:25.91 before crashing out of the lead. Sadly, Germany's rising star driver was killed in a 956 at Spa in 1985 when he collided with Jacky Ickx' 962C at the daunting Eau

Rouge corner. The 956 and its successor, the 962, scored more than 200 race wins for Porsche over a 12-year period.

Porsche 944

From the launch of the 924, Porsche knew that the supply of engines to be produced by Volkswagen would eventually run out. Given the intention to keep the basic platform in production for years into the future, Porsche had to explore future engine options. They also hoped to improve performance, as always, and to smooth out other flaws inherent to the 924.

In the fall of 1981, Porsche surprised the automotive world with the announcement of the 944. Although based on the 924 platform, the new car was dramatically improved. The shapely exterior was based on the 924 Carrera GT. The general shape from the racing version of the 924 had to be altered slightly for mass production in steel, but it improved the car's visual presence as well as allowing for a wider track to further improve the handling. At first, the 944 inherited the basic interior from the 924, which continued in production for several years beyond the introduction of the 944. By 1985, the 944 received its own attractive, modernized interior.

The engine decisions were all-important to the success of the 944. Porsche considered several alternatives before electing to build their own inline four-cylinder unit. The design was based on one bank of 928's V8. Part of the strategy was to share as many parts as possible with the sister car, as well as sharing materials (silicon-aluminum alloy block) and a common manufacturing process. Building the engine in-house would potentially be more profitable for Porsche as opposed to the sourcing arrangement from VW. It was also seen as maintaining Porsche's independence from Germany's largest car maker, although final assembly of the four-cylinder cars would continue at Audi's Neckarsulm plant. The connection to the 928 engine enhanced the perceived Porsche-ness of the 944. The 924 suffered (like the 914) from a perception in some minds that it was not a 'real' Porsche.

The central problem for building an inline-four to Porsche's

Porsche CEO Peter Schutz with the 944.

high standards was vibration, especially at high rpm. Under the leadership of Paul Hensler and Gerhard Kirchdorffer, the engineering team explored both internal and external solutions for smoothing the engine's running character. In the end, Porsche overcame any 'not invented here' prideful instinct and licensed existing patents from Mitsubishi. The Japanese design for rotating balance shafts was deemed to be the most economical solution for the 944. Of course, Porsche carefully fine-tuned the use of the technology for the 944, with results that became a major success in the marketplace. Porsche also designed the 2.5-liter unit with room to expand both capacity and power, through the use of turbocharging.

The balance shafts were positioned on either side of the engine, one low and to the right, one higher up and on the left side. The belt-driven shafts rotated at twice the speed of the crankshaft and in opposite directions to each other. The 944 temporarily lost some commonality with the 928 when it was introduced with a 100 mm bore size for the pistons. Although heavier than the base 924 engine, the 944 exceeded its design intent by producing more than 160 horsepower in Europe (143 horsepower for the US version running on unleaded fuel). The 944 inherited its basic suspension layout from the 924 Turbo, with a further 'sport' suspension option. It also carried on with the Audi-sourced transaxle as used in the 924.

The combination of engine power, handling and a racy shape drew near-universal praise from the automotive press and the

> **PORSCHE POINTS**
>
> » The seventh-place car at Le Mans in 1981 was a 924 GTP with the Type 949 engine, a turbocharged race engine using the forthcoming 944's engine block. The car was dubbed the '944 Le Mans'.
>
> » To test the concept of a V6 engine for the 944, Porsche used a Renault-Peugeot-Volvo engine in a 924. This is the engine that powered the DeLorean.

TAG Turbo F1 Engine

In 1983, Porsche designed and built a new engine for the McLaren Formula One team, sponsored by the Saudi company Techniques d'Avant Garde. The Ojjeh family business provided financing and a Porsche team led by Hans Mezger worked with McLaren designer John Barnard to tailor the engine to his chassis concept. The engine itself was a 1.5-liter, 80-degree V6 with twin turbos, managed by a version of the Bosch Motronic system. Initially making 550 horsepower, development over the life of the engine eventually resulted in 960 horsepower on race boost (1,060 horsepower with maximum boost for qualifying).

For 1984, the engine powered the new McLaren MP4/2 chassis. Drivers Nikki Lauda and Alain Prost dominated the season, winning 12 of 16 races. McLaren won the Constructors' Championship and Lauda won the Drivers' title by a scant half point over Prost. In 1985, Lauda retired, but Prost was World Champion and he repeated in 1986 despite rising competition from Williams with Honda engines. Over four full seasons, the engine took McLaren cars to 25 victories in 64 starts.

Alain Prost in the TAG-McLaren Formula One car.

PORSCHE POINT » The TAG-McLaren F1 engine was first tested in a modified 956 at Weissach, receiving an enthusiastic first impression from Jürgen Barth. McLaren boss Ron Dennis was miffed about the potential connection of the engine to a Porsche car.

buying public. The 944 was priced well relative to its competition, including sports cars from Japan. Given the strength of the dollar at the time, the 944 was an excellent value in the USA where initial 944 sales exceeded 50% of production volume. Few, if any, customers, dealers or journalists had a problem accepting the 944 as a real Porsche. No surprise, the 944 also found its way into racing almost immediately, winning the 'Longest Day' 24-hour showroom stock endurance race at Nelson Ledges, Ohio in 1982.

As noted above, in 1985 the 944 received a new interior and dashboard along with improved climate control and ventilation. It also got power-assisted steering. These cars are generally referred to as 1985½ models. ABS (anti-lock braking system) became an option in 1986 models, airbags optional for model year 1987. Also, for the 1987 model year, Porsche introduced a mid-range 944, the 16-valve, twin-cam 944 S, positioned between the base 944 and 944 Turbo. The 944 S benefited from parallel development of the 928 S4, also using four valves per cylinder. It borrowed suspension tuning and components from the faster 944 Turbo. The 944 S had no automatic transmission option although an early version of the PDK (twin-clutch) concept was tested. See sidebar on page 215 for more on PDK.

In 1989, for 1990 models, Porsche introduced the 944 S2 and base 944 with larger engines. The 16-valve S2 grew to 3.0-liter capacity with the base 944 up to 2.7. The S2 received a Turbo-style nose cover with slatted intakes. At the same time, a cabriolet top became an option for the 944 S2. The contract for production was awarded to ASC (American Sunroof Company)

924S

Porsche continued to sell 924s in the years after the introduction of the 944. These later 924s were sold in markets other than the United States and used up the supply of engines that Porsche had 'banked' prior to production being discontinued by Volkswagen. However, the 924 found new life in 1985 with a 150-horsepower 944 engine installed and featuring 928-style telephone dial aluminum wheels. Known as the 924S, these cars were sold as 1986 through 1988 models (1987 and 1988 in the US). Nimble and fun, although hampered by some standard 924 flaws, these cars presented a lower priced alternative to the 944. The 14-year run of the 924 came to an end in 1988 with about 150,000 cars built. Although it was not particularly profitable for Porsche, the 924 did provide enthusiasts with entry-level Porsche ownership for many years.

who partnered with Karosseriewerke Weinsberg to build the convertibles. Production was complicated and expensive as cars had to be transported between Neckarsulm and the Weinsberg facility twice before completion. A little over 4,000 944 Cabriolets were built, including the Turbo S versions.

944 Turbo

This high-performance 944 came to market in 1985 after extensive testing and development work. Offering 220 horsepower, the Turbo featured a new nose styling with long, horizontal intake slats plus a lower rear skirt. These changes helped to reduce aerodynamic drag, resulting in the lowest figure for any production car at the time. The Turbo also benefited from sharing the new interior design from the base 944.

Porsche took on an extensive engineering effort aimed at cooling the turbocharger (apart from the engine itself) for long-term reliability. The turbo was packaged to maximize spacing from the hot exhaust manifold. At the same time, the plumbing was engineered to reduce turbo lag and to integrate an air-to-air intercooler. Adding to the technical challenge, Porsche committed to producing a single, global model with a catalytic converter for emission control. Years of experience with turbo engines and the engineering effort specific to the inline four-cylinder resulted in performance numbers very similar to the 911 3.2 Carrera, although the 944 Turbo was heavier and more expensive. The automotive press generally gave rave reviews but Porsche also began to hear some criticism of overlap in its model range.

As with the first 944, the Turbo was tested as at Nelson Ledges,

BELOW: 944 Turbo promotional photograph.

Helmuth Bott

Helmuth Bott spent a glorious career at Porsche that touched on nearly every important model and racing success from 1952 until 1988. Born just south of Stuttgart in 1925, he served as a tank crewman near the end of World War II. He worked as a mechanic at Daimler-Benz and studied engineering at the University of Stuttgart. Hired by Porsche in 1952, Bott started as a factory assistant. One of his early tasks was creating test equipment and testing new versions of the synchromesh transmission for the 356. In the early 1960s, he was involved in chassis development and road testing the 901 as well as enthusiastically championing the growing test facilities at Weissach. During the Ferdinand Piëch era, Dr. Bott was second in command for Porsche's engineering team, succeeding Piëch in 1972 as head of R&D. He overcame a loss of testing and development work for Volkswagen by obtaining new business from clients like GM, Volvo and Mercedes. He also had an early vision for what today are Porsche's Experience Centers.

Dr. Bott was known as a patient, approachable man who could also achieve tremendous productivity with his engineering team. Writer Kieron Fennelly quoted Peter Schutz in *Excellence* magazine, regarding Bott's management skill: *"He reorganized them every day to keep them focused on the task and the customer. He developed people to handle specific tasks, he reconfigured teams regularly, and it kept his engineers creative as he was always feeding them challenges."*

Dr. Bott served on Porsche's management board from 1979 until 1988. His career ended in some controversy after criticism of the cost of the 959 program and concerns about the lack of parts commonality in Porsche's three main product lines. After leaving Porsche, Dr. Bott worked as a consultant but died of cancer much too early, in 1994, at the age of 69.

Helmuth Bott (center) with Peter Falk (left) and Norbert Singer (right) at Le Mans in 1985.

running as a prototype in the 1983 and 1984 Longest Day races (winning the event in 1984). To further the competition credibility for the 944 Turbo, cars were entered in the Playboy-sponsored showroom stock series. Porsche also built competition-modified 944 Turbo Cup racers for one-make series' in Canada and Europe. In 1988, Porsche offered the special edition 944 Turbo S, a 944 Turbo powered by the 250-horsepower 'Cup' engine with seven-spoke Design 90 wheels. These cars also had upgraded suspensions, limited-slip differential and a transaxle oil cooler. The 'S' engine then became standard in 1989 944 Turbos. A small run of 944 Turbo Cabriolets were built in 1991 after the car had been discontinued for the US market.

Over 160,000 cars of all types were produced in the 944 range before the introduction of the 968. As Porsche was expanding the 944 range, the decline in the value of the US dollar became a major problem. Porsche's dependence on US sales had grown beyond 50% and a declining dollar caused a pricing disadvantage against Japanese competitors from Mazda and Nissan. This was further complicated by terms with VW for assembly of the four-cylinder cars at Audi's Neckarsulm plant. The exchange rate issue and a slowing global economy in the later 1980s put significant pressure on Porsche's financial health.

Porsche 959

Along with Group C, for which Porsche built the 956, the FIA also introduced Groups A and B as new motorsport categories at the beginning of the 1980s. Group B was mainly intended to become the top category for the World Rally Championship. It required building 200 cars for homologation and numerous interesting and exotic cars resulted, from the Ford RS200 to the Ferrari 288 GTO, MG Metro 6R4, Audi Sport Quattro, Lancia 037 and others. Group B ultimately produced some spectacular, super-fast rally racing but the speeds proved to be incredibly dangerous. Group B was canceled after the 1986 season. For Porsche, the pursuit of a Group B entry resulted in what became the most technologically advanced car of the 1980s. The 959 became a 'supercar', as the term is used in the 21st century (for ultra-high performance, very limited production and very expensive).

Porsche was relatively late to the game in considering and designing a Group B car, having concentrated at first on Group C. Rather than build a competition 'special' like Lancia's 037, Porsche's head of engineering, Helmuth Bott insisted on Porsche's Group B car being a test bed for future developments applicable to the 911 (or other Porsche cars). Dr. Bott also felt pressure to explore the growing interest in four-wheel drive for passenger cars (as evidenced by Audi's success, as one example close to home).

The first public showing of the future 959 was the *Gruppe B* styling exercise, star of the September 1983 Frankfurt Show. It was a futuristic vision of the 911, presented on the 20th anniversary of the original car. It retained the existing 911 greenhouse,

ABOVE (BOTH): Gruppe B show car on display at the Porsche Experience Center in Carson, California. (Jay Gillotti)

PORSCHE POINT
» Bodies for the 959 were made in the former Reutter building space in Zuffenhausen while final hand-assembly took place in the second story of a former bakery (!) on the Adestrasse.

doors and basic structure. From there it proposed much wider, flared fenders and an integrated basket-handle rear wing. The concept envisioned four-wheel drive powered by a twin-turbocharged race engine. Just to prove that Porsche was serious about adapting four-wheel drive to the 911, the Type 953 won the desert raid Paris-Dakar Rally in northern Africa in January 1984. The 959 project was led by Manfred Bantle but many of Porsche's key engineers had a hand in its development.

The production version of the 959 incorporated a prodigious list of new or advanced technologies. The engine was partly derived from the 935 and 956 race motors. At 2,849cc, the flat-six featured twin overhead cams with water-cooled, four-valve cylinder heads and oil-cooled pistons. Making approximately 450 horsepower, controlled by the Bosch Motronic engine management system, the 959 had two turbos working in sequence (rather than parallel) to reduce lag. Although catalytic converters were considered, the production 959 lacked the emission controls needed for markets such as the US. The car was originally conceived with the PDK transmission concept but ended up using a six-speed manual version of the Type 950 (G50) gearbox.

The four-wheel drive system was designed with help from Helmut Flegl. Numerous options were explored and tested. The final drivetrain used multi-plate, oil pressure clutches for electronically-managed, variable torque split. Drive to the front differential could be as much as 100% or as little as 20%. For normal driving, the split was approximately 40/60 front-to-back. The driver could select from four drive programs: dry, wet, ice/snow or traction (for the most severe conditions).

The suspension was double-wishbone front and rear, hydraulically adjustable and self-leveling with two shock absorbers at each corner. The 959 had ABS with upgraded brake calipers working on vented and cross-drilled discs descended from Porsche's racing cars. Race-developed tire pressure sensors were installed in magnesium alloy, hollow-spoke, center-lock wheels made by Speedline. Initial tire development was conducted with Dunlop, but production cars had Bridgestone tires (using Dunlop's Denloc technology) designed to run flat if necessary and capable of handling speeds over 200 mph.

The production bodywork closely resembled the *Gruppe B* show car. It was a team effort from Tony Lapine's styling department with input from Wolfgang Möbius, Dick Söderberg and Hermann Burst. It featured a combination of aluminum

959 promotional photograph.

Walter Röhrl testing the 959 at Weissach.

or composite Kevlar body panels with polyurethane bumper covers. The final 0.32 coefficient of drag was much lower than the standard 911. Despite the car's hefty weight, at 3,500 pounds, the zero-to-60 mph time was under five seconds and the 959 was capable of a top speed near 200 mph. It was no surprise that journalists widely praised the 959 as the best car in the world at the time.

Because of currency fluctuation, sale prices ended up well over $200,000. 338 cars were built including prototypes and pre-production vehicles. Standard 959s were known as *Komfort* models while 37 stripped-down 959S 'Sport' models were built without the hydraulic suspension. Pre-production 959s were not ready for testing until mid-1985 and given the depth and complexity of the development process, the first customer cars were not delivered until the first half of 1987.

The rally-prepared version of the 959 took first and second places in the 1986 Paris-Dakar Rally, with René Metge driving the winning car ahead of Jacky Ickx. The 961 circuit racing version of the 959 only ran a few times, with a notable seventh-place finish at Le Mans in 1986. A mildly modified 959S achieved a speed of 210.7 mph, just slightly slower than tuning firm Ruf's 911 Turbo, when *Auto Motor und Sport* magazine conducted a top speed test at the Nardò test track in 1988.

Emission and other regulations prevented the import of the 959 to the US when the cars were new. Years of political maneuvering resulted in a 'show and display' exemption starting in 1998 that allowed for limited use of cars like the 959 on US roads. Californian Bruce Canepa's Canepa Design began modifying and updating a trickle of 959s for use in the 21st century. One of the main modifications was the conversion to parallel rather than sequential turbocharging. Over the years, numerous 959s have been modified by the Canepa team. The later 959SC, for 'Sport Canepa', involved complete rebuilding of the car

> PORSCHE POINTS
>
> » In 1984, Porsche built the 965 prototype. Separate from the 959, this was a test 911 looking at concepts to replace the 911 Turbo (930) and follow-on from the 959. Porsche even tried a water-cooled V8 sourced from Audi in the 965.
>
> » In 1986, the 961 version of the 959 became the first four-wheel-drive car to compete at Le Mans, finishing seventh overall.

1980s

THESE TWO PAGES: Canepa Design 959 SC – a modernized, higher-performance 1980s supercar. (Canepa Design)

with substantially upgraded components, resulting in a much more modernized driving experience with 825+ horsepower available. Only 50 959SCs will be built.

Although Porsche is thought to have lost a significant amount of money on each car sold, numerous 959 innovations found their way into future Porsche cars. In hindsight, this R&D effort and the 'halo' effect of the 959 help to justify the investment. Fair or not, criticism of the cost of the 959 program likely played a part in ending Dr. Bott's illustrious career with Porsche in 1988. As of this writing, 959s are highly prized by collectors with values sometimes exceeding $2 million.

Porsche 962

The Porsche 962 was a development of the 956 built to conform to IMSA GTP rules for racing in the US. The wheelbase was lengthened by moving the front wheels forward so the driver's feet would be behind the centerline of the front wheels. The cars also received integrated steel roll cages. IMSA called for production-based engines even in prototype racers. As a result, the 962 was equipped with a fully air-cooled, single camshaft, single turbo version of the 2.8-liter flat-six, similar to the engines used in the 935. Bodywork from the 956 had to be modified to comply with IMSA rules and to accommodate airflow to the intercoolers and single turbo arrangement. The design and construction of the first car were completed with great speed in the fall of 1983. After a bare minimum of testing, the 962 made its debut at the 24 Hours of Daytona in February 1984. Driven by Mario and Michael Andretti, 962-001 sat on the pole and ran well before dropping out with transmission failure after only four hours.

Later in 1984, customer 962s began to arrive and race in IMSA for entrants Bruce Leven, Al Holbert and Bob Akin. These teams were eventually joined by Jim Busby, Preston Henn, Rob

Dyson and others. By 1985, perhaps to the dismay of IMSA, the 962s were dominating the competition, winning most of the races and the season championship for Holbert (who became Porsche's North American representative for motorsport). The 962 continued to be the dominant car in US prototype racing in 1986 and 1987. The Holbert Racing team, with co-driver Derek Bell, won two additional season championships. Under IMSA's rules to balance performance, some 962s eventually ran with larger engines, twin-plug cylinder heads or twin turbos (carrying intake restrictors and a weight penalty). With no fuel consumption restriction, the larger engines proved advantageous at certain tracks, even with the weight penalty.

As the FIA transitioned to safety rules similar to those of IMSA, Porsche developed a Group C version of the 962 for racing in the World Endurance Championship. The 962C first appeared in 1985 (1986 would be the last year of eligibility for the 956s). Over the next three seasons, Porsche carried on in the Championship with a factory team and continuously improving the 962. This was necessary as Group C saw greater competition for Porsche from the Tom Walkinshaw TWR-Jaguar team as well as Peter Sauber's team (with increasing involvement from Mercedes). The FIA as well as the Le Mans organizers reduced the fuel allocation for races starting in 1985, requiring engine and aerodynamic improvements to stay competitive. For Porsche, this effort was aided in 1986 by the completion of Porsche's own wind tunnel at Weissach.

Among many interesting developments in the 962C was the first use of a fully water-cooled flat-six Porsche engine. This 3.0-liter, twin-cam engine made use of greater cooling capacity to produce something over 700 horsepower. The additional power was helpful for some of the shorter-distance races that were included in the 1986 World Championship season. Along with the PDK transmission concept (see sidebar), Porsche also began experimenting with ABS in the 962. Although first tried in the 917 as early as 1971, anti-lock braking had become a viable option for street cars and was deemed worthy of trying at the race track. For the drivers, who simply wanted to race (and win), these development exercises were often troublesome. However, Dr. Bott patiently reminded his men of the need to justify the expense of racing as a benefit to Porsche's street cars.

At Le Mans, always the main event for Porsche, 1985 proved an inauspicious first attempt for the 962C. The best the factory team could manage was third place for Derek Bell and Hans Stuck, son of the 1930s Auto Union ace, who had replaced Stefan Bellof in the factory team. Results were much better for the factory in 1986 and 1987. In 1986, the Jaguar challenge was growing with the big, normally aspirated V12s having beaten the Porsches at Silverstone before Le Mans. However, at Le Mans, the Jaguar team was no match for the might of Porsche. Derek Bell, Al Holbert and Hans Stuck had a trouble-free run to victory, with Stuck getting his first win in the 24-hour classic. Porsches filled the top seven spots in the final results.

LEFT: Bayside 962 driven by Bruce Leven at Riverside, 1985. (Kurt Oblinger)

> » The first win for a 962 came on June 10, 1984 at Mid-Ohio (Holbert/Bell).
>
> » In 1985, Hans Stuck put his 962C on the pole at Le Mans, breaking the speed record set in 1971 by the 917 LH. Stuck's average speed was 156.48 mph on a slightly longer, slower version of the circuit.

962C pit stop at Le Mans, 1987. Derek Bell ready to take over the winning car.

The 1986 race was marred by the death of Jo Gartner who crashed in a 962C on the Mulsanne Straight. It was the third fatal accident for a Porsche prototype pilot in a 12-month period. German drivers Manfred Winkelhock and Stefan Bellof, driving privateer-entered 956s had died in crashes during 1985. The losses were a shock to Porsche, especially in the case of Bellof as the young German was a rising star in Formula One who might have gone on to challenge Ayrton Senna. Bellof had won the WEC drivers' championship in 1984.

In 1987, Bell, Stuck and Holbert won Le Mans again after several of the top Porsche teams were knocked out by a fuel problem. The fuel supplied by the organizers was slightly too low in octane level compared to what the Motronic system was programmed for. This caused engine failures as the cars were set to run as lean as possible to stretch their fuel. After the Mass/Wollek factory car dropped out with a holed piston, quick thinking in the Porsche pit changed the fuel mapping and ignition timing to run slightly richer in their only remaining car (the third team car had been heavily damaged in a practice crash). The Bell/Stuck/Holbert car also had to overcome a challenge from Jaguar, who had won the first four races of the season. The three Jaguars were not affected by the fuel problem and matched the race pace of the Porsches until a huge crash on the Mulsanne and mechanical failures took them out. Beyond the fuel concerns, the rain, extended running behind the pace car and a few small repair issues couldn't stop the surviving factory car.

The only race for the factory team in 1988 was at Le Mans where the TWR-Jaguar team finally triumphed. However, the 962C could have chalked up a third consecutive win had Klaus Ludwig not miss-timed a fuel stop. Running dry and limping back to the pits resulted in a second-place finish for that car, co-driven by Bell and Stuck, which was on the same lap with the Jaguar XJR-9 at the end. This ended a seven-year streak of Porsche victories at Le Mans.

Similarly to the 935, the 956 and 962 chassis were eventually built by other manufacturers using Porsche's design. Between Porsche and other fabricators, approximately 200 Group C and IMSA GTP monocoques were built.

928 in the Eighties

The 928S was introduced in Europe for the 1980 model year (1982 for the 1983 model year in the US). The most significant change was a revised 4.7-liter engine approaching 300 horsepower in the European version. The 928S had a revised chin spoiler at the front and grew a small spoiler integrated at the

Peter Falk

As the son of a German archeologist, Peter Falk was born in Athens. After World War II, he started as an apprentice with Mercedes-Benz before studying mechanical engineering. Gravitating to sports cars and lighter-weight vehicles, he moved to Porsche at the age of 26 in 1959. Falk worked on the early development of the 901 and he co-drove a 911 to fifth place in the 1965 Monte Carlo Rally with Herbert Linge. He became head of vehicle testing in 1969 and his career continued to blend work on racing and production cars during the 1970s.

During the 1980s, Falk became a familiar face after being named Porsche's head of competition in 1981. Known as an excellent and enthusiastic test driver, Falk was considered a master of driving on the skid pad. In 1984, he broke Willi Kauhsen's 917/10 g-force record on the Weissach skid pad, driving a 956. He also took on desert test-driving duties during Porsche's Paris-Dakar program in the mid-1980s. In 1988, he became head of all chassis development. One of his significant duties was writing the design requirements for the 993 version of the 911. This document set the tone and guidance for one of Porsche's most successful cars. Peter Falk retired from Porsche in 1993.

PDK Transmission

The PDK, for Porsche *Doppelkupplungsgetriebe*, 'double clutch transmission', is a concept explored by Porsche since the early 1960s. A version appeared in 1963 based on the patented work of Imre Szodfridt and Richard Hetmann. It was considered a possible automatic transmission option for the first 911. Helmut Flegl picked up on the idea in the late 1970s for the Type 995 design study for a 'car of the future' sponsored by the German government.

Development work continued in the 1980s intended for production cars but actively tested and used in racing. The first appearance of a PDK-equipped 956 was in practice for a race at Kyalami in South Africa, in December 1983. Its first use in a race was at Imola in September 1984. The concept had some appeal for the racers, given the potential to change gears faster and optimize engine speeds to overcome turbo lag. As with any radically new technology, race successes were slow to materialize during the 956 and 962 era. A PDK first appeared at Le Mans in 1986 and won for the first time in a sprint race that year at Monza. However, the computer and electrohydraulic control technologies took two additional decades to mature before the PDK made it to a production car. Prompted by Ferdinand Piëch, the first production use of the PDK came in the Mk5 VW Golf of 2005 that was branded DSG for Direct Shift Gearbox.

The PDK uses two clutches with alternating gears running on two separate drive shafts within the transmission. Switching from one gear to the next (also from one drive shaft to the other) can be accomplished within milliseconds in the current version. This can be done automatically or manually by the driver through the use of paddles on the steering wheel. The majority of Porsche sports cars sold today use this transmission concept.

PORSCHE POINTS

» As part of the ongoing 928 development program, Porsche experimented with fuel efficiency ideas that came into production decades later. These included cylinder shut-off, 'stop/start' (engine shut-off at stoplights), use of aluminum for the chassis and body panels and using a smaller, four-cylinder turbo engine in the 928 (from the 944).

» The Porsche 984 concept was a design for an entry-level sports car that looks a bit like a cross between a 924 and a 928, but as a mid-engine roadster, it also foreshadows the Boxster. It was inspired by development work on small car engines for SEAT. The project was canceled in 1987 when the car couldn't meet its production cost target.

THIS PAGE: Porsche 928S postcard with brochure photos (below).

928 S4 shown in rare Cassis Red.

base of the rear glass. Numerous other improvements (including transmissions, clutch, and suspension) made for a more sporting 928, while the base car featured improved fuel economy. The 4.5-liter base model enjoyed improvements designed for the 928S, but was then phased out during the 1982 model year.

For the 1985 model year, Porsche introduced the 928 S4 with twin-cam, 32-valve cylinder heads for increased power and better fuel efficiency. Fuel economy was an important consideration due to the loss of the CAFÉ (Corporate-Average-Fuel-Economy) umbrella from VW in the US and the 928 was in danger of being awarded a 'gas guzzler' tax. The new engine also featured an innovative but complex induction and exhaust system to improve engine breathing. Ongoing improvements to Bosch's engine management system also helped to balance performance with consumption. The S4 was still capable of figures close to 300 horsepower even with improved fuel economy. Depending on the market, compression ratios were adjusted to the available fuel quality.

928 production was enhanced in 1986 by completion of a paint facility in Porsche's Werk II complex. This allowed the old paint shop to be repurposed for special-order paintwork as well as a separate line for painting plastic bumper covers (previously an outsourced function). Painting the plastic material to color-match metal bodies had been a difficult and error-prone process

Special Wishes

Porsche has had some version of a special customer request department, formal or informal, for many decades. After first joining Porsche in 1967, Rolf Sprenger was promoted in 1969 to run the factory *Reparaturewerkstatt* 'repair shop', servicing customer street cars as well as race cars. In the 1980s, the *Sonderwunsch*, 'special wishes' department under Sprenger was separated from the customer racing department which moved to Weissach under Peter Falk. Sprenger continued to lead the team responsible for cars like the *Flachbau* 'slant nose' 930s, some 2000 of which were built over the years. The team continued the tradition of creating special interior and exterior color and trim requests along with some performance modifications for customers who could afford them. In 1989, the *Sonderwunsch* was renamed Porsche 'Exclusive' and continued the mission. 'Paint-to-Sample' exterior colors became a treasured mark of collectability in the Porsche world, often resulting in one-of-a-kind examples. Rules for testing and approval of custom colors changed over the years along with testing requirements for modern paints on non-metal bodywork. Rolf Sprenger retired from Porsche in 2006 and passed away in 2021.

928 S4 in Guards Red. (Adrian Johnson)

since it involved different paint formulation for the plastic material. Also in 1986, ABS became standard in the 928. The telephone dial wheels were dropped in favor of flush-faced, slotted alloys first introduced on the 928S.

In 1987, the 928 received an exterior styling update with a new nose and tail shape with a larger spoiler. After being produced in multiple versions for different countries based on the availability of higher-octane fuels, the 928 S4 received a standard 320-horsepower engine in 1987 (altered slightly for Australia only). Although hardly dated, this version of the 32-valve engine was extensively redesigned. The manual transmission cars also received an improved, single-disc clutch. 1987 represented peak production for the 928 at 5,400 cars even though US pricing zoomed above $60,000.

Although competition was never a high priority for Porsche with the 928, the car did uphold tradition by winning a 1983 24-hour race at Snetterton in England for production sports cars. Privateers in Europe and North America tried racing with the 928 at times although not with great success. The Porsche factory carried on another tradition by using the 928 to set speed records. The first came in 1982 on the circular track at Nardò in Italy. It was a 24-hour run averaging 156.22 mph. In 1987, on the Bonneville Salt Flats in Utah, Al Holbert set production car records for the flying mile and kilometer.

Although hampered by the car's high price and a weak dollar, the 928 would soldier on well into the 1990s. The S4s received generally high praise from the press, and the cars were highly capable sports or GT machines. Perhaps they were too good, in the sense of lacking driver involvement relative to the 911. It is also possible they suffered from a lack of clear identity; sports car versus grand touring car. After 1987, annual production of the 928 never exceeded 4,000 cars.

Porsche CART/IndyCar Program

As Porsche's Group C program wound down in the late 1980s, Peter Schutz decided to go single-seat, open-wheel racing in

> **PORSCHE POINT** » Porsche's Type number, 2708, for the ill-fated IndyCar project, was chosen to signify the approximate engine size, 2.7 liters, and number of cylinders, 08.

1980s

THIS PAGE: Teo Fabi in the March/Porsche IndyCar at the New Jersey Meadowlands, 1988. (Harry Hurst)

Porsche 989

When Ulrich Bez was hired from BMW and replaced Dr. Bott as head of engineering, he revived the concept of a four-door as a way to move Porsche upmarket. Porsche had experimented with front engine and four-door 911 concepts over the years. The 989 was a front-engine, rear-drive car styled by Bez' fellow ex-BMW employee, Harm Lagaaij who worked at Porsche in the early and mid-1970s. Some of the 911 visual references in the 989 survived to be applied to 993 and 996 styling. The engine design was an entirely new V8, separate from the 928. After an investment of 100 million dollars (or more), the 'Learjet for the road' project was canceled in 1992. In part this was due to the projected build cost for the car which came in far above target. Opposition from Ferdinand Piëch may have been influenced by Audi's plans to introduce the A8. Customers would have to wait until 2009 for a four-door luxury sedan from Porsche.

THIS PAGE: 989 prototypes (ABOVE) and the near production-ready car (TOP).

America. The CART (Championship Auto Racing Teams) IndyCar series was highly popular and offered the enticing possibility of adding the Indianapolis 500 to Porsche's list of glorious race wins. Porsche's first attempt, in 1980, had come to nothing. Norbert Singer was in overall charge of the Type 2708 project with an engine designed by Hans Mezger and the chassis designed by 956 and 962 author Horst Reitter. The first engine was ready to test by the end of 1986.

The program was marked by disagreement between Porsche and Al Holbert in his role as head of competition for North America. Holbert preferred to use the Porsche engine in a March chassis (March had been building successful IndyCars for many years). There was also increasing competition from Lola as a constructor and Porsche eventually raced the engine mainly in the March chassis. The program was also affected by Holbert's tragic death, in a plane crash, in September 1988. Management changes at Porsche in 1988 may also have played a role in the lack of traditional Porsche cohesion in this racing program.

The 2708 had its first race at Laguna Seca in 1987 with the goal of contending for a championship by 1989. Despite good engine performance, the only win for a Porsche-powered IndyCar came at Mid-Ohio in 1989, with Teo Fabi driving a March 89P for the Derrick Walker-managed team. This was cause for some optimism, but the best result in 1990 was a third place for Fabi at the Meadowlands in New Jersey. The program was discontinued after the 1990 CART season.

Porsche Tuners

The 1980s saw a rise in popularity of small firms known for customizing and enhancing the performance of Porsches (and other marques). Some focused on wildly expressive, even outlandish looks, including individual takes on Porsche's *Flachbau* treatment on 911s.

Gemballa was known for wide-body kits and Ferrari-like side strakes. The brand still exists despite the mysterious murder of founder Uwe Gemballa in South Africa, in 2010. Koenig was known for their creative takes on Ferraris and Mercedes as well Porsche 911s and 928s. They were among a small number of companies that created road-legal 962s in the 1990s. Rainer Buchman's B&B also modified other cars, but their fanciful take on Porsche's are classically representative of 1980s style. The Kremer brothers were well-known for their customer racing efforts with Porsches, especially the 935 K3 and K4. They produced road-going versions of the 935 and body kits for 930s. Alois Ruf's firm focused on ultimate performance 911s, including the famous CTR 'Yellowbird'. Ruf carries on to this day under Alois Ruf, Jr. making cars (including an electric 911) and performance parts. Like Gemballa, Ruf has manufacturer status for its Porsche-based cars. Other tuner names like Rinspeed, Strosek and DP may jog the memories of those who followed Porsche in the 1980s. DP had supplied the bodywork for the Kremer Brothers' legendary K3.

In some ways, these companies were forerunners to today's customizers and high-end tribute builders like Singer and Gunther Werks.

1990s

1989	»	Porsche 964
1991	»	Porsche 968
	»	928 in the Nineties
		Mercedes and Audi Outsourcing
1994	»	1994 Le Mans
		Porsche 993
		Kaizen
		C88 - China People's Car Project
1995	»	WSC 95
		Porsche Engineering Services
1996	»	Porsche Boxster
		Porsche 911 GT1
		Wendelin Wiedeking
1997	»	Porsche 996

It had more of a high-performance look about it. It was more muscular with its big fenders, and everyone loved it. – *Harm Lagaaij*

The 1990s represent Porsche's most challenging and perilous decade since the 1940s. As noted in the previous chapter, cost disadvantage, partly due to the strength of the Deutschmark relative to the US Dollar, was a major problem. This continued into the early 1990s. Porsche was also hampered by aging product designs that shared few components in common. Inefficient production methods added to the problems stemming from lack of parts commonality. Although it took time, Porsche eventually tackled these problems with consulting help from Japan. At the same time, Porsche had to be concerned about competition from Japanese sports cars like the Acura NSX, Nissan 300 ZX and Toyota Supra. With the reunification of Germany in 1990, Porsches would no longer be products of 'Western Germany' and within ten years, a city in the former East Germany would become Porsche's second home.

Management changes in the late 1980s and early 1990s worked against stability in Porsche's programs. Finance head Heinz Branitzki was a temporary CEO after Peter Schutz left at the end of 1987. Porsche then hired Arno Bohn to lead the company into a new decade. Bohn's background was in the computer industry with Nixdorf AG. He started at Porsche in November 1989, becoming a board member in January 1990. Unlike Schutz, Bohn was a Porsche customer and enthusiast before joining the company. His was a bumpy tenure at Porsche and he eventually left to run GE's medical business in France in 1992. Porsche's second CEO hire from outside the automotive industry would be its last.

There was understandable turmoil at the board level during this same period. Butzi Porsche became Chairman of the supervisory board in 1990 as his father eased further into retirement as Chairman Emeritus. The Porsche and Piëch families had to deal with a troubled business and did not always agree on major decisions. At the same time, with a portion of Porsche shares publicly owned, there was a risk of unwanted takeover attempts. Amidst the storm, Wendelin Wiedeking returned to Porsche in 1991 as board member in charge of production. He replaced his mentor, Rudi Noppen, who had clashed at times with Helmuth Bott and Ulrich Bez from the engineering and development side. Wiedeking had led two very important projects for Porsche in the 1980s. First was the completion of the new paint shop. Later in the decade, Wiedeking took on the design and construction of Werk V. This was a new facility for producing body shells making use of robotics for welding. When Arno Bohn fell out of favor, Wiedeking was chosen as Porsche's next CEO. This was not long after Horst Marchart replaced Bez as head of engineering and development. Although certain key projects were already underway, Wiedeking and Marchart ultimately led Porsche's resurgence in the later 1990s.

As if the early 1990s weren't difficult enough, Porsche's sterling competition record was tarnished by two ill-fated ventures in open-wheel racing. The CART/IndyCar program was discontinued in 1990 after only one victory (see Chapter Eight). The Indy engine had been developed into a respectable competitor, but the partnership with March for the chassis was questionable both in terms of performance on track and PR and advertising value. It will never be known if continuing with Porsche's in-house chassis would have produced better results (the cost of that type of program was hard to justify at the time). 1990 results were certainly impacted by the CART organization disallowing the full carbon fiber monocoque version of the March 90P. This forced March and Porsche to redesign the car, leaving inadequate time for testing and development.

Worse yet was Porsche's ill-fated 3.5-liter V12 F1 engine designed for the Footwork/Arrows team. The engine was perhaps the only notable failure during Hans Mezger's otherwise illustrious career designing engines for Porsche. Concern about the length of the engine, which started in concept as two TAG V6s mated together, led Mezger to design a central power take-off (not unlike the 917 in concept). In this case, the drive shaft sat above, rather than below the engine's crankshaft. Sadly, the engine proved to be too heavy and never produced competitive levels of horsepower. It also suffered some reliability problems.

To be fair, the Arrows team did not put their best chassis forward to receive the Porsche engine. Perhaps the chassis was too ambitious with its early use of the high-nose concept and single-post rear wing. The Arrows-built transaxle also had its share of problems (Porsche had considered a version of the still-developing PDK to pair with the V12 F1). Another challenge, experienced also during the IndyCar program, was coordination between

> **PORSCHE POINT** — The first Porsche bicycle prototype was built in 1990 under cycling enthusiast Ulrich Bez. After Bez' resignation in 1992, he and Thomas Menzel produced a small run of the first 'Bike Spyders' without the Porsche name. Later, official Porsche bikes have been produced by Votec and Rotwild. Current Porsche eBike offerings, suitable for commuting, are built by Storck and Greyp. In 2021 Porsche released new Sport and Cross versions of the eBikes.

Porsche's engine work and the chassis design from another firm. In this case, Arrows' owners Jackie Oliver and Alan Rees were also dealing with the pressure of having sold the team to the Japanese owner of Footwork, Wataru Ohashi.

In the end, the Type 3512 engine did less than half a season in 1991 before the project was paused and ultimately canceled. This costly episode not only added to Porsche's financial woes, but also factored in the dismissals of Ulrich Bez and Arno Bohn.

Porsche 964

The Type 964 was developed under Peter Schutz during the 1980s as an effort to modernize the 911. By this time, the original layout of the 911 had been developed incrementally over 20-plus years. The basic product concept document (*Lastenheft*) for a new 911 was approved in April of 1984.

With the 964, Porsche launched a car that was significantly different in all but exterior appearance. The basic 911 structure, or 'tub', remained constant except for alterations required for the new suspension design and the planned implementation of four-wheel drive. Although the stylists were not allowed to change the basic shape above the top of the existing bumpers, the car was still said to be 85% new. The main change to exterior styling was the integrated plastic and fiberglass bumper covers. New rocker panels changed the look of the profile and a new wheel design moved away from the classic five-spoke Fuchs. The appearance of the 964 was controversial when the cars were new and remains somewhat polarizing to 911 fans even now.

Without doubt, the new shape of the 911 improved aerodynamic performance. The front valence, bonded windshield and inclusion of an engine undertray reduced the drag coefficient to 0.32. This was dramatically lower than previous 911s. The 964 also featured an automatic rear deck wing needed to defeat lift at higher speeds. Integrated with the engine cover, the wing extended at 50 mph (an early form of 'active' aero). It should be noted that many 964s now run without the undertray in the interest of better engine cooling.

The list of mechanical changes to the 964 was lengthy, to say the least. First, Porsche needed more power to move the significantly heavier (especially 4WD) 964. The M64 air-cooled flat-six engine increased to 3.6 liters. It also featured twin-plug

THESE TWO PAGES: The 964 shows controversial new bumper covers, revised side sills and 'jelly bean' colors.

ignition for the first time in a regular production 911. Porsche redesigned the combustion chamber to accommodate twin spark plugs along with new induction plumbing and redesigned lubrication and cooling functions. The goal was a global target of 240 horsepower minimum in an engine that could run leaner (for fuel savings) and meet California emissions standards. This was achieved with help from an upgraded Motronic engine management system.

Also new for the 964 was the Tiptronic transmission option. Porsche had not offered a two-pedal 911 since the last of the Sportomatics. The Tiptronic could be shifted manually or could be run in a fully automatic mode with multiple electronic shift 'maps' that self-adjusted based on parameters indicating the way the car was being driven. Although Porsche had worked on an update to its old Sportomatic, and was still developing and racing the PDK transmission, ultimately the decision was to adopt the ZF technology that was in development for Audi. The new automatic became available starting in 1990, with four forward speeds but offered only in the rear-drive Carrera 2. Porsche found the majority of average drivers could accelerate faster using the 'Tip' than they could using the manual transmission. Tiptronic added about $3,000 to the price of the car.

The suspension of the newest 911 was completely new with coil-over-struts and anti-roll bars. The redesigned front suspension and power assist improved steering, especially at lower speeds. The rear suspension maintained traditional trailing arms but adapted the 928's self-correcting 'Weissach' axle design to control wheel movement. Porsche's venerable and sacrosanct torsion bars finally fell victim to the need to accommodate a drive shaft for the front wheels. The rear of the 911 body structure had to be redesigned to accommodate the struts and to regain rigidity lost with the absence of the torsion bar tube. ABS brakes became standard on the 964. Airbags also became standard starting with the 1990 model year. The new five-gauge interior looked familiar but rationalized controls and offered a new HVAC system.

The first 964s became available in early 1989 and only the four-wheel drive Carrera 4 was offered at first. It featured a simplified four-wheel drive system (compared to the earlier 959). It was more like the 953 desert racer with center differential control.

The standard torque split was 31/69 front to rear but adjustable for road conditions by a hydraulic, multi-disc clutch reacting to computerized input from the wheel sensors. Production of the 3.2 Carrera continued in parallel with the 964 C4 until the end of the 1989 model year. The rear-drive Carrera 2 became available in the autumn of 1989 as a 1990 model. The 964 was designed from the beginning to be offered with cabriolet or Targa top options.

Press reaction to the 964 was generally favorable although there were some concerns that the 911 was somehow less 'lively' even though it was certainly faster than its predecessors. The fact that 're-imagined' 911s by Singer are based on the 964 says a lot about the capabilities of the platform.

The 965/969 project, which envisioned a new Turbo similar to the 959 but separate from the standard 911, was ultimately scrapped by Ulrich Bez. The 911 Turbo based directly on the 964 appeared in October 1990. At first, it used a carryover but improved version of the 3.3-liter *Sportpaket* engine from the late-1980s 930. For the 1993 model year the 964 Turbo S arrived with a 360-horsepower, turbocharged version of the 3.6-liter. It was a modified, single spark plug version of the base engine, retaining Jetronic rather than the Motronic fuel injection and engine management system. The larger engine brought zero-to-60 mph times under five seconds and top speed up to 180 mph. That performance came at a price well over $100,000 at the time.

As racing for production-based cars returned to popularity in the 1990s, high-performance and competition versions of the 964 were prolific. The first step was the 964 Carrera Cup which replaced the 944 as Porsche's one-make series racer in 1990. This was a stripped-down but performance-upgraded 964 that pointed toward a new 'RS'. The official 964 RS arrived as a car similar to the 'Cup' but with some comfort features added back. It was available in numerous versions depending on intended use. Because the RS was not legal for the US, Porsche devised the RS America. This was a lower-price, lightweight Carrera 2 with 17-inch diameter Cup wheels, sport suspension, similar interior and no power steering or cruise control. The RS America also featured a classic 'whale tail' spoiler in place of the standard 964's electric rear wing. Available with optional radio, air conditioning, sunroof and limited-slip differential, the RS America proved more popular than expected, with sales of 701 cars during its lifespan.

Numerous other special versions based on the 964 were offered. For some, the ultimate 964 is the rare Turbo-look, wide-body RS 3.8 which was available only as a special order from the motorsport department. Although most higher-performance versions of the 964 were rear-drive cars, Porsche did build a small series of Carrera 4 'Lightweights' for the US market, although these were not officially announced as a catalog option. Other 964 'specials' included the Turbo-look Cabriolet (America Roadster), the Speedster and 30th Anniversary 911 (Turbo-look coupe but with a standard rear deck wing and four-

PORSCHE POINTS

>> For 1991, Porsche became the first manufacturer to offer ABS brakes standard in all their models.

>> In 1991, PCNA (Porsche Cars North America) built 964s for a US version of the Carrera Cup racing series. Lack of sponsorship caused cancellation of the series in May. Most of the 45 'Cup' cars had to be re-converted for street use and were sold as Carrera Cup US Edition models.

>> North American customers were disappointed at not being able to buy a 964 RS. However, with prodding from PCNA CEO, Fred Schwab, and enthusiast lobbying channeled through Porsche insider (and US resident since 1984) Vic Elford, the RS America was created as a slightly lighter, sharper handling version of the Carrera 2. An added bonus was Porsche's decision to offer it at a base price significantly below a standard Carrera 2.

>> The rare 964 Turbo S used carbon fiber for the doors as well as the engine cover/spoiler and trunk lid. The car also featured lightweight side and rear glass, deleted air conditioning, electric window lifts and central locking, removal of some soundproofing material and the rear seats. In all, the cars were about 400 pounds lighter than a standard 964 Turbo. Only 86 of these very expensive cars were built.

>> In 1994, the French Larbre Competition team won the first international motor race in China, using a factory Porsche 911 Turbo S Le Mans GT. In the BPR GT Championship finale at Zhuhai, the car sat on the pole and the experienced French crew of Jean-Pierre Jarier, Jacques Laffite and Bob Wollek won the three-hour race.

The Type 968 version of the 944 shows new nose and headlight treatment, looking more like its bigger stablemate, the 928.

wheel drive). 76 very special 3.6-liter Turbo S 911s were built with the *Flachbau* or slant nose front bodywork (a $60,000 option at the time). Less than 20 were built with the standard Turbo nose bodywork.

From the racing department, Porsche offered the Group N race version of the 964 RS for GT racing. The 964 achieved great success in IMSA Supercar racing with the 911 Turbos winning the championship for Brumos in 1991 and taking second place in 1992 with Hurley Haywood driving. Hans Stuck was the winner in 1993 for Brumos. Porsche also won the SCCA World Challenge series in 1994. The Turbo S LM-GT and 3.8 RSR 911 racing cars were competitive in FIA competitions, winning their class in numerous major races including Le Mans in 1993 and 1994.

The 964 had a lifespan of six model years although some of its technical content carried over to the restyled Type 993 version of the 911.

Porsche 968

The final version of the 944 was originally thought of as the '944 S3'. In the end, given the substantial revisions to the car and the need for a marketing boost, the decision was made to give the car a new name, '968'. With the end of the 924-era VW contract, assembly of the 968 moved to Porsche at Zuffenhausen. The first model year was 1992 and it gradually phased out of production by 1995.

To stay competitive, more power was an obvious goal. Porsche explored various options for upgrading the engine, including six- or eight-cylinder non-turbo versions of the IndyCar engine. Another possibility was sourcing BMW's new M50 inline six, designed by ex-Porsche employee, Karlheinz Lange. Porsche also considered a version of the inline six they had designed for Volvo. After exploring the other options, Porsche proceeded with a heavily revised version of the 944's four-cylinder. One of the main improvements was the introduction of VarioCam technology. The concept was to vary the timing for engine inlet valves by moving the camshaft to a slightly different lobe position automatically. This could help to optimize performance across the entire rev range. In typical Porsche fashion, horsepower was also improved by improving the induction airflow. Because of the power increase, about ten percent greater than the 944 S2, various components of the engine were redesigned for greater strength and to increase cooling capacity.

The 968 received two new transmission options. For manual shifting, there was a six-speed, Audi-designed, Getrag-built transaxle. For automatic shifting, the 968 received a version of the Tiptronic (the PDK concept was still under consideration, but not ready for production). Airbags and ABS were standard equipment.

The most obvious change for the 968 was its exterior styling. One of Harm Lagaaij's first projects upon returning to Porsche from BMW in 1989 was the styling refresh, which nearly ended any resemblance to the car's 924 ancestors. The new nose had exposed, round, pop-up headlights, and a new bumper cover. The mirrors were modernized along with the door sills and the new tail received a 928-like plastic bumper cover (unlike the exposed bumper on all previous four-cylinder transaxle cars). For the first time, the design team assisted with the look of the engine, styling the inlet and cam cover assembly. With new 'Cup'-style five-spoke wheels, the 968 presented a much greater family resemblance.

Opinions were mixed on the 968 when it was new. Many felt the cars were too expensive relative to performance, especially when compared to the competition from Japan. There was general agreement on the spectacular handling afforded by basic chassis and suspension design. A particular favorite among current enthusiasts is the stripped-down, 968 Clubsport. Porsche followed the tried-and-true formula of reducing weight, further improving the handling and reducing the price to create a real driver-focused automobile. 1,371 CS editions added to the relatively modest sales total for the 968. An extremely rare version of the 968 was the Turbo S, inspired by the ADAC GT Cup race series. The cars used a single-cam, eight-valve head with a single water-cooled turbocharger to produce over 300 horsepower. Only 17 were made including four extra-ferocious, 337-horsepower Turbo RS race cars.

By the end of production in 1995, Porsche had sold over 325,000 four-cylinder transaxle cars over two decades. Although the profitability and branding benefit of these cars might have been questionable, they did provide significant volume for dealers including the service and parts business over the years.

928 in the Nineties

Although Porsche was clearly committed to the future of the 911, they continued to improve the 928 as it approached its final years of production.

For 1990 the 928 motored on in either S4 or GT configuration, now with airbags as standard equipment. The 928 GT was a slightly lighter, more sporting 928 similar to the limited-production Clubsport and SE versions offered in 1988 and 1989. In 1991, less than a thousand 928s were sold. During the late 1980s, Porsche had considered various options to upgrade the 928 engine. The final upgrade decision was to build a 5.4-liter for the 1992 model year. Along with the new, 350-horsepower engine, the latest 928 received improved front brakes, a new six-speed manual gearbox from Getrag, and larger wheels. The wheels were 'Cup'-style five-spoke units. The width of the wheels called for a subtle flaring of the fenders, especially notable at the rear. The rear spoiler was now body-colored and the car received a new tail light treatment along with more aerodynamic rearview mirrors (replacing the traditional 'flag'-style units from the earlier cars). The new car was named the 928 GTS.

The GTS became available as a 1992 model in Europe and for 1993 in North America. It was praised for its immense capability, especially at high speed, and as usual for the 928, it may simply have been too good, in the sense of lacking driver involvement compared to the 911. However, given its sometimes harsh ride and sporting nature, it could be considered an uncompromising, 'driver's' GT car. It was certainly a car that could cover huge distances in relative comfort (especially on good roads) and at tremendous speed. It can also be thought of as a private jet for the

> **PORSCHE POINTS**
>
> » In 1992, nine additional 'continuation' 959s were built using spare parts. They received special approval to be registered as 1988 models (given that emissions requirements for new cars had changed).
>
> » The rarest of all RS Porsches is the 968 Turbo RS, with only four built.
>
> » 1993 was a low point for Porsche's North American car sales at 3,713 for the year. US sales today exceed 70,000 annually. Total production of Porsche sports cars in the fiscal year 1992/93 was only 11,763.

1990s

ABOVE AND OVERLEAF: 928 GTS promotional photographs.

highway, best suited to high-speed travel in European fashion. The 928 GTS was expensive when it was new, most costing over $90,000 at the time. They are now prized by collectors as the last and best-performing 928.

The 928 ended its 18-year run in 1995, quietly selling only 328 cars. Over its lifespan, Porsche produced a little over 61,000 928s. It remains an enigmatic car in Porsche's production history. One wonders how it might have fared if Porsche had given it a more luxurious, grand touring character once the decision was made to continue the 911 as a sports car.

Mercedes and Audi Outsourcing

In the early 1990s, with sales in decline, Porsche had spare production capacity. Some of that capacity was used to build cars for other companies. First came the Mercedes-Benz 500E W124. This was a higher-performance version of Mercedes' mid-size E-Class sedan. Porsche assisted with engineering and development, adapting the 5-liter V8 to the W124 and tuning the handling to match the performance potential of the engine. Production began in 1990 and final assembly was handled by Porsche through 1992. Between Porsche and Mercedes, over 10,000 were built by 1995.

A second project was the Audi Avant RS2 Sportwagen. Every aspect of the car's performance was improved and both interior and exterior styling separated the RS2 from a standard Audi 80 wagon. Production by Porsche started in 1994 and ran through July 1995 with 2,891 cars built. Unlike the Mercedes contract arrangement, Porsche and Audi shared profits on the RS2 through a joint venture.

Mercedes-Benz 500 E in the Porsche Museum.

Porsche 993

Unlike other generations of the 911, the 993 never experienced a period of being out of favor. When new, it achieved near-universal acclaim and excellent sales results, helping Porsche on the road to recovery from the problems of the early 1990s. Although some considered the conversion to water cooling overdue, the decision was made to hold that change off until the 996. As a result, the 993s had the added cachet of being the last air-cooled Porsches. Production was limited to just four model years, from 1995 through 1998 (US). The 993 also had to 'save' the 911 in concept, as some Porsche employees thought that issues with the 964 were a sign the 911 should not survive beyond its third decade.

The 993 was championed by Ulrich Bez. As head of R&D, he wanted to advance and enhance the special qualities of Porsches in general and the 911 specifically. In 1989, he commissioned Porsche veteran Peter Falk to study, in detail, the attributes that a future 911 should achieve. The result was Falk's 'agility' brief (similar to but broader and more philosophical than Porsche's traditional design document or *Lastenheft*). Agility was defined in empirical terms and general guidelines that the engineers could work toward in the design and development of the 993. There were statements on direct agility ("quick, direct, perceivable and visible vehicle reaction"), along with indirect agility ("impressions of lightness, speed and controllability").

	964 1989-1994	993 1995-1998
ENGINES	New 3.6-liter engines, horsepower range 250 up to 380 (Turbo)	3.6 to 3.8-liter, improved engines, horsepower range 270 to 400+ (Turbo)
CHASSIS	All-new suspension, 4WD, Tiptronic and automatic rear spoiler were introduced	Redesigned rear suspension and 4WD system, six-speed manual transmission
STYLING	Controversial bumper cover design, interior redesigned	Unique 911 style that was widely appreciated. New front hood, fender and headlight profile
OTHER	Approximately 85% new compared to 3.2 Carrera	Sliding roof Targa introduced

1990s

Dauer 962s finished first and third at Le Mans in 1994.

1994 Le Mans

Customer and racer Jochen Dauer approached Porsche with the idea of making a street-legal version of the 962 in 1993. At first, Porsche was not interested. That was until Norbert Singer realized that a single street-legal car could open the door for a Le Mans entry in the GT1 category. Although the ACO said that converting a race car to a street car was outside the spirit of GT1, the Dauer 962 was eventually allowed a one-time entry at Le Mans in 1994.

Porsche and Dauer worked to prepare a two-car team with an all-star driver lineup. The car was different from a standard 962 because it was required to have a flat bottom, narrower rear wheels and a minimum weight of 1,000 kilograms. In a hodge-podge year at Le Mans, Porsche hoped to win the LM-GT category. However, when the leading LMP1 Toyota C90 SARD team suffered gear selection problems in the final 90 minutes, the 962 driven by Hurley Haywood, Mauro Baldi and Yannick Dalmas won the race overall. Haywood became the only driver to win Le Mans with and without chicanes on the Mulsanne Straight (the chicanes were added in 1990 when the FIA limited straight sections on approved circuits to two kilometers maximum). Dauer Sportwagen went on to sell a small number of street-going 962s.

PORSCHE POINTS

» Early 1990s assembly of the Mercedes 500E and Audi Avant RS 2 Sportwagen made use of the facility originally used to build the 959.

» When Porsche won Le Mans in 1994, Mauro Baldi became the 100th individual driver to achieve overall victory in the 24 Hour race.

» In addition to Dauer, ex-Porsche factory driver Vern Schuppan and tuning firm Koenig also built street-legal versions of the 962.

» Both the 993 and 996 Type numbers were intended to indicate the year in which production would begin for those versions of the 911.

A big part of the acclaim for the 993 resulted from the exterior restyle. Its roots dated back to the controversial Panamericana show car of 1989, but there was also a 959 and 965 styling influence. Head of design at the time, Harm Lagaaij has said: *"In the early 1990s, sales figures showed that customers were becoming bored with the shape of the 911 – certainly in the form of the then-current 964 model. It was okay for the real enthusiasts, of course, but overall sales showed that it just wasn't exciting enough. That was why we went in the direction with the 993. It had more of a high-performance look about it. It was more muscular with its big fenders, and everyone loved it."*

Primarily the work of Englishman Tony Hatter (with veteran Dick Söderberg), aided by Porsche's experienced modeling team, the 993 was an instant hit. The lowered front fenders and laid-back headlights modernized the face (although at the risk of violating Ferry Porsche's traditional view of the fender tops as a driver aid for positioning the front wheels). The shapely rear fenders recalled earlier high-performance 911s but carried over the 964's retractable spoiler. Only the roof shape could not be changed (as a result of limits on the development budget). Although there was no budget to redesign the 911 interior, the 993 did receive a few incremental improvements to the overall layout.

Perhaps the most important change in the 993 was at the rear suspension. The goal was to eliminate the noise and harshness experienced in the 964. Based in part on work from the 989 project, the new design was multi-link with parallel wishbones, along with the now-standard coil springs over shock absorbers. The 911 finally lost the trailing arms that had been used since the beginning in 1963. Porsche considered using active rear-wheel steering for the 993 but cost and technical difficulties meant it would employ a form of passive rear steering using the 'Weissach axle' concept from the 928 (see Chapter Seven). A lightweight aluminum subframe connected all the pieces in what was dubbed the 'LSA' (for light, stable, agile). The entire rear suspension could be produced separately and installed as a unit during final assembly of the car. Attaching to the tub with rubber mounts helped to further reduce noise. Front and rear anti-roll bars, shocks and springs were offered in both standard and 'sport' configurations. Although challenging to fit within the limited development budget, the 993 rear suspension design influenced the future of Porsche suspensions in general.

For the 993 Carrera 4, Porsche designed an all-new drive system meant to increase agility and eliminate the 964 version's

ABOVE: 993 Cabriolet, note the wind deflector behind the seats. (Dean Holbrook)

OPPOSITE: 993 promotional photograph. Note lowered front fenders and laid-back headlights.

PORSCHE POINTS

- The lone 993 Speedster was built in 1995 as a 60th birthday present for Butzi Porsche. Porsche also converted a 993 4S Cabriolet to similar 'Speedster' bodywork for Jerry Seinfeld in 2001.

- For the 1997 model year US 911s, Porsche introduced an ignition interlock, requiring that the clutch pedal be depressed in order to start the engine.

- The 993 Turbo featured a new design for the outlet of the oil coolers ahead of the front wheels. The outlet shaped the exiting air to improve flow around the front wheels. This technique continued with the 996 version of the 911.

- 14 narrow-body 993 Turbo Cabriolets were created by Porsche Exclusive for Munich dealer Fritz Haberl.

- In 1997, the Ruf CTR2 version of the 993 went to the Pikes Peak Hill Climb. Brothers Steve and David Beddor had 702 horsepower to play with and finished second and fourth overall.

- The last narrow-body 993 produced went to Ulrich Bez in recognition of his vision for a better 911.

tendency to understeer. The new system was simpler, relying on viscous control of the drive to the front differential. Power to the front wheels could vary from five to forty percent based on rear-wheel traction. Although the multi-disc, temperature-driven silicone oil coupling to the front wheels was outside technology not invented by Porsche, the performance of the system justified swallowing a bit of pride. The new four-wheel driveline was lighter, less expensive and was combined with a limited-slip rear differential and automatic braking at the rear wheels. ABD for 'Automatic Brake Differential' was an early form of traction control for a Porsche street car, using the wheel speed sensors from the anti-lock braking system.

The 993 inherited the basic 3.6-liter engine layout from the 964, however, in typically exhaustive Porsche fashion, nearly every part was redesigned and upgraded. For the first time, the 911 received hydraulic rather than mechanical valve lifters. Initial power for the base engine was up to 272 horsepower. In 1996, the engine was further improved with the VarioRam induction system which varies the length of the intake ducts in the manifold based on engine speed. This improved low-end torque and increased horsepower to 285. In the 993, the G50 transmission was redesigned and upgraded to six speeds. Tiptronic was retained as an option for those preferring a semi-automatic 911.

The 993 first appeared in 1993 for the 1994 model year in

Customers loved the aggressive profile of the 993, with its more steeply raked windshield.

Europe (1994 in the US as 1995 models). It benefited from improved production processes (see Kaizen sidebar) which allowed Porsche to sell the car for a lower price and realize an improved profit margin. Labor input for a 911 fell from 120 to 76 hours between 1991 and 1995. The influence of the new CEO Wendelin Wiedeking (who came from a production background at Porsche and then led bearing supplier, Glyco) was critical to Porsche's survival during a period when the 911 would account for the bulk of the company's car sales. In model year 1996, the 993 911 was the only Porsche for sale. Calendar 1996 production of 911s exceeded 20,000 cars. This was an outstanding result at a time when Porsche had no other product line (Boxster production would ramp up late in 1996).

The 993 was available as a coupe and a cabriolet from the beginning. The cabriolet was noted for its more stylish appearance with the top up. It also received an optional wind deflector that popped up behind the seats. A new Targa debuted in 1996 with two large glass panels in the roof (the motorized front panel slides under the rear window). In the open position, the roof rails recall the look of the Panamericana concept car. The Targa roof was built as a unit by Webasto and installed on the 993 Cabriolet chassis. The stylish Carrera 4S was a wide-body, four-wheel drive 911 with lowered suspension and brakes from the 993 Turbo.

Press reception for the 993 was almost universally positive and filled with 'best 911 ever' kudos. Falk's 'agility' memo as championed by Bez and implemented by the 993 development team under Bernd Kahnau paid off handsomely.

As usual, Porsche produced several special and higher-performance versions of the 993. The 993 Carrera RS marked the first appearance of the VarioRam induction system. These lightweight 911s also featured 3.8-liter engines with numerous performance upgrades good for 300 horsepower. Suspension and brakes were also upgraded for track duty and the cars were identifiable by their large, fixed rear wing. A further, even lighter version of the car was referred to as the RSR or Clubsport version, wearing an even larger rear wing with two horizontal elements. Produced in 1995 and 1996, the RS 993s were not legal for the US market.

Porsche made an interesting decision relative to the turbocharged version of the 993. Sporting 400-plus horsepower and now with two turbochargers, Porsche decreed that all would be made with four-wheel drive. Unlike the sequential arrangement in the 959, the 993 turbochargers worked in parallel, one feeding each bank of cylinders. As with the 964, the 993 Turbo featured a fixed spoiler housing the intercooler (now made from much lighter plastic), plus wider rear fenders and door sills. Suspension, brakes and wheels were upgraded to match the additional performance and a hydraulic assist was added to the clutch action given the need for higher clutch spring pressures in the transmission.

The 993 Turbo was showered with glowing press compliments and awards, although a few pundits wondered if it might be somehow too perfect, lacking a bit in the category of driver involvement. The Turbo's performance put it in supercar territory, exceeded only by the likes of the McLaren F1 (a much more expensive car at five to six times the price of the Porsche). However, the Turbo was usable as an everyday car while capable of sub-four second times from zero-to-60 mph.

Porsche had offered a performance upgrade kit for the 993 Turbo (allowing higher boost pressure) and this prompted a run of Turbo S cars from the factory. They were built with the engine upgrades at the factory and sported some 450 horsepower (425 for the US market). These 1998 model year cars are rare, with less than 350 built, and highly prized by collectors. Being the last air-cooled 911 Turbos makes them extra special.

As always, racing was integral to Porsche's development and promotion for the 993. The rise in popularity of GT racing in the mid-1990s led to the 993 GT2. Named for the GT2 category in the BPR race series, this was a lightweight 993 Turbo built specifically for racing or track duty and with rear-wheel drive only. The pure race versions had success including winning the SCCA World Challenge series in 1995. Porsche also found success in Japan and closer to home in the ADAC GT Cup. A small number of GT2s were built for street use while a still smaller number were upgraded to near-GT1 specs, with 600 horsepower, for racing against the likes of the McLaren F1 and Dodge Viper. The last GT2s built for the street were based on the 450-horsepower Turbo S and had road car amenities like air conditioning and electric windows.

In July 1996, a 911 Carrera became the one millionth Porsche to be produced. It was configured as a police car and presented to Porsche's home state of Baden-Württemberg. 993 production ended in 1998 and the very last air-cooled Porsche produced, a Mexico Blue 911 C4S, was sold to comedian and Porsche enthusiast, Jerry Seinfeld. A little over 410,000 air-cooled 911s were built between 1963 and 1998. It was nearly an even 50 years between production of the first and last air-cooled Porsche cars.

Ferry Porsche with the 1,000,000th Porsche – a 993 for police duty, built in 1996.

Kaizen

Kaizen is the Japanese word referring to 'continuous improvement', specifically in manufacturing processes and productivity. Under Wendelin Wiedeking, Porsche worked with both the Kaizen Institute and the Shin-Gijutsu organization to improve overall production efficiency and quality. Although the aggressive, action-first methods of the Japanese were a shock to the cautious Swabian culture at Porsche, the improvements played a significant part in modernizing Porsche's approach and ensuring financial independence. Wiedeking also reorganized Porsche's management and culture around end-to-end process responsibilities. These efforts were critical to selling a lower-priced but more profitable 911.

WSC 95

After the success at Le Mans in 1994, Porsche was prompted to look at options for the 1995 World Sports Car category. Competition would be formidable from the new and very fast Ferrari 333 SP (Ferrari's first sports racer in 20 years). Porsche's North American motorsport head, Alwin Springer, received an inquiry from Tom Walkinshaw Racing about adapting a chassis from their inventory to use a Porsche powertrain. The chassis started life as a TWR-Jaguar XJR-14 in 1991 and the basic design was subsequently used with a Judd V10 engine by Mazdaspeed in 1992. TWR's man in Valparaiso, IN, Tony Dowe made a workable proposal and the carbon fiber monocoque chassis was modified to accept a Porsche engine and transmission. Open-cockpit spyder bodywork was created to suit the WSC rules and Porsche drivetrain. The goal was to

C88 - China People's Car Project

In 1994, Porsche produced a family car prototype for the Chinese government. It was a compact sedan designed for one child seat under China's one-child-per-family policy at the time. With a range of 47 to 67 horsepower from a 1.1-liter flat-four, the spec was reminiscent of the VW Beetle, although the C88 had front-wheel drive. It was designed for multiple body configurations and for both domestic as well as export sales. China invited some 20 auto companies to submit proposals and Wendelin Wiedeking gave his Beijing conference presentation speech in Mandarin.

The 'C' stood for China while the '88' signifies good fortune. Ironically, the Chinese government never officially selected a design for a locally produced 'people's car'. The C88 prototype can be seen in the Porsche Museum.

C88 prototype in the Porsche Museum.

enter the car at Daytona in 1995, however, it was significantly off the pace in pre-race testing and IMSA thought Porsche was sandbagging. (Dowe disclosed in recent years that there was an element of sandbagging by manipulating the throttle settings to show full throttle telemetry at less than true wide open.) IMSA imposed both a weight and air intake restrictor penalty on the car. Porsche elected not to race but ironically the Daytona 24 Hour was won by the Kremer Racing K8 (a spyder that also had a 962-based engine).

While Porsche worked on building a 911-based car for the GT1 category, the two WSC 95s were set aside at Weissach and spent months in limbo. However, Reinhold Joest pressed Porsche to use the cars for a Le Mans entry in 1996. After wind tunnel testing to solve the problem of front-end lift, the two cars were entered at Le Mans in the LMP1 class. It was a superb race for Joest and Porsche, with the WSC 95 Spyders finishing first and third. Davy Jones, Manuel Reuter and Alex Wurz drove the winning car, finishing just a lap ahead of the factory-entered 911 GT1. Part of Joest's reward from Porsche was the purchase of the winning car "at a special price" according to Norbert Singer.

After helping Porsche to sweep the podium in 1996, Joest entered just one car at Le Mans in 1997. One was enough as the Joest team won the race again, this time by a single lap over a McLaren F1 GTR. The drivers were Ferrari F1 veterans Michele Alboreto and Stefan Johansson along with Le Mans rookie Tom Kristensen. After running a conservative pace early in the race, Kristensen drove a blistering quadruple stint during the night, setting and re-setting fastest lap, recalling Jacky Ickx'

TOP: WSC 95 on the way to winning at Le Mans in 1996.

ABOVE: WSC 95 wins again at Le Mans in 1997.

1977 performance. This would be the first of nine Le Mans wins for Kristensen who became 'Mr. Le Mans' for a new generation. The WSC 95 Chassis 001 became the fifth car in history to win Le Mans in two successive years (and the second time the Joest team managed this feat). It also sat on the pole and set fastest lap in the 1997 race. At this point, the Joest team was responsible for four of Porsche's 15 wins at Le Mans.

The aging WSC Spyders came out for one last go at Le Mans in 1998. This time they were part of the Porsche factory team entry but managed by the Joest crew (designated LMP1-98). The cars were extensively reworked, now with 3.2-liter, fully water-cooled engines, sequential six-speed gearboxes, revised suspension and new bodywork (mainly eliminating the large airbox behind the driver). Third time was not the charm as both cars retired, one with mechanical failure and the other as a result of crash damage that caused the rear bodywork to fly off (taking the rear wing with it).

PORSCHE POINTS

- The first WSC 95 Spyder, Chassis 001, was derived from an original Jaguar XJR-14, believed to be Chassis 691/01. The second WSC 95, Chassis 002 which resides in the Porsche Museum, was built new for the 1995 season by TWR.

- Porsche's design, engineering and consulting clients are too numerous to list but they have included names like Yamaha, Volvo, Scania, SEAT, Rolls-Royce, Opel, Mercedes-Benz and Audi in addition to Volkswagen.

- In 1994, Porsche Consulting GmbH became an organization separate from Porsche's engineering and design business. Porsche Consulting has helped clients in numerous general management and strategic functions.

Porsche Boxster

The original 986 version of the Boxster was the result of several years of work on a new entry-level Porsche. An interesting precursor was the mid-1980s Type 984. This was a roadster concept with a high beltline and high rear deck. Porsche explored it both as a front- and rear-engine design. Although the 984 never got close to being produced, the roadster concept came back to life around 1991 and ultimately became part of the 'New Generation' product direction for Porsche. Horst Marchart, working under Ulrich Bez, was leading the effort to create a future 911 (beyond the 993) and hit upon the idea of an entry-level car that would share significant content with the Type 996.

When Ulrich Bez left Porsche in 1991, Horst Marchart took over as head of R&D. He had the advantage of three decades' experience working in Porsche engineering, including years on the services side working with Porsche's outside clients. He also had an easy-going, familial style well suited to the Porsche culture. Marchart also got along well with new CEO Wendelin Wiedeking. As Wiedeking pushed to improve Porsche's efficiency, Marchart reorganized Porsche's approach to project management including the development of its own new products. Although this included painful reductions in staff at Weissach, it helped in the creation of the 'New Generation' 996 and 986.

Grant Larson is an American who began his design career at Audi but was hired by Porsche in 1989. The Boxster *Studie* show car was partly inspired by Larson's visit to the Tokyo Motor Show in 1991 and noting the volume of 'concept' cars from competitors (whereas Porsche had none in the pipeline). During 1992, multiple design teams were working on roadster concepts for a new Porsche. Larson hit on a theme that had the right combination of sexiness and Porsche's historic design language.

Porsche Engineering Services

In October 1996, Porsche Engineering Services GmbH was set up as a separate legal entity for Porsche's long-established customer development work. With the Porsche Development Center at Weissach and the development of teams for customers worldwide, the amount of activity needing to be coordinated has been growing since the 1990s. In 2001, with increasing new demands on project management, Porsche placed the customer development area of Weissach on a new footing with Porsche Engineering Group GmbH (PEG) and since then it has been expanding internationally:

2001: Founding of Porsche Engineering Prague

2012: Porsche Engineering takes over Nardò Technical Centre

2014: Founding of Porsche Engineering (Shanghai) Co., Ltd.

2016: Founding of Porsche Engineering Romania SRL

2018: Founding of Porsche Engineering Ostrava (Czech Republic)

The concept car had an attractive interior design done by Stefan Stark and was realized by a team of seven people based on Larson's drawings and clay modeling by Peter Müller.

The Boxster concept car debuted at the Detroit Auto Show in January 1993, proudly unveiled by new CEO Wendelin Wiedeking and head of design, Harm Lagaaij. The concept caused quite a sensation and drew near-universal praise, especially in light of Porsche's stated intent to sell such a car in the entry-level, $40,000 price range (far less than the 968). Unveiling the car in Porsche's most important export market was also noteworthy. The production version of the Boxster was already being worked on at this time, so it was known internally that some features of the show car would not make it to showrooms. Boxster styling was intended to pay tribute to Porsche's roadster heritage from the 1950s and early 1960s. Although often compared to the 550, designer Larson has noted that the Boxster was more influenced by the 718 RSK, RS60 and 61 (when the FIA required full-height windscreens for sports racers).

February of 1992 was the date of board approval for the 'New Generation' cars. The 986 and 996 would move forward in conjunction for design, development and manufacturing. The momentous decision was to share nearly everything from the doors forward between the 996 and 986. This allowed for major efficiency in the design and production with the cars sharing the forward structure, bodywork, doors, windshield, front suspension and steering. They could also share expensive systems like the dashboards and HVAC. Another expensive item is lighting and the Boxster would share the single headlight unit, containing all the lighting elements, with the 996. Although the aesthetics were somewhat controversial at the time, the logic of sharing the headlights is obvious in hindsight. Every aspect of the Boxster was challenged to minimize cost without detracting from target performance. Some 36 sub-assemblies were provided by outside suppliers and the overall number of suppliers was reduced to 300 from 950 as compared to the 993.

The Boxster was always intended to be a mid-engine car like its distant ancestor, the 914. There was also a thought that the 986

The Boxster study with the responsible designers, 1993. From left to right: Grant Larson (exterior designer), Jörg Kirschbaum (model designer), Otto Geffert (engineer in the Design Studio), Dorothea Müller-Goodwin (responsible for colors and equipment), Klaus Ziegler (model construction), Stefan Stark (interior designer) and Peter Müller (chief modeler). In the Boxster is Harm Lagaaij (head of the Design Department).

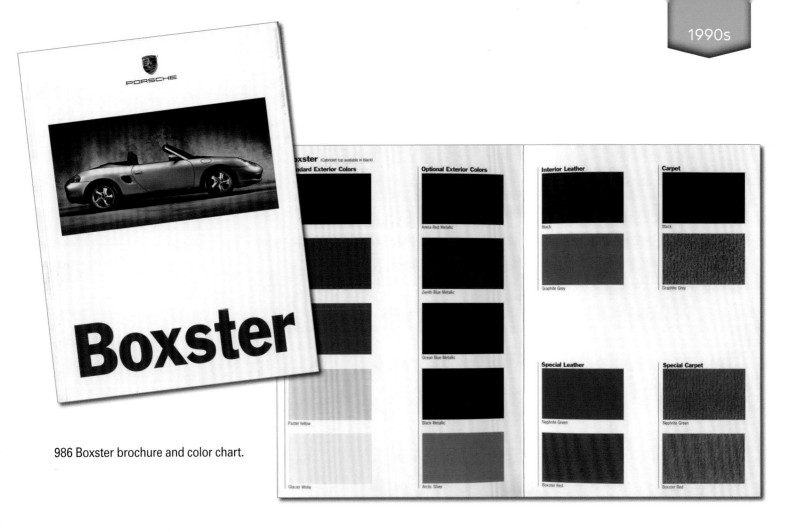

986 Boxster brochure and color chart.

and 996 could share an engine, or at least a family of engines. After years of considering the options for water cooling, Porsche settled on a redesign of the familiar flat-six, boxer engine. The water-cooled version was coded 'M96' and the architecture was meant to serve both the Boxster and the 911. The complete redesign was meant to reduce component costs, weight, and assembly time (which came down from seven hours for the 993 engine to one hour for the M96). Water cooling allowed for the long-awaited move to four valves per cylinder. For the Boxster, the capacity was 2.5 liters with just over 200 horsepower in the first cars. This provided adequate power for a car weighing approximately 2,800 pounds with a manual transmission (and depending on other equipment options). The Boxster made use of an Audi-sourced five-speed manual or a new version of the familiar Tiptronic that now offered five speeds as well.

One thing the new cars could not share was the rear suspension. This was a thorny problem at first, but Marchart and team eventually realized that the front suspension could be reversed and used at the rear of the Boxster (sharing several of the major components, with other parts such as the Bilstein strut and spring units adjusted for size). This helped the Boxster's lower price point by being much less expensive than the 911's more complex rear suspension. The Boxster featured innovative one-piece aluminum brake calipers from Italy's Brembo (co-designed by Porsche).

Grant Larson fought for the Boxster to get its own front bumper cover with differently shaped and arranged air intakes (rather than sharing with the 996 as originally mandated). The low tail shape led to development of a movable, vertical spoiler that provided needed downforce at speeds above 75 mph and remained upright until the car slowed below 50 mph. The production Boxster managed to retain the concept car's signature central exhaust, after some negotiation with the engineering team. The innovative and fast-folding soft top not only looked good but managed to fit around the cockpit and engine, leaving usable room for storage in the rear compartment. The Boxster was also available with a removable hardtop. Instead of the traditional five-gauge dash, reserved for the 911, the Boxster featured a similar layout but with only three gauges ahead of the driver. The interior was functional and stylish, especially in the swooping door panel design.

Production of the Boxster began on August 18, 1996 for the 1997 model year. The reception from the press, dealers and customers was extremely positive. So much so that Porsche was flooded with orders and had to explore outside assembly capacity. To

986 Boxster promotional photograph.

avoid long wait times, Porsche began a relationship with Valmet in Finland to produce Boxsters (starting in the 1998 model year). The new roadster benefited from its unmistakable and uncompromised Porsche character while avoiding any threat to the venerable 911. The Boxster also nearly met its attractive original price target (most selling in the low $40,000 range) while giving Porsche a share of the growing roadster market (as evidenced by the BMW Z3 and Mercedes-Benz SLK). The mid-engine layout gave Porsche's new entry-level car a distinct advantage in track tests over its German rivals.

Porsche 911 GT1

The 911 GT1 was not a singular car, but rather a series of racing cars and homologation street cars that gave Porsche a chance to win overall at Le Mans and other major sports car races of the era. Under motorsport Director Herbert Ampferer

standard 911. 993 designer Tony Hatter was mainly responsible for the exterior styling. Under GT1 rules, the car had to be homologated as a street car and 21 *Strassen* version supercars were sold in addition to the race cars (some of which were sold to private teams). In the first two attempts at Le Mans, Porsche's first mid-engine 911s ironically lost out on overall victory to a Porsche-built Spyder (the WSC 95).

In 1996, factory-entered GT1s finished second and third at Le Mans, winning their class. The second-place car of Hans Stuck, Bob Wollek and Thierry Boutsen was only one lap back from the Joest-Porsche WSC 95. 1997 was something of a disaster for the 'Evo' version of the GT1s at Le Mans. Bob Wollek crashed on Sunday morning when a driveshaft broke and the other car broke an oil line while leading the race with only two hours remaining. The line broke at full speed on the Mulsanne and caused a spectacular fire. Driver Ralf Kelleners had to get the car stopped and abandon ship in great haste.

To have any chance of competing with Toyota and Mercedes in 1998, Porsche decided it needed an all-new, lighter-weight car. The solution was the first all-carbon fiber Porsche monocoque. As with the earlier GT1s, the chassis was designed by Horst Reitter. The Porsche team benefited from the presence of Dutch engineer Weit Huidekoper who had designed the composite chassis Lola T92/10 sports racer. An advantage of the all-new chassis was the ability to design a no-compromise racing suspension. Tony Hatter led the styling, working with Norbert Singer on the aerodynamics. The bodywork looked even less like a 911 and more like a pure sports prototype racer. The new car also featured a sequential gearbox for quicker, more efficient shifting. To homologate the GT1-98, Porsche only had to build one street-legal car.

and engineer Norbert Singer, Porsche sought to compete head-on with the McLaren F1. The program started in July 1995 with the idea to build a mid-engine 911 using the steel forward structure from the 993 combined with a roll cage and a racing-specific tube frame at the rear. The rear structure would carry a fully water-cooled, turbocharged flat-six derived from the unit used in the Dauer 962s. The mid-engine configuration and hybrid structure made for a longer wheelbase than the

> *Uusikaupunki* means "new town" in Finnish. Porsche cars built there by Valmet are identified by a 'U' in the 11th digit of the VIN. Production started on September 3, 1997 (Ferdinand Porsche's birthday) and by spring of 2000, the Finns had built some 30,000 Boxsters. The author's own Cayman was built there in 2007.

> Porsche's 911 GT1 cars were very competitive at Le Mans but had a relative lack of success in the BPR/FIA GT Championship series against the likes of McLaren and Mercedes. Jochen Rohr's team did win the 1997 IMSA GTS-1 championship with a GT1.

911 GT1 wins at Le Mans in 1998.

Le Mans in 1998 looked like a titanic battle in the classic mold. Porsche took on a highly competitive field with entries from Toyota, Nissan, AMG-Mercedes, BMW, McLaren and Ferrari. The Porsches were not the fastest cars, qualifying fourth and fifth. However, as Porsche proved many times, to win Le Mans requires speed combined with reliability and solid race craft. The heaviest competition came from Mercedes and Toyota, but all of their cars had significant mechanical trouble. In spite of rain during the night, the Porsches ran a mostly trouble-free race and staged a perfect one-two photo op at the checkered flag. The third-place Nissan R390 was three laps behind in the end.

Laurent Aiello, Allan McNish and Stéphane Ortelli drove the winning car, finishing a lap ahead of Jörg Müller, Uwe Alzen and the ever-unlucky Bob Wollek (certainly one of the best and longest-tenured drivers never to win at Le Mans). Porsche's 16th victory at Le Mans came on June 7, 1998, just one day short of the 50th anniversary of the registration of Porsche's first road car in 1948. After 1998, Porsche withdrew from top-level competition at Le Mans and did not return until 2014. Expansion of the Porsche product line would take priority over the next decade.

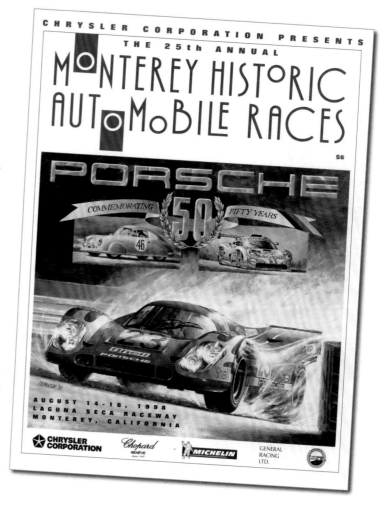

RIGHT: In 1998, the Monterey Historic Races celebrated a 25th anniversary. Porsche was the featured marque, celebrating their 50th as a manufacturer.

Wendelin Wiedeking

Wendelin Wiedeking was born in Ahlen, Westphalia in 1952 and became a Doctor of Engineering before his first stint at Porsche. Unlike Peter Schutz or Arno Bohn, Wiedeking brought extensive automotive industry and Porsche experience to the CEO role, plus he was a Piëch family choice. He was responsible for returning Porsche to profitability in the mid-1990s, thanks in part to embracing production methods recommended by the Japanese. This also involved painful job cuts to management, engineering and production labor, however, with an emphasis on saving seven of ten jobs rather than losing three of ten. He was also responsible for Porsche's expansion with products beyond sports cars. The Cayenne and Panamera appeared during his tenure, in the 2000s, along with the Cayman. By 2008, Porsche AG was valued at 80x its 1990s low point.

Wiedeking's highly successful run at Porsche ended after the failed takeover of VW in 2009. This bold, if overreaching strategy put a black mark on his tenure (see Chapter Ten). Outside of Porsche, Wiedeking has invested in a shoe business, a restaurant chain and technology companies.

Porsche 996

Ferry Porsche lived to see the launch of the New Generation 911, the Type 996. The car was unveiled at the Frankfurt Auto Show in September 1997, the same month that Ferry Porsche turned 88 years old. Happily, the company was showing strong improvement in terms of financial performance. However, as the introductions of the 996 continued, the son of Ferdinand Porsche was in declining health. He passed away on March 27, 1998 just after the Geneva Auto Show (where the first Porsche cars had been shown in 1949). His wife, Dorothea had passed away in 1985. Ferry's death was a somber moment in a year when Porsche celebrated its 50th anniversary as a manufacturer. His sister Louise, a force in company affairs behind the scenes for more than 50 years, passed away at age 95 in February 1999.

Like its stablemate, the Boxster, the 996 was all-new but shared significant common parts and systems. As a new, ground-up design, 'packaging' for a water-cooled engine and new heating/air conditioning system could be engineered more easily. The challenge for the styling department was to make the new car look like a 911 and differentiate it from the Boxster. The designers took a traditional approach to *formsprache* or 'form language'.

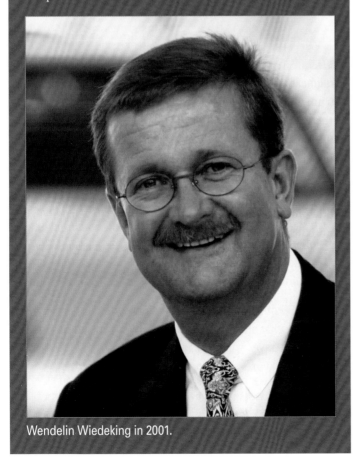

Wendelin Wiedeking in 2001.

1998 *Autoweek* cover on the passing of Ferry Porsche.

996 Cabriolets showing soft top and hard top.

of suspension and braking were upgraded to suit the power of the engine. Porsche also offered an even less-compromising, track-oriented GT3 'Clubsport'. The 911 GT3 series became the base for an ever-evolving series of Porsche GT racing cars. Unfortunately for the American market, Porsche did not produce a US-legal GT3 until the 2004 model year.

In 1998, Porsche Cars North America moved from Reno, Nevada to Atlanta, Georgia. Among other things, this offered better time-zone compatibility with Stuttgart. Porsche's growing financial strength allowed it to consolidate ownership of several distributors in Europe during the later 1990s. Porsche also formed Porsche Japan to control distribution in that important market. Speeding into the new millennium, Porsche had even bigger changes in store. Most shocking to the traditionalists was the prospect of Porsches with four doors.

A 986 and 996 promotional photograph in 1998 shows the New Generation product line. The 986 Boxster sits between a 996 coupe and cabriolet, showing the commonality of design from the doors forward.

2000s

2002	» Porsche Cayenne
	Rennsport Reunion
2000s	» Porsche 996 in the 2000s
2003	» Carrera GT
2000s	» Boxster in the 2000s
2005	» Porsche Cayman
	Norbert Singer
	RS Spyder
	Porsche 997
2008	» Porsche 997.2
2009	» Panamera
	Porsche versus Volkswagen
	New Porsche Museum

We will never allow our design to be 'hip' or trendy, only to lose its appeal after a single season. – *Wendelin Wiedeking*

On February 7, 2000, Porsche broke ground for a new plant in Leipzig. This city in the former East Germany would be home to final assembly of an all-new Porsche. The Cayenne was Porsche's first departure from building sports cars and provided entry to the market for Sport Utility Vehicles (a market approximately eight times larger than the market for sports cars at the time). Porsche did, however, take the 'Sport' in 'SUV' seriously. In 2001, Porsche began vehicle sales into China, first through Hong Kong into the mainland. Porsche then established direct imports in 2005 through Guangzhou. There was a missing element from the traditional program during the 2000s. Porsche stayed away from frontline sports car racing although they eventually developed the RS Spyder for the second tier (P2) class in the ALMS. However, celebrating Porsche's racing heritage took a major step up in the 2000s. Ex-factory driver Brian Redman and Porsche Cars North America PR manager Bob Carlson worked to found the Rennsport Reunion in the US. The decade concluded with Porsche launching a luxury four-door sedan into a global financial crisis and a risky attempt to take over Volkswagen AG.

Porsche Cayenne

An off-road capable Porsche was not a new idea. Dating back at least to the mid-1950s Type 597 (the *Jagdwagen* or 'hunting vehicle'), Ferry Porsche had thought that Porsche could compete with the likes of Land Rover. Porsche built a small number of 597 prototypes for the West German military but ultimately did not receive the contract to supply the army. Another study took place in the early 1970s under Type number 926. Fortunately, Ferry Porsche lived to see the work begin in the mid- to late-1990s on a Porsche SUV. Part of that work was deciding that Porsche had to partner with another manufacturer for such a large and complicated undertaking. That partnership was very nearly established with Mercedes-Benz. Mercedes was interested in updating its G-Class (*Geländewagen* or 'terrain vehicle') and the deal only foundered when Mercedes asked for an ownership position in Porsche.

A more traditional Porsche partner was also lacking an SUV in their product mix at the time. Porsche shareholder and board member, Ferdinand Piëch, had risen to become CEO at Volkswagen AG. The combination of relationships, talent, resources and purchasing power now seems like an obvious winner, however, it took prodigious work to create a central design that could work for vehicles to be sold by both companies and have individual brand identities. Rather than work at Weissach, where space was limited and VW employees might catch a glimpse of projects in progress for other Porsche clients, the companies set up shop in office space at Hemmingen (located between Stuttgart and Weissach). The agreement between Porsche and Volkswagen was announced in 1998, although design work was already quite far along at that point.

Cooperation with Volkswagen on the 'E1' (the internal code before the car was officially named) allowed Porsche to produce a vehicle with its own styling, engine and suspension tuning. At the same time, the basic structure and many of the unseen components could be shared and therefore built at much lower cost. Rather than have vehicles assembled in a VW plant (as with the 924 and 944), Porsche elected to handle final assembly of the Cayenne. There was no room in the Stuttgart facilities, so a new location was found in Leipzig. The site had plenty of space for future expansion and a 2.3-mile test and demonstration track was eventually built there. Partnership with Volkswagen allowed Porsche to finance development of the Cayenne and the Leipzig facilities without taking on any debt.

Creating an appropriate Porsche 'look' for an SUV was the responsibility of a team under Harm Lagaaij. Steve Murkett was the lead designer and despite the need for large front air intake openings, the project succeeded in making the Cayenne look like a Porsche. Extra help came from Butzi Porsche, an enthusiastic supporter of the SUV idea. Butzi's independent Porsche Design studio put forward a styling proposal that ultimately lost out to the work of Lagaaij' team. However, Butzi Porsche's ideas are credited with influencing the simplicity of the final Cayenne form. The Porsche look was helped by incorporating traditional elements such as the front hood line sloping slightly below the fender tops and the curving rear quarter windows. A version of the 996 headlight aided family resemblance. This new unit was developed with an automated,

> **PORSCHE POINTS**
>
> » During the design stage, the code name for the Cayenne was 'Project Colorado'.
>
> » In 2002, when the Cayenne was unveiled at the Geneva Auto Show, Wendelin Wiedeking offered a quote that crystallizes Porsche's approach to styling: *"Our designers faced the challenge of creating a vehicle that didn't try to follow current fashions or trends that would soon be out of date. This is the true secret behind the fascination and emotion of Porsche. We will never allow our design to be 'hip' or trendy, only to lose its appeal after a single season."*

movable low beam for cornering, synchronized with steering angle. The Cayenne Turbo was distinguished by power bulges along the edges of the hood.

The dashboard and interior of the Cayenne were graced with Porsche styling elements similar to the 996 and 986. Following Porsche tradition, the ignition key was on the left for left-hand drive cars. Three overlapping gauge housings presented the important information directly ahead of the driver. As with most modern cars, more control functions moved to the steering wheel. Sophisticated front and side airbags were engineered into the Cayenne. Porsche's first four-door vehicle was equipped with five seating positions (room for three across the back row).

To power the Cayenne, Porsche designed a modern, 90-degree V8 that was a spiritual successor to the 928 engine. As in the 928, it was an aluminum alloy engine starting at 4.5 liters in capacity. The Cayenne had twin, chain driven overhead cams with four valves per cylinder (sharing cylinder dimensions with the Boxster). VarioCam was used to alter timing of the inlet valves to suit performance needs while maintaining fuel efficiency. The latest version of the Bosch Motronic system took care of engine management and fuel injection. The engine was designed for future expansion in capacity and for turbocharging. Specific attention was paid to the use of weight-saving materials in the engine to help lower the center of gravity in a tall vehicle.

For the Cayenne S, the new engine produced 340 horsepower with 310 foot-pounds of torque. For the ultimate in sporty SUV performance, customers could opt for the twin-turbo with 450 horsepower and maximum torque of 457 foot-pounds. The four-wheel drive system used a multi-disc clutch to allocate power to the axles. This was reminiscent of the 964 and unlike the viscous coupling used in the 993 C4. A version of the Tiptronic transmission was developed with numerous features intended specifically for the Cayenne, tuned for off-road duties and heavy-duty hill climbing. A relatively small number of Cayennes were built with a ZF six-speed manual transmission.

Porsche employed several new technologies to give their SUV sports car handling capability. Air suspension, which had been considered for the 964, was employed in place of conventional springs as an option in the base Cayenne and standard for the Turbo. This allowed for adjustability of ride height for different conditions. PASM (Porsche Active Suspension Management) controlled the shock absorbers to improve stability in cornering

Cayenne promotional photograph – happy couples could now combine a Porsche sports car with an SUV in their garage.

2009 Cayenne S Trans-Siberia Edition celebrated competition success for Porsche's SUV.

and adjust ride comfort. Further to the growing list of acronyms, PTM (Porsche Traction Management) controlled the 4WD clutches to optimize power to the wheels. An innovation meant for off-road use was optional disconnect for the anti-roll bars. This allowed for greater wheel movement on rough terrain (limited to speeds below 30 mph).

Famous for its braking innovations, Porsche's Cayenne development was challenged with stopping such a large and heavy vehicle capable of speeds above 150 mph. Vented, 13.8-inch diameter discs were employed up front with 13-inch at the rear. To avoid collecting stones off-road, the discs were not cross-drilled. Six-piston Porsche 'monobloc' calipers acted on the discs which were over an inch in thickness. Optional tire pressure sensors offered continuous monitoring at each wheel, programmed to allow for change in temperature of the tire itself. The all-new, foot-operated parking brake (a first for Porsche) was designed to cope with heavy weight and hillside parking.

Test vehicles were built at Leipzig as early as August 2001, but the official start of Cayenne production was in August 2002. Sales in Europe began late in 2002 with US sales starting in the spring of 2003. At the time, the idea of a Porsche 'truck' seemed like heresy to some of the sports car faithful (the author included). However, in hindsight the business and marketing logic was spot on. There was significant doubt whether Porsche could survive

PORSCHE POINTS

» Putting the 'sport' in SUV, the first Cayenne Turbo's best lap time at the Nürburgring, 8:42, was just ten seconds slower than the 1968 1000 KM race pole time, set by Rolf Stommelen in a 907 at 8:32.8. This new SUV benchmark came from a vehicle weighing a little over 5,000 pounds.

» After the launch of the Cayenne, Porsche considered alternative body styles including a coupe roof and a convertible.

Rennsport Reunion

Porsche's periodic Rennsport Reunion events have grown to celebrate the racing and brand heritage in a way far beyond any other manufacturer. The idea grew out of the 1998 Double-50 event celebrating Porsche's 50th anniversary as a manufacturer and the 50th anniversary of racing at Watkins Glen, New York. Brian Redman and PCNA's Bob Carlson then worked to put on Rennsport Reunion I in August of 2001 at Lime Rock in Connecticut. That first event was a fairly intimate three-day gathering with the focus on vintage racing and a superb gathering of historic Porsche racing cars. The next two events were held at Daytona in 2004 and 2007. The 2007 event featured no less than 17 of the existing 917s. In 2011, Rennsport moved to Laguna Seca and Monterey, California. Since then it has grown into a festival of all things Porsche. The 2018 event attracted more than 80,000 spectators.

Brian Redman in the Collier Collections 908/03 at Rennsport Reunion I. (Jay Gillotti)

by producing only sports cars. And what family wouldn't enjoy having a Porsche sports car and a Cayenne in the garage?

Press reception for the Cayenne was quite positive, particularly on its driving merits. Some reviewers clearly had trouble getting past the potential impact on Porsche's brand image. Fuel consumption was a concern, while most reviewers acknowledged the Cayenne's game-changing blend of performance and capability both on the street and off-road. Although most Cayennes would never see a dirt road or top speed on the autobahn, buyers responded in traditional Porsche fashion to owning vehicles with such high potential.

In 2004, Porsche introduced a 'base' Cayenne with a V6 from Volkswagen, tuned by Porsche's engineers. This was a unique, narrow-angle 'V' with cylinders set at only 15 degrees apart and sharing a common head across all six cylinders. Although criticized for its leisurely nine second sprint to 60 mph, the base Cayenne was some $12,000 less expensive than the Cayenne S and offered better fuel mileage. Fuel economy was a factor in mid-decade as gasoline prices increased and Cayenne sales fluctuated.

For the wealthier buyers, Porsche introduced a Turbo S Cayenne in 2006. With more than 500 horsepower on tap, the Turbo S had a top speed over 160 mph and 5.5-second zero-to-60 time. Naturally, Porsche employed improved tires and braking to match the increased performance. The line was further expanded with the extra-sporty GTS model in 2007. This was a lowered Cayenne running on steel springs combined with PASM. It had a six-speed manual transmission (Tiptronic optional) for enthusiast driving. It also featured the new, lighter-weight 4.8-liter engine designed by Porsche in part to address fuel economy concerns without sacrificing performance.

The GTS was part of the Cayenne 'refresh' for the 2008 model year. The Cayenne received a mild restyling and the engines were improved across the board (including the base VW V6). In addition to weight-savings, the engines benefited from direct fuel injection (where fuel is injected directly into the combustion chamber, beyond the intake port). The top-of-the-line Cayenne Turbo now raced to 60 mph in less than five seconds. Porsche also introduced PASM with PDCC (Porsche Dynamic Chassis Control). In the Cayenne, PDCC divided the anti-roll bars with a high-strength, hydraulic controller. This allowed for a range of handling options, from super firm and flat cornering on the street to nearly free-moving suspension for off-road use.

Cayennes upheld Porsche tradition in competition by winning the Trans-Siberia Rally in three consecutive years, from 2006 through 2008. In 2009, Porsche offered the Transsyberia edition of the Cayenne S. It featured both standard and optional off-road equipment and a livery inspired by the actual rally-prepped vehicles.

Porsche 996 in the 2000s

At the start of the new decade, Porsche was well into development and expansion of the 996 lineup. By the early part of the decade, the 911 was available in nearly all the body style, engine and drive configurations that have become standard up to the present day. In 2000, the 911 Turbo returned to the lineup as a 2001 model. However, Porsche had made the decision to create the Turbo's engine using the dry sump 'Mezger' architecture. This grew out of Porsche's need to continue with the dry sump for competition 911s, starting with the Porsche Cup 996 and subsequent GT3s. In order to create economies of scale for an alternate engine, Porsche planned to adapt the Mezger-style engine to all of the high-performance street-going 911s as well as the competition cars.

2001 996 Turbo. (Allan Caldwell)

For the 911 Turbo, the dry sump, water-cooled, twin-turbo flat-six could trace its origins back through the Le Mans-winning 911 GT1 and Dauer 962 to the 959. However, this new engine featured Porsche's VarioCam 'Plus' valve gear. The system introduced electronically controlled, adjustable tappets atop each valve stem. Managed by the Motronic engine control unit, the tappets allowed for changing both the timing of the valve as well as the lift. Induction could be adjusted to the power demand and driving conditions. In addition to better performance, the engine achieved significantly improved fuel economy and lower emissions. This newest 911 Turbo offered a maximum of 420 horsepower and powered the car to a top speed near 190 mph.

As with the 993, the 996 Turbo was built with four-wheel drive but there were several new features. For the first time the Turbo was available with an automatic transmission, a five-speed unit sourced from Mercedes. Newly-developed PCCB Carbon Composite (ceramic) Brake discs were available as an option (signified by yellow-painted brake calipers). A product of advanced research by Porsche, the carbon ceramic discs were much lighter, better in performance and expected to last far longer than cast iron.

The 911 Turbo was a distinctly different looking 996. It featured strikingly large front air intakes and a chin spoiler for front downforce. At the rear, the Turbo had fender inlets and rear bumper outlets for the intercoolers. A ducktail spoiler with a separate wing extending up at 75 mph provided downforce at the rear. Restyled headlights introduced Bi-Xenon lamps, using a single bulb for both high and low beam. Even with wider fenders to accommodate bigger tires, wider rear track and the additional air intakes, the 996 Turbo managed only a negligible increase in drag thanks to cooperation between the styling and aerodynamics teams.

In 2002, the 996 received a mid-cycle upgrade. The mild restyle included headlights from the 911 Turbo and a new, sleeker-looking front bumper cover. The rear bumper cover was also restyled. The chassis was strengthened to double the torsional rigidity of the original 996 (causing a minor increase in overall weight). Engine size increased to 3.6 liters and 315 horsepower for the base engine. Adopting VarioCam Plus improved fuel efficiency despite increased power output. At this same time, the Turbo-look option returned with the Carrera 4S. In addition to its curvaceous appearance, the car was lowered with Turbo suspension tuning, brakes and wider tires. In 2004, Porsche offered an upgrade to the 911 Turbo itself, the 450-horsepower Turbo S.

996 GT3 promotional photograph shows fixed rear wing.

Promotional photograph of the fearsome 996 GT2. The Nürburgring lap record time was comparable to the 908/03 in 1970/71.

The 996 received a Targa-style, sliding glass roof option for the 2002 model year. Unlike the 993, with the Targa roof installed on a cabriolet body, the 996 Targa roof system was engineered for robotic installation into a reinforced coupe body. The unit was produced by CTS, supplier of tops for the Boxster and 911 Cabriolet. The rear glass panel could now be tilted up when the sliding portion of the glass was closed. This glass hatch gave easier access to the rear seat storage area. In 2003, Porsche produced the 40th Anniversary Edition 996, building 1,963 Metallic Silver 911s with 345-horsepower engines, lowered suspension and front valence similar to the 911 Turbo.

The 996 GT3 had already been introduced before 2000, but it continued to develop both as a street car and for racing. Porsche's motorsport department had great success selling GT3-based cars for competition. This was highlighted by the GT3 RS which made 420 normally aspirated horsepower for track duty. In 2001 at Le Mans, all finishers in the GT class were 911s with the GT winner achieving sixth overall. In 2003, at the 24 Hours of Daytona, a GT3 RS outlasted the Daytona Prototypes and the other GT class entries to take the overall win. The triumph was not unlike the 1973 911 victory by Brumos. This time The Racer's Group team of Kevin Buchler, with Michael Strom, Timo Bernhard and Jörg Bergmeister won by nine laps over the Ferrari 360 Modena GT from Risi Competizioni.

For the 2004 model year, the GT3 became available as a road car in North America. As with the 911 Turbo, it featured an engine separate from the regular M96 found in standard 996s. Nicknamed the 'Mezger' engine, the traditional 3.6-liter, dry sump, seven main bearing, aluminum engine case was adapted for water-cooled cylinders. As originally introduced in the late 1990s, the cylinder heads had been redesigned to match the bore of the new cylinder layout. For this second-generation GT3, the engine produced 380 horsepower and sub-5-second zero-to-60 times. Ever expansive with high-performance options, Porsche made the GT3 in standard and RS form for the street, with a racing-only RSR version. The RS version was built to

PORSCHE POINTS

» Porsche worked on engine design projects for Harley-Davidson as early as 1978. A 1990s project went into production in 2001. The water-cooled V-twin design came to market in Harley's V-Rod motorcycles.

» Original, air-cooled Volkswagen Beetle production finally ended at the last site, in Puebla, Mexico, in 2003. Total production came to 21,529,464 cars over the life of the model.

homologate the RSR and harkened back to the 1973 911 RS. Unfortunately, the 996 RS was not available in North America, but Porsche remedied that in later generations.

In 2001, Porsche produced a 996 GT2. A spiritual successor to the 993 GT2, this ultimate motorsport-derived 996 Turbo never achieved significant competition success, but it did provide wealthy enthusiasts the opportunity to own the ultimate high-performance 911. In addition to its 460-plus horsepower, the GT2 was built with rear-wheel drive only and a competition suspension. It had a larger front spoiler and adjustable rear wing. As with the GT3, the center radiator vented air upward over the hood, adding to frontal downforce. Zero-to-60 times were under four seconds and top speed knocked on the door of 200 mph. Walter Röhrl set a production car record with the GT2 at the Nürburgring, at 7:47 (Porsche 908/03 territory). The GT2 Clubsport option included a roll cage, fireproof seats and fire suppression system for track duty. These pricy Porsches sold well, even at $180,000 and up. This sales success demonstrated a healthy market for ultra-high performance and special edition Porsches.

Carrera GT

The Carrera GT is generally viewed as the second 'supercar' from Porsche, after the 959. Although similar in some ways, the Carrera GT was also quite different from the 959 in that it was not necessarily a showcase for new technologies and it was clearly required to earn a profit.

Also, rather than growing out from an existing street car like the 911, the Carrera GT was based on a racing car. Porsche stayed out of top-level sports car racing in 1999 to develop a new Le Mans prototype, the LM 2000 (Type 980). However, late in 1999, Wendelin Wiedeking canceled the racing program in favor of applying maximum engineering effort to the Cayenne. At the same time, Porsche was on a parallel path for an entry in the early 21st century wave of supercar offerings. In typical fashion, Porsche adapted some elements from the canceled racing car into the Carrera GT.

An important decision for the LM 2000 was to discard turbocharging in favor of a large, normally aspirated engine (the better to accelerate out of slower corners on more modern race tracks). Porsche's head of motorsport, Herbert Ampferer, based a new 5-liter, pneumatic valve V10 on his own earlier design that might have replaced the 3.5-liter V12 in the failed Footwork F1 project. A larger capacity, conventional valve version of the V10 found its way into the Carrera GT. In production form, it ended up at 5.7 liters producing more than 600 horsepower.

Like the LM 2000 racer, the Carrera GT was conceived as a mid-engine, open-cockpit spyder with stark differences from the 959. Design proposals were submitted by Porsche's then-new California design studio and the year 2000 concept car was designed by Grant Larson. Shape and styling were finalized for production by Tony Hatter. An important interior feature was the steeply sloping center console with a high-mounted shift lever, carrying a 917-inspired wooden knob. This console treatment

RIGHT: Carrera GT promotional photograph.

would later appear in the Panamera and 991-generation 911. The interior also sported a three-spoke steering wheel similar to the Cayenne design. Tradition was upheld with the instrument cluster ahead of the driver. Since a pure spyder is impractical for a street car, Porsche designed small, very lightweight, double 'bubble' roof panels which can be stored in the front trunk and quickly installed if needed.

The Carrera GT was Porsche's first carbon composite chassis for a street car, in this case produced by ATR Composites in Italy. The Carrera GT also featured a newly designed, racing-style transverse gearbox, with synchromesh gears, helping

PORSCHE POINTS

» The LM 2000 Spyder is a fascinating 'what if' in Porsche's racing history. Before the project was canceled, the V10 racer was tested by Bob Wollek who pronounced it *"an excellent car."*

» Employees assembling the Carrera GT in Leipzig wore white overalls, rather than red, hence being nicknamed the 'angels' of Leipzig.

to shorten the wheelbase. The small, racing-derived ceramic composite clutch caused some difficulty with low-speed engagement and take-off, requiring owners to master a specific technique. The racing-style suspension with pushrods acting on inboard, chassis-mounted coil spring and shock absorbers was partly inspired by the GT1-98. To reduce weight and give the Carrera GT a pure sports car aura, it was not offered with four-wheel drive, PASM, or other electronic driver aids (except ABS). The Carrera GT was the first Porsche designed to run larger diameter wheels at the rear, putting more rubber on the road. Large, spiral vented and cross-drilled ceramic brake discs were installed to stop the Carrera GT, a car capable of speeds above 200 mph. The carbon underbody pan was shaped to create most of the car's downforce at high speed.

With engines built in Stuttgart, Carrera GTs were assembled at Leipzig and every car was tested on the Leipzig track before release to a customer. US pricing started at $440,000. Deliveries began in 2003 and by 2006, 1,270 Carrera GTs were sold (somewhat less than the originally planned 1,500). The performance of the car was spectacular, including a 7:28 time at the Nürburgring. The legacy of the Carrera GT is somewhat controversial as a result of its goal to bring near-race car capability to the street.

In 2001, during Carrera GT development, Horst Marchart retired after 40 years with Porsche. Wolfgang Dürheimer took over as head of R&D. Dürheimer would lead new car development over the next ten years.

TOP: Workshop photograph shows larger diameter rear wheels on the Carrera GT. (Jason Tang)

ABOVE: Carrera GT interior with 917-inspired gear knob. (Jason Tang)

Boxster in the 2000s

Porsche followed tradition by developing and improving the Boxster throughout its first full decade. The first major step was the introduction of the Boxster S in 2000. While the base Boxster was upgraded to 2.7 liters and 220 horsepower, Porsche introduced the 3.2-liter, 250-horsepower 'S'. Both models were granted 'E-gas' drive-by-wire throttle systems. The Boxster S received a 911 transaxle to handle the additional power while braking and suspension were revised and re-tuned to match the engine's potential output. In addition to its tail script, the 'S' was recognizable by a twin-outlet exhaust. Performance and handling of the newest Boxster put to rest any complaints about the roadster not being a 'real' Porsche. For 2001, PSM (Porsche Stability Management) became available as an option and the Boxsters received effective cup holders, a critical improvement for the US market.

Porsche refreshed the Boxster for the 2003 model year. As usual, Porsche looked at all areas of performance as well as appearance. VarioCam technology was adopted to improve performance in the engine room. With help from revised Motronic engine management, both base and 'S' engines achieved mildly increased horsepower and torque, but more usable torque and improved fuel mileage. Suspension tuning was adjusted and the Boxster received both larger and lighter wheel options. The interior was restyled and a glovebox was added below the passenger airbag. The exterior refresh gave the Boxster a slightly more muscular appearance and a less pointy nose. The new face with its revised air inlets improved cooling airflow. The rear fender air inlets, now body-colored, were also moved to improve airflow. The rear bumper cover and spoiler were reshaped in coordination with a revised convertible top (now featuring a glass rear window).

M96-M97 Engine Issues

As with most cars, Porsches have their known problem areas. Potential purchasers of the Boxster, Cayman and 911s with early-generation water-cooled engines should become familiar with engine issues that could affect these cars. Problems with the intermediate shaft (IMS) ball bearings are relatively well-known, as are the potential solutions. The IMS controversy resulted in a class-action lawsuit, details of which can be found online. Some of these engines also developed problems with cylinder surface cracking or 'scoring' of the bore surface material.

In addition to internet searching, the Porsche clubs as well as experienced Porsche service specialists can be good sources for information, recommendations and inspection of cars from the M96 and M97 generations. This is equally true for other generations and Porsche models. Proper and specific maintenance is also important for all generations of Porsches, especially as they transition to 'vintage' status.

Porsche product line including the 2002 Boxster S (left).

987 Boxster RS60. (Brent Jones)

Still bigger changes for the Boxster came along for the 2005 model year. In coordination with the 997-generation 911, the newest Boxster received the '987' designation. Porsche claimed the car was 80% new compared to the previous model. Without changing the size of the engines, their structure, intake system (including larger rear fender inlets) and exhaust were reworked and produced another bump up in horsepower. The 987 also offered improved cooling and a revised six-speed manual transmission. Base and Tiptronic transmissions were revised in coordination with larger wheel sizes. As usual, faster cars require better brakes and Porsche now offered larger, improved discs and improved brake cooling on both the base car and Boxster S. PCCB ceramic brake discs became an option for the first time. Thanks to new castings, the 987 rear suspension was now stronger and lighter while steering improved with variable-ratio hydraulic assist. The suspension tracks were wider, playing into the restyling of the bodywork.

The underlying chassis was again lighter, but more rigid. Lightness was added with the use of aluminum for the front and rear deck lids as well as careful management of materials and design choices throughout the package. Styling of the 987 Boxster was led by the Boxster's original designer, Grant Larson. The shape changed to accommodate new details, such as the oval headlights which replaced the somewhat controversial 996/986 units. The front bumper cover looked a bit more muscular with larger intakes that now shared space with the signal and fog lights. The rocker panels, fenders and wheel openings were bigger and the tail subtly reshaped with a tighter rear fender profile. Below decks, the 987 featured a new, more complete undertray which helped overall improvement to the drag coefficient.

The new interior featured a range of seat options from the 997, with better seat position and adjustability. The steering wheel was now adjustable for height as well. A further option affecting the interior was the Sport Chrono performance package which included a dashboard-mounted stopwatch.

Press and customer reception for the 987 Boxster was positive with multiple 'best Boxster yet' accolades. However, this first-generation 987 (sometimes referred to as the 987.1) Boxster was in for more improvements in conjunction with a new stablemate, the Cayman. The Boxsters received further engine upgrades in 2007, including VarioCam Plus and the upgraded (M97) engine for the Boxster S. For the 2008 model year, Porsche produced 1960 special Boxsters as tributes to the RS60 Spyder racing car. Based on the Boxster S, these metallic silver cars had a unique front spoiler treatment and dashboard (with instrument cover removed), special 19-inch wheels, and a bump to just over 300 horsepower. For the 2009 model year, Porsche introduced the next generation, 987.2 Boxster and Cayman (see Chapter Eleven).

Porsche Cayman

In *Excellence Was Expected*, author Karl Ludvigsen neatly summarized the Cayman project: *"Elaboration and elevation of the 911 Carrera left a significant price gap to the Boxster, holding fort at the bottom of Porsche's (price) range. How best to fill it? The answer was a Boxster-based coupe. Adventurous styling and engineering gave the Cayman a powerful personality that many considered the most scintillatingly sporting of Porsche's road cars."*

The Cayman, named for a South American alligator, became Porsche's first regular production hardtop, mid-engine two-seater (after having produced many racing cars in this configuration). It recalled the 904 in some ways, as well as the 916 (the hardtop 914/6 that never reached production). From a styling point of view, the team led by Pinky Lai drew inspiration from the 1956 550A coupe that Porsche raced at Le Mans. This is especially clear in the roof line, rear fenders and tail treatment. Unlike a 904 or the 916, the Cayman was designed as a hatchback, to maximize rear storage space. At the front, the Cayman had new-style, angled oval headlight/signal light units but with a unique bumper cover and air intakes also housing the round fog lights. Along the sides, the Cayman had wide sills sweeping up to rakish fender intakes different from those on the Boxster. At the rear, the Cayman had a movable rear wing programmed to raise at 75 mph. Like the 987-generation Boxster, the Cayman would benefit from a more complete undertray, helping to reduce overall drag.

The Cayman S made its official debut at Frankfurt Auto Show in September 2005, after some months of sneak previews and 'coming soon' announcements. Although sharing approximately 40% of its content with the Boxster, the 2006 Cayman S had a number of unique features, including the engine. This new 3.4-liter engine shared some dimensions with the 996 engine and was the first Porsche outside the 911 range to receive VarioCam Plus technology. Rated at 295 horsepower and revving to 7,300 rpm, the Cayman S engine provided an exciting blend of Boxster and 911-derived elements. Porsche even went so far as to give the Cayman S a unique crankshaft and Motronic-controlled induction system.

Adding a roof to the basic Boxster structure allowed Porsche to increase rigidity. Porsche made extensive use of aluminum to save weight in the Cayman's body, chassis and sub-assemblies.

TOP: 987 Cayman promotional photograph.

ABOVE: 987 Cayman in Macadamia Metallic. (Jay Gillotti)

This included some improvements already engineered into the Boxster. The end result was a car slightly lighter than the Boxster but with more than twice the calculated rigidity. The Cayman chassis allowed for suspension tuning different from the Boxster, giving the coupe its own handling and ride character. For higher-level performance, the Cayman could be ordered with PASM, PCCB and Sport Chrono packages. The Cayman S had several optional wheels beyond the standard 'surfblades' which were styled similarly to the Carrera GT. One wheel not available in the Cayman was the spare. To improve luggage space, the spare was eliminated in favor of fix-a-flat sealant and an onboard air pump.

For the 2007 model year, the base Cayman was introduced with a 2.7-liter, 245-horsepower engine. This model could be equipped with a VW-produced five-speed transmission, a Tiptronic or the superb six-speed manual from the Cayman S. Far from being 'cut rate', the base Cayman was praised by many for its elemental, sporting qualities and it further opened Porsche's 987 range to new customers. The counter to affordability in the 987 line was option pricing. Porsches of all types were increasingly customizeable, but often at a hefty price per option selected.

Both versions of the Cayman were showered with praise by the automotive press and won a plethora of awards during the first years of production. To produce the Cayman, Porsche expanded its relationship with Valmet, a company that by 2005 had produced over a million cars for various manufacturers. Valmet in turn increased investment in their facilities to further the relationship with Porsche.

RS Spyder

While Porsche concentrated its resources in other areas during the early 2000s, racing activities were limited to the GT categories with 911-based cars. In the US Grand Am series, teams including Brumos had some success employing Porsche engines in chassis built by Fabcar and Riley. By the middle part of the decade, Porsche was ready to return to prototype sports car racing and selected the LMP2 category as their target. This would allow for entries in the American Le Mans Series and potentially in the 24 Hours of Le Mans itself. The LMP2 category was largely intended for privateers and Porsche planned to offset some of their cost by selling cars to select teams. Rather than enter any cars directly, Porsche selected Penske Racing to field a team in the ALMS series. This recalled Porsche's partnership with Penske in the Can-Am from 1972/1973. The cars were sponsored by Porsche partner DHL and appeared in the bright yellow and red livery of the logistics company's trucks and planes.

Le Mans Prototype 2 rules allowed teams to choose one of several possible engine configurations, including production-based engines. Porsche settled on developing a pure racing engine, a 3.4-liter unit designed from scratch for LMP2. Performance was primarily governed by an air inlet restrictor, visible on the RS Spyder in the cockpit at the driver's left. Initial output from this new Porsche V8 was 480 horsepower at 10,000 rpm (high revs for a sports racer of that era).

The RS Spyder used a couple of vintage ideas from Porsche's past. One was the use of torsion bars for springs in the suspension. Another was the 908/03-style gearbox/transaxle arrangement. It featured a longitudinal gearbox in front of the final drive. The gearbox was electronically controlled through paddle shifters using technology from Megaline. The clutch pedal was only used for starting from a stationary position. Departing from tradition, the drivetrain used non-synchromesh gears and a differential from Xtrack. Given the compact size of the engine and the position of the gearbox, engine exhaust was at the sides of the car ahead of the rear wheels. The chassis was a carbon composite monocoque with Kevlar-reinforced carbon fiber bodywork. Norbert Singer, who retired officially in 2004, helped with early design work and was then called back to consult on aerodynamic development for the RS Spyder during 2005.

The first outing for the new racer was at Laguna Seca at the end of 2005. In an auspicious debut, the car finished fifth overall and won the LMP2 class, at times lapping within two seconds of the victorious LMP1 car. For the full 2006 season, the Penske-entered cars, driven by Sascha Maassen, Lucas Luhr, Timo Bernhard and Romain Dumas, had an excellent

> ### Norbert Singer
> Like Dr. Porsche, Norbert Singer was born in Bohemia, an area that is now in the Czech Republic. He began his illustrious career as a Porsche engineer in 1970 and worked on every important Porsche racing car from the 917 up to the RS Spyder. In addition to being a brilliant race engineer, he was also a master at interpreting the racing rule book and coming up with creative solutions within those rules to maximize Porsche's opportunity to win. Trained as an engineer at the University of Munich, Singer also had a talent for aerodynamics and body design. He retired from Porsche in 2004 and served as an advisor to the rules-making body, the FIA. He remains an enthusiastic ambassador for Porsche's racing history.

run in the ALMS. They finished first and third in the LMP2 championship with seven class wins. The first win for the RS Spyder came appropriately at Mid-Ohio, where the 962 and the Porsche Indy car had also achieved maiden victories. With the LMP1 R10 Audis absent to prepare for Le Mans, the RS Spyder won the race overall from the LMP2 category, Bernhard and Dumas driving.

In 2007 there was new LMP2 competition from Honda's Acura. Porsche also made the RS Spyder available to an additional team in the ALMS series, Dyson Racing (who had achieved significant success with their 962s in IMSA racing). The revised 'Evo' cars for 2007 were built in Porsche Motorsports' new facility near Flacht, at the south end of the Weissach complex. The new cars could produce more than 500 horsepower with improved downforce and no penalty in terms of aerodynamic drag. As ever, Porsche's development effort was first-rate and complimented Penske's industry standard race preparation and management. This was important as the rules for fuel and air intake were tweaked by the ALMS to balance performance. Penske's Ryan Briscoe replaced Lucas Luhr in the driving team

RS Spyders at Laguna Seca in 2007.

when Luhr moved to Audi. Despite additional competition, the DHL-sponsored Penske-run Porsches really delivered. They won 11 of 12 races in the LMP2 class and took a shocking eight overall wins against LMP1 cars from Audi. This even included wins at high-speed circuits like Road America and Miller Motorsports Park in addition to several shorter and twisty tracks like Lime Rock and again at Mid-Ohio (where the RS Spyder Evo was more than four seconds per lap quicker than the previous year).

In 2008, the Penske team even managed to win the fast and grueling Sebring 12-Hour race overall, with Emmanuel Collard joining Bernhard and Dumas. In a thrilling race, the lead changed hands some 27 times. At the end, three cars were on the lead lap with the Penske team heading the Dyson RS Spyder by 62 seconds. Audi's LMP1 R10 diesel was third. It was Porsche's first overall win at Sebring since 1988. The rest of the 2008 season was a bit less glorious than 2007, with only five class wins and two overall ALMS victories, at Miller Motorsports Park in addition to the Sebring triumph. American driver Patrick Long joined Penske, starting at Sebring, and helped the team win the LMP2 season championship for a third time. This was the last season for Penske's second Porsche program.

RS Spyders were entered at Le Mans and took LMP2 class wins in 2008 and 2009, finishing only tenth overall in those races as large fields of LMP1 cars dominated the top ten.

Porsche 997

By the end of its run, the 996 generation had sold nearly 175,000 cars. For the 2005 model year, Porsche introduced the Type 997 version of the 911. Once again, Porsche's engineers managed to continue the trend dating at least to the 993. They produced a better, faster, more rigid and lower drag 911. The 997 shared very little with its 996 predecessor, in part because of the adoption of larger, 19-inch diameter wheel sizes and also a wider track. This required redesign of the suspension and the chassis itself, not to mention the fenders and general shape of the car. Of course, Porsche did continue the strategy of sharing substantial content with its 987 Boxster and Cayman stablemates.

The looks of the 997 were an immediate hit and remain popular with 911 enthusiasts. The styling was muscular and aggressive with more shapely fenders to accommodate the larger wheel sizes. The 911 got a new 'face' with traditional-looking oval headlights and separate signal/running lights integrated into the front bumper cover. The swoopy rear lights were carved into the higher-cut bumper cover, giving the rear end a somewhat cleaner, more integrated look. Four-wheel drive 911s would receive even more shape at the rear fenders to cover still wider rear wheels. Frontal area for the 997 was 2.6% greater, yet the engineers and stylists still managed a slight reduction in drag compared to the 996. The 997 turned out to be the last Porsche design completed under Harm Lagaaij as head of styling. He retired in 2004 and was replaced by Michael Mauer who came to Porsche from GM of Europe. The main styling theme for the 997 was credited to Grant Larson.

Engineering for the 997 was led by August Achleitner and Bernd Kahnau. The basic structure of the car was extensively reworked using new, high-tech steel and aluminum elements. The result was a huge 40% increase in bending stiffness over the 996 and much-improved crash safety. Overall weight of this new 911 was less than 2% above its predecessor. In a return to the 1976/77 911 strategy, the 997 was available with two engine sizes, 3.6 or 3.8 liters. The base Carrera carried on with the M96

	996 1999-2004	997 2005-2011
ENGINES	First water-cooled 911, 300+ horsepower, note IMS issues, 996.2 VarioCam Plus introduced, 'Mezger' engine Turbo and GT cars	997.1 325+ horsepower, two engine size options, 997.2 all-new direct injection MA1 engine, no IMS issue
CHASSIS	All-new chassis, rear suspension and six-speed transmission, longer wheelbase, PSM introduced	More rigid, larger wheels, wider track, PASM and Sport Chrono introduced, new six-speed manual, PDK semi-automatic transmission introduced in 997.2
STYLING	Controversial headlights, now beginning to take on classic appreciation status	Traditional headlights, slightly more curvaceous 911 form
OTHER	Carbon brake disc option introduced, safety improvements over 993 including side airbags	'Mezger' engines retired during 997.2 run, GTS option introduced

997 promotional photograph shows a return to a more traditional 911 'face' and slightly flared fenders.

engine from the 996 while the 911 Carrera S featured the new M97 3.8 liter. This largest-yet 911 engine produced a maximum of 355 horsepower. The base engine was tweaked up to 325 horsepower.

The 997 received a new six-speed transmission from Japanese supplier, Aisin, rather than Porsche's traditional supplier, Getrag. Aisin was already a Porsche supplier (building automatics for the Cayenne). The new box was completely different from that of the 996 in terms of construction, gear arrangement and ratios. The Tiptronic was updated to work with the 997's increased power and larger wheel sizes. A welcome feature was that the automatic could be operated with fully manual shifting for the first time (when the PSM was turned off). Porsche Stability Management was standard on all 997s. Porsche's Active Suspension Management feature (PASM) was standard on the Carrera S, optional on the base 997 Carrera. This system allowed for variable spring rates to suit the driver's preference and road conditions. For the more performance or track-oriented driver a traditional sport suspension combined with a limited-slip differential was also an option. Sport Chrono

> ▶▶ In 2005, Porsche sold CTS, the convertible top supplier that had been co-founded with Daimler-Benz, to auto parts giant, Magna International.
>
> ▶▶ The 2008 24-hour race at the Nürburgring was the first appearance for a full, 4.0-liter (3996cc) version of the 911 flat-six. Use in the 997 RSR became standard in 2009.

was an option to further integrate and coordinate the high-tech systems for serious performance driving. PCCB brakes were also an option across the 911 line.

Like its 987 stablemates, the 997 achieved near-universal praise from the automotive press and the first-generation cars sold very well in the years leading up to the Great Recession. The 997 was offered in an ever more bewildering number of configurations during its first four years. Buyers could select between two-wheel and four-wheel drive with base, or 'S' engines, manual or Tiptronic transmissions. The coupe and cabriolet versions were available with these drive options, joined later by a sliding roof Targa (curiously available with four-wheel drive only). 911 Turbo and GT car options were introduced using the 997 platform but continued with dry sump, Mezger-based engines until the 2009 model year.

To celebrate the 50th Anniversary of the Porsche Club of America, PCNA and Porsche worked to create a special 997 'Club Coupe'. 50 cars were built with one retained for the Porsche Museum and one given away in a PCA lottery. The remaining 48 Azurro California blue cars were offered by lottery for purchase by PCA members. In addition to their unique appearance, these cars also featured 25 extra horsepower thanks to special cylinder heads, modified intake and exhaust plus a retuning of the Motronic engine control.

Porsche 997.2

For the 2009 model year, Porsche introduced the second-generation 997. This was far more than a typical mid-cycle refresh. Porsche not only designed an entirely new engine but also finally brought the PDK transmission into production. The Porsche *Doppelkupplungsgetriebe*, 'double clutch' automatic became viable with advances in computing power and hydraulic control making coordination of gear and gear-shaft changing reliable and extremely fast. Other manufacturers, including Volkswagen, had already begun offering forms of twin-clutch automatic transmissions. However, Porsche aimed for a much higher level of performance including readiness for track duty.

997 in Nordic Gold at PCA Werks Reunion Monterey, 2019. (Jay Gillotti)

The PDK featured seven speeds driven by five-disc 'wet' clutches using oil for lubrication and cooling. The outer clutch drove the odd number gears, inner clutch drove the even numbers. Based on driving conditions, the PDK can sense and pre-engage the next gear to be selected (such as 2nd gear while running in 1st) with a high degree of accuracy. Numerous automatic program functions were built in, such as detection of uphill or downhill driving and adjusting the shift points appropriately. The PDK unit itself even managed to be 22 pounds lighter than the Tiptronic. Operating modes selected by the driver were Normal, Sport and Sport Plus. At first, manual shifts were executed by buttons on the steering wheel, a choice influenced by Wendelin Wiedeking. However, this was one feature of the 997.2 which drew criticism from the press and customers. Not long after Wiedeking's 2009 departure from the CEO role, the PDK 997 received paddle shifters similar to other semi-automatics of that era. The 997.2 also became available with launch control as an option (having first appeared in the 997.1 GT2).

Porsche was not overly aggressive in advertising the all-new 9A1 engine. However, this engine was a total redesign featuring major changes to structure and systems. Construction of the engine case was simplified as was the drive to the camshafts. Similar to early 911s, the 9A1 cylinders and crankshaft were carried in a two-piece block bolted together. The traditional 911 intermediate shaft was eliminated. Instead, the camshafts were driven by belts directly from the crankshaft. No intermediate shaft meant no troublesome bearing issues. Although not a pure dry sump engine, the 9A1's sophisticated oil pumping system allowed for steady pressure throughout the engine even in extreme driving with high lateral g-forces. To supply fuel, Porsche employed direct injection in the new 911 engine (first used on the Cayenne in 2008). This system supplies fuel directly into the combustion chamber rather than the intake port. Although requiring sophisticated controls, direct injection helped the new engine produce approximately 8% greater power with a 7% improvement in fuel efficiency. The engine itself was lighter, had fewer parts and could sit slightly lower in the chassis.

The 9A1 became the basis for 997.2 Turbo and Turbo S models. Carrera and Carrera S 911s continued with different engine capacities at 3.6 and 3.8 liters respectively. The second-generation 997 carried on through the 2011 model year with all of the same body style, drivetrain and high-performance options as had now become customary for the 911 line. In addition, Porsche introduced the 3.8-liter 911 GTS as an enthusiast option slotting between the Carrera S and the GT3. Another special model, conceived by Porsche Exclusive and built toward the end of the 997.2 run, was a 911 Speedster. 356 examples were made as a tribute to the original 356 Speedster. The 997 Speedster used the 408-horsepower 3.8-liter engine from the Carrera GTS with the seven-speed PDK and PCCB brake package. In Speedster tradition, the windscreen was lower and more steeply raked. When stowed, the soft top was covered by a 'double-bubble' shell similar to the one found on the 1989 911 Speedster.

Panamera

In addition to the 997.2, the 2009 model year saw the debut of a four-door, luxury-performance GT sedan from Porsche. Porsche had explored the four-door sedan concept several times over the preceding 50 years with the 989 of the early 1990s getting closest to production. The Panamera finally gave Porsche an entry into the executive sedan class, competing directly with German rivals from Mercedes, BMW and Audi. The name 'Panamera' was a contraction referring to the classic Mexican road race (see Chapter Five).

Unlike the Cayenne, which had been funded jointly with VW, the 'G1' (code name for the Panamera) development project was entirely funded by Porsche. This included significant costs to expand the assembly facilities at Leipzig and for engine capacity in Zuffenhausen. Porsche's financial success in the mid-2000s made the finances workable for Porsche to expand the product range with a second four-door vehicle. VW was involved in the Panamera program as the builder of the bodies at their Hanover plant. Porsche worked with Deutsche Bahn rail to bring the components of the Panamera (and other Porsches) together for final assembly in an efficient, environmentally friendly manner.

Styling of the first-generation Panamera was credited to Grant Larson and newcomer Peter Varga working under Michael Mauer. The challenge was significant as the designers had to blend Porsche's historic form language into a large, modern

---

PORSCHE POINTS

>> According to Porsche transmission engineer Rainer Wüst, racing driver Hans Stuck was the first to suggest using gear shift paddles on the steering wheel in conjunction with the PDK.

>> In 2009 Porsche produced the 997 Sport Classic. Inspired by the 911 RS, it marked a return of the ducktail spoiler and also featured a unique 'double-bubble' roof form. Only 256 were made. A similar 911 was offered in 2022.

sedan. This task was further complicated by the decision to make the car a hatchback with a large volume of storage space. Space for the rear seats was also an issue, especially with a very tall CEO, Wendelin Wiedeking, testing head and leg room. The result was successful from the front, with a low, purposeful and very Porsche-like visage ready to show up in rearview mirrors on the Autobahn. However, the back half of the car had an awkwardness resulting from the demand for rear seat headroom and the hatch. Overall, the looks of the Panamera were judged rather harshly by the press and many of the Porsche faithful.

However, in terms of performance, the Panamera was every bit a Porsche. Development was led by R&D head Wolfgang Dürheimer and project lead Michael Steiner. Steiner and other hires from Daimler-Benz brought years of experience working on luxury sedans. The Panamera S, four-wheel drive S4 and the Turbo used a V8 engine related to its Cayenne sibling but extensively redesigned for the specific needs of the sedan. The S engine was 4.8 liters, normally aspirated, producing 400 horsepower with the Turbo spinning up to 500. Naturally, with such a large car Porsche was concerned with weight-saving. Much of the engine design work was concerned with making the desired power from lighter packages. The engines featured all of Porsche's current innovations, including direct injection, VarioCam Plus, electronic throttle and the Sport Chrono option. Engine controls were mapped to optimize cold-start functioning and the Panamera was the first Porsche to combine the PDK with an engine start-stop feature (for fuel economy). In 2010, Porsche introduced a 3.6-liter, V6 version of the engine for the 'base' Panamera.

A huge amount of work went into the basic structural design of the Panamera with the intent of saving every ounce possible. 75% of the structure was steel, but used various high-tech formulations. The remaining 25% was made of lighter alloys and plastics. Extra attention was paid to the front, to offset the weight of the engine, and the upper cab (such as the use of magnesium window frames) to keep the center of gravity low. As a result, the big sedan had a center of gravity only .6 of an inch higher than a 997. The Panamera S weighed just under 4,000 pounds.

Aerodynamics was an important consideration for speed, stability, interior sound and fuel economy. The Panorama featured a variable wing, hidden at low speed but designed to deploy for reduced drag at moderate highway speeds. As speed increased, the spoiler could angle upward to provide downforce. Above 127 mph, aerodynamic lift was near zero thanks to the increased wing angle. For the Turbo, the spoiler elements spread outward for extra downforce (helpful in a car capable of exceeding 180 mph).

THESE TWO PAGES: The Panamera gave Porsche an entry in the German executive-class sedan market.

Unlike earlier front-engine sports cars from Porsche, the Panamera did not use a rear transaxle. This was due in part to the intention for a four-wheel drive version. The four-wheel drive system featured an unusual arrangement with an angled bevel gear driving forward from the transmission to a differential near the right front wheel. A drive shaft for the left front extended across the car through the engine sump. The seven-speed PDK was designed and configured specifically for the Panamera in collaboration with supplier ZF. The Panamera was also available with a six-speed manual.

Like the engine, the Panamera's suspension was derived from the Cayenne but extensively reworked for lighter weight and the needs of a high-performance sedan. Steel springs were standard with air springs optional (air standard on the Turbo). Similar to the Cayenne, air springs allowed for variable spring rates depending on the setting but for the Panorama, variable ride height was also possible. PDCC worked with the anti-roll bar mounts and differential locking to stabilize the chassis depending on cornering load.

The interior met the mark for luxurious high-speed travel. It featured a dramatically sloping center console similar to the Carrera GT. The console continued between the rear seats. Rather than a bench, the Panamera featured two individual seats for the rear passengers. Numerous leather color and interior trim options could be combined for interior style. Another option was a Burmester stereo system with 16 speakers on individual channels and 1,000 watts of output. A host of other Porsche Exclusive options were available along with dealer-installed Tequipment.

Panamera Turbo numbers were sensational, challenging those of its vaunted cousin, the 911 Turbo even with a vehicle weight over 4,000 pounds. Acceleration to 60 mph was well under four seconds. Unlike its rivals, the Panamera was not artificially speed-limited and the Turbo could reach speeds upwards of 180 mph. The press reception was very positive, particularly on the driving experience. Aside from the concerns about its appearance, reaction to the Panamera can be summed up by the quote from Motor Trend's Angus Mackenzie: *"Relax, it's a real Porsche."*

PORSCHE POINTS

>> Available only in Europe, a manual transmission Panamera is a rare car with only 146 built. 50 were V6s, 96 with V8 engines.

>> In April 2009, the Panamera was revealed to the press on the 94th floor of the Shanghai World Financial Center. The car had to be stood on its tail to fit in the freight elevator.

Porsche versus Volkswagen

As if the 2008/2009 period weren't stormy enough, with a severe global recession and financial crisis, Porsche ended up in a tumultuous bid to take over Volkswagen AG. The twists and turns of this complex business and family battle would play out over seven years.

By the mid-2000s, CEO Wendelin Wiedeking had led Porsche to astounding success as measured by vehicle sales, revenue, profitability and stock price. Earnings exceeded $1 billion on sales approaching $10 billion. Porsche had also invested in a major expansion of its dealer network, created the Porsche Classic business to support vintage cars, and seen Porsche clubs expand to 60 countries around the world. Porsche also acquired the Porsche Design personal products business started by Butzi Porsche in the 1970s.

For reasons of partnership as well as investment, Wiedeking and his finance chief, Holger Härter, began to pursue a strategy of Volkswagen AG stock purchases by Porsche. In 2005, this was justifiable in part to block any attempt by other entities that might try to take over Porsche's partner in the SUV market. September 2005 was the date for Wiedeking's announcement that Porsche intended to acquire 20% of Volkswagen (roughly equal to the state of Lower Saxony's ownership in the company).

By 2007, Porsche's share of VW exceeded 30% and the increase in VW's stock price had been hugely profitable for Porsche. The situation in 2007 led to the creation of Porsche Automobil Holding SE (for *Societas Europaea*). Porsche SE would be the holding company controlled by the Porsche and Piëch families and representing the other shareholders of Porsche AG. Also in 2007, Wolfgang Porsche became Chairman of the Supervisory Boards of both Porsche AG and SE. Another important event in 2007 was a European Court of Justice decision that could have invalidated Germany's 'Volkswagen Law'. The law limited any single entity to owning no more than 20% of the voting shares in Volkswagen AG.

In March 2008, the Porsche SE board approved a plan allowing for an increase in ownership of VW beyond 50%. Although Porsche stated it did not want to interfere with day-to-day management of VW, nor merge any operations, Porsche did want to influence VW management at a strategic level to achieve better performance. This likely fed resentment on the VW side and made things more awkward for Ferdinand Piëch, serving as Chairman of Volkswagen AG, but also a major shareholder in Porsche SE.

By October 2008, Porsche's ownership stood at 46% of VW with the announced intent to increase the stake to as much as 75% in 2009. This caused a severe shortage of VW shares available on the market and the price of VW shares skyrocketed by a factor of five times over two days on October 27/28, (very briefly making VW the most valuable company in the world). Short-sellers of VW incurred severe losses in the squeeze. Porsche SE announced it would work to increase availability of VW shares and resolve the 'distortion' of VW market value. By January 2009, Porsche SE had reached just over 50% ownership in VW. However, Porsche's debt was now dangerously high, at about 170% of its own market value.

The financial crisis of 2008/2009 not only affected Porsche's sales negatively, it also made access to capital increasingly difficult. Porsche's own share price had declined significantly.

The months of April, May and June of 2009 involved a series of meetings and maneuvers attempting to resolve a struggle for ultimate control of VW and Porsche, complicated by global financial conditions and Porsche's debt situation. One could argue that Porsche's heavy use of debt to finance the purchase of VW shares ultimately made the strategy of acquiring VW impossible to sustain. David would not, in the end, take over Goliath.

In July 2009, the situation was finally resolved with investment from the sovereign wealth fund of Qatar, as well as Volkswagen and the Porsche/Piëch families agreeing on the basics of a new structure. Volkswagen AG would own the Porsche brand while Porsche SE would own just over 50% of Volkswagen's voting shares (and approximately 30% of total shares). Wendelin Wiedeking and Holger Härter lost their jobs in the process. Wiedeking had been CEO for 16 years by this time. Michael Macht became the new CEO of Porsche. Wiedeking and Härter were later prosecuted for illegal stock market manipulation, but both were ultimately acquitted in 2016.

Were it not for the severe economic downturn, Wiedeking's strategy for Porsche to become majority owner of VW might have succeeded. Instead, the merger of Volkswagen and Porsche was finally completed in 2012 with Porsche SE holding the majority of voting shares in Volkswagen AG (which, in turn, owned the Porsche brand and operating businesses). A further consequence of the merger was the absorption of the Porsche-Salzburg distribution business into Volkswagen.

> **PORSCHE POINT**
> ≫ German billionaire Adolf Merckle committed suicide in January of 2009 partly as a result of gigantic losses from short-selling VW shares during Porsche's takeover attempt.

New Porsche Museum

Porsche began assembling a small collection of its important cars during the 1950s. Prior to 2009, the growing collection of historically important cars and prototypes was displayed in a modest area of Werk II with numerous other cars in storage. In 2009, Porsche opened a world-class museum in Stuttgart to rival that of any other manufacturer. It was designed by the Vienna firm, Delugan Meissl. In addition to gleaming, multi-level display space for approximately 80 cars, the new building also became home to Porsche's archive of historical documents, photos, film, books and other memorabilia. A repair and restoration workshop is included in the facility and visible to visitors. The Museum Collection keeps most of the cars in running condition and they travel the world to promote the brand at significant events. The Archive also helps to promote the brand by supporting internal marketing efforts and external media sources.

RIGHT: Porsche's Archive space in the new museum facility.

BELOW: The Porsche Museum opened in 2009.

2010s

- 2009 ›› Super 911s
  987.2, Boxster Spyder and Cayman R
- 2010 ›› Boxster E
  Cayenne E2 and E22
  Diesels and Hybrids
  Porsche 918
- 2012 ›› Porsche 991
  Porsche 981
- 2014 ›› Porsche Macan
  Porsche 919
  Passing of Butzi Porsche and Ferdinand Piëch
  Porsche Customs
- 2016 ›› Porsche 718 Boxster and Cayman
  Four-Door Fortitude
  Special 911s
  Wolfgang Porsche
- 2019 ›› Porsche 992
  Porsche Taycan

One can easily imagine Dr. Porsche smiling down on these all-electric vehicles proudly carrying the Porsche name…

In 2010, Porsche had an excellent lineup of cars to sell. The 997.2 911 and 987.2 Boxster and Cayman were superb sports cars and are considered highly desirable right up to the present day. The Panamera was the new product and the Cayenne held its position as a benchmark luxury performance SUV. Affluent buyers could fill their garage with a Porsche for nearly any purpose or occasion. The difficulty was selling these cars in a global economy that was very slowly recovering from the Great Recession of 2008 and 2009. That slow, but steady recovery extended across the decade, a time during which Porsche continued to introduce new products, including a 'crossover' SUV, the Macan. Hybrid Porsches became available as the trend to electrification gathered momentum. In racing, Porsche made a triumphant return to Le Mans with the hybrid 919. The 918 was Porsche's supercar entry for the decade, utilizing a sophisticated gasoline/electric drive system. The decade concluded with the start of production for the full EV Taycan and Porsche racing in the open-wheel, pure electric Formula E series. One can easily imagine Dr. Porsche smiling down on these all-electric vehicles proudly carrying the Porsche name.

Super 911s

The later 2000s and early 2010s were a difficult time for automotive journalists. Finding new and proper superlatives for high-performance 911s was ever more of a challenge. The first-generation 997 continued the proliferation of 911 Turbos, GT3, GT3 RS and GT2s and they would only get better in the 997.2 generation.

The 997 Turbo featured Porsche's newest four-wheel drive system, PTM for Porsche Traction Management. Developed in conjunction with BorgWarner, the new system employed electromagnetic rather than viscous coupling for the drive to the front wheels. This allowed more precise control of the drive characteristics. Porsche also worked with BorgWarner, who had purchased the KKK turbo business, on new turbos. The 997.1 was the first Porsche to employ variable turbine geometry. VTG allowed for changing the angle of the turbocharger vanes to optimize performance. Electronic control of the turbos allowed Sport Chrono-equipped cars to vary the boost pressure for increased torque at lower rpm. Zero-to-60 times were well under four seconds and the top speed could approach 200 mph.

Overhead view of a 2012 gathering of 997 special editions, top to bottom: PCA Club Coupe, Edition 918 Spyder, GT3 RS, RS 4.0, Sport Classic, Speedster and Turbo S. (Jason Tang)

The 530-horsepower GT2 was so fast that it could meet or exceed the performance of Porsche's original supercar, the 959. The GT2 remained a lightweight, two-wheel drive, highly tuned version of the standard 911 Turbo. It featured handling and aerodynamic improvements suited to a car capable of speed over 200 mph. The GT3 was the non-turbo but track-focused, uncompromising driver's 911, although the 997 was less raw and somewhat more civilized than its 996 predecessors. The GT3 RS was a homologation stepping stone to the full race RSR and thankfully the RS was now available to United States customers. The 997.1 GT3 was the basis for Porsche Supercup racing cars as well as taking the fight to Ferrari in GT racing, with notable class wins at Le Mans and Sebring. In addition to winning Sebring in 2008, the Flying Lizard Motorsports team also won the ALMS GT championship for that year.

The 997.2 911 in any form is considered a superb sports car. In the second-generation, 997.2 era, the 911 Turbo moved from the Mezger-based engine architecture to the new 9A1 shared with other 997s. The 9A1 had been designed from the beginning to serve in the 911 Turbo and Turbo S models. In production, the Turbo version featured design and construction modifications specific to the needs of its performance goals. This included features to cool the engine itself, the turbos and to maximize cooling for the intake air. Like its normally aspirated stablemates, the Turbo featured direct injection of the fuel to the cylinders although modified for the Turbo's appetite. This latest 911 Turbo could be combined with a PDK specifically developed for use in this car, or run with a traditional manual transmission. PTV, for Porsche Torque Vectoring, was made available along with the other handling assistants including PADM (see below). PTV allowed for braking effect on individual wheels if needed in high-speed turns.

The 997.2 Turbo was lauded as a civilized, more user-friendly alternative to its GT cousins but one that gave away very little in terms of performance. Porsche also offered the fully loaded Turbo 'S' package, including all the highest performance features but limited to PDK for the transmission given the engine's higher torque and 530 horsepower.

The 997.2 versions of the GT3 were the last to use classic Mezger-based engine architecture that was so well suited to high rpm track driving and racing applications. The 997.2 GT3 used a larger bore 3.8-liter (3777cc) version with VarioCam applied to both inlet and exhaust valves. In addition to a round of other engine performance updates, the GT3 received the first Porsche Active Drivetrain Mount (PADM adding to the acronym list). Magneto-rheological engine mounts allowed for the movement of the engine itself to be managed in real-time based on the driving situation. PADM was coordinated with PASM and PSM, suspension and stability management controls, for ever-better handling. This latest GT3 was five seconds faster at the Nürburgring in the hands of ace Porsche test driver, Walter Röhrl.

The GT3 RS as usual upped the ante with a higher downforce rear wing, wider track and considerably modified engine tuning to reach 450 track-ready horsepower. This in a package more the 50 pounds lighter than the already-lightweight GT3. However, in 2011 the Porsche Motorsports team led by Andreas Preuninger topped itself. The 4.0-liter GT3 RS was inspired by and based on the very successful 2009 RSR racing engine. Modified and massaged for street use, this car still offered 500

2010 24 Hours of Daytona

As noted in Chapter One, Dr. Porsche never designed a car without some consideration of competition potential. Although Porsche did not directly support the idea of using a Cayenne for racing, its engine did find a way to the track. The Texas-based Lozano brothers worked with the Cayenne V8 to create a race engine for Daytona Prototype, Grand Am competition. In 2010, the Action Express team, using a Riley chassis, made it to victory lane at Daytona with the Cayenne-based engine. The drivers were João Barbosa, Mike Rockenfeller, Terry Borcheller and Ryan Dalziel. The car was prepared and partially crewed by the Brumos team who only entered one car for the 2010 race rather than their usual two. Barbosa had to hold off Chip Ganassi Racing's BMW/Riley driver, Scott Pruett, with the Cayenne-engine car winning by just 52 seconds.

>> Porsche Motorsport's internal project number for the 997.2 GT2 RS was 727. According to Andreas Preuninger, this number was chosen as a reference to the Nissan GT-R lap time at the Nürburgring, a time the GT2 RS was meant to beat (and did so by nine seconds).

>> In 2012, Hurley Haywood raced in the 24 Hours of Daytona for the last time, marking his 40th year driving in the race. He won it five times, all in Porsches.

non-turbo horsepower, a significant increase from the previous 3.8-liter GT3 RS. Suspension tuning was borrowed in part from the GT2 and aerodynamics were specifically tweaked to the needs of the 4.0 liter. 600 were made excluding test and pre-production cars. These were the last of the 'Mezger' line of engines tracing their history back to the 1960s. They were also the last manual shifting GT3s until the advent of the GT3 R model in 2016. The combination of features, performance, rarity and historical significance enhanced the collectability. Journalists again had to ratchet up the kudos, with several finding ways to say 'No, *really*, this is the greatest 911 ever made....'

Yet another special model made to close out the 997 era was the 997.2 GT2 RS produced in 2010 and 2011. This outrageous performer was the last turbocharged Porsche to use Mezger-based engine architecture. Careful weight-saving brought the car down to just 3050 pounds with engine output raised to 620 horsepower. Naturally, Porsche combined all that power with steering, handling and braking capability that garnered many 'instant classic' reactions from the press. At the Nürburgring, the GT2 RS posted a 7:18 lap time, 14 seconds faster than the previous GT2 and faster than a Carrera GT around the 'Ring.

987.2, Boxster Spyder and Cayman R

For the 2009 model year, Porsche introduced the second-generation 987 Boxster and Cayman. These cars featured many improvements in common with the 997.2 911s. Most significantly, the Boxster and Cayman received their versions of the new 9A1 engine. For the 'S' versions, the engines were just over 3.4 liters, with base engines at 2.9 liters. The 987 S models also received direct fuel injection. Following a familiar pattern, Porsche challenged the airflow on both sides of the engines, making improvements on the intake side and providing a completely new exhaust system. In addition to increasing power, the new exhaust was designed to improve sound quality as well. The Boxster S was set to 310 horsepower, with the Cayman S at 320. The base Cayman was also rated ten horsepower above the Boxster, at 265 versus 255.

THIS AND FOLLOWING PAGE, TOP: 2010 Boxster Spyder promotional photographs.

For the 987.2 the PDK replaced the Tiptronic as the automatic transmission option and the manual shift cars all had six-speed gearboxes. A new option was the limited-slip differential on either Boxster or Cayman. The Sport Chrono package now included launch control for extra-fast zero-to-60 runs. As the reader might predict, Porsche was able to improve fuel economy even though the 987.2s were slightly heavier than their predecessors. Core styling was unchanged, but designer Mitja Borkert drew new nose and tail details giving the cars a more angular and aggressive look. New halogen and LED lighting packages were incorporated into the restyle. Before moving to VW sister company Lamborghini in 2016, Borkert worked on other 2010s Porsches such as the E2 Cayenne, Macan and Mission E.

During the 987.2 era, both the Boxster and Cayman had significant special editions. The first was the Boxster Spyder in 2010. Perhaps more of a roadster than the standard Boxster, this special car harkened back to the original Speedster. It had a fabric 'emergency' top that could be manually installed (in five minutes or less, with practice). The aluminum rear deck was a separate piece, different from the standard Boxster, featuring twin humps behind the seats and individual roll hoops. The Spyder was also a performance package, featuring a lowered and re-tuned suspension, plus the higher horsepower engine from the Cayman S. The lighter top, rear deck and aluminum doors contributed to performance with an overall 176-pound weight savings (depending on options such as sport seats, CD changer and climate control system). The optional PDK transmission was now operated with paddle shifters rather than steering wheel buttons (a change migrating gradually across the Porsche product line).

For the 2011 model year, the Cayman R joined the Boxster Spyder as the 987.2s approached the end of production. Borrowing the hallowed 'R' designation from the 1960s 911 R was a little risky, but this higher-performance Cayman did follow the 'R' ethos of lighter weight and more power. Weight saving was achieved with aluminum doors and front hood, but the overall calculation depended on choice of seats and whether the new owner could live without a radio or air conditioning. The Cayman R took after the Spyder with a re-tuned suspension and lower ride height. The Cayman's growing reputation for superb handling was only enhanced with these changes and the addition of the limited-slip differential as standard equipment. A revised exhaust raised the horsepower figure slightly, to 330, but responsiveness felt greater thanks to more rigid engine mounts. Those who selected the paddle-shifting PDK enjoyed revised, sportier program settings for both upshifts and downshifts. The Cayman R received a fixed rear wing and revised front air dam for reduced lift at those sporty, higher speeds. Like the Boxster Spyder, the R featured a vintage-style Porsche side graphic above the rocker panels. Many Cayman Rs are instantly recognizable, wearing the launch color, Peridot Green.

Boxster E

Porsche had designed electric cars before 2010, but the Boxster E was the first Porsche EV constructed and tested. It was part of a program subsidized by the German Ministry of Transport. Three test cars were built with the first on the road by May 2011. Porsche tested two motor configurations; a single-motor 122-horsepower version and a dual-motor 254-horsepower car. The dual-motor car had four-wheel drive. The motors were supplied by Volkswagen while the batteries were designed and developed by Porsche Engineering. The dual-motor car made a zero-to-60 time of 5.5 seconds with the vehicle weighing about 400 pounds more than a regular Boxster. Practical range of the cars was about 90 miles with a full charge. The project gave Porsche a chance to test many aspects of EV construction and technology including features like Active Sound Design which produced simulated engine noise. The cars also allowed for real-world testing of driving character and usability. Porsche CEO Michael Macht said: *"We will definitely be offering an electric sports car in the future. But such a concept only makes sense if it offers product qualities typical of a Porsche."*

OPPOSITE: Experimental EV Boxster in 2012.

Both 987.2 specials garnered high praise from the press. As sharp as the Cayman R was, the Boxster Spyder seemed to win the highest scores for real-world enjoyment. While the Cayman R received its share of 'best Cayman ever' reviews, the Spyder was described as easier and more fun, even for track driving in some cases. Pete Stout, editing *Excellence* magazine, said the Spyder was the best-handling factory Porsche he had ever driven.

As with the 997, Porsche closed out the 987 run with highly-optioned Black Editions of both the Boxster and the Cayman.

Cayenne E2 and E22

The first major redesign of the Cayenne was known internally as 'E2' and first appeared in 2010 for the 2011 model year. By this time, Porsche had sold more than a quarter million Cayennes around the world and had introduced a diesel option (see below). The E2 was larger, with a longer wheelbase, but managed to look smaller than the original Cayenne. Thanks to the work of designer Mitja Borkert, it was certainly more curvaceous and perhaps more Porsche-like in appearance. No surprise, given the concerns about energy, it was also lighter and more fuel efficient. The weight saving was due to complex blending of the metals forming the body/chassis. Working with their partners from VW, the Porsche engineers created a new, lighter but stronger steel structure featuring aluminum skin on the doors, hood and tailgate, plus aluminum fenders. Porsche also found weight savings in components like the four-wheel drive system and even the engines.

For the sportier Cayenne S and Turbo, Porsche adapted computer-controlled traction management from the Panamera and 911. This controlled drive to the front wheels as needed and worked in combination with the optional torque vectoring (PTV) to control wheel spin side-to-side. PASM and PDCC were now options as well for those who might want ultimate handling in an SUV. The base Cayenne could still be ordered with a manual gear shift, but most Cayennes featured a new eight-speed automatic with additional overdrive gears to minimize fuel consumption at highway speeds. This new Tiptronic also managed the start-stop feature and provided 'Normal', 'Sport' and 'Off-Road' modes. Redesign of both the V6 and V8 engines provided greater maximum power but combined with

significant fuel and emissions savings (in the region of 20%). The stylish interior of the newest Cayenne was more similar to the Panamera with a rising console between the front seats and featured a 911-like grouping of the instruments ahead of the driver.

For 2013, Porsche took a further page from the sports car book and introduced a GTS version of the Cayenne. Suspension tuning, 420 horsepower and more aggressive styling made for a high-performer that some reviewers felt could rival the Turbo and Turbo S Cayenne models.

For the 2015 model year, the Cayenne 'E22' represented much more than a typical mid-cycle update. In addition to another aggressive restyling of the Cayenne, there was significant new engineering not least of which was under the hood. The engine family was reconfigured with the diesel becoming an alternative to the 'base' V6 Cayenne. For the Cayenne S, the E22 received the V6 turbo shared from the Panamera and the Macan S. At 3.6 liters with two turbochargers, this engine could produce 420 horsepower. The new Cayenne also added a plug-in hybrid option (see below). The sporty GTS option package continued in the E22 along with the range-topping Turbo and Turbo S. Suspension was redesigned with new mounting points and continuing with three spring options, standard steel, PASM shock absorbers or PASM with air springs. In styling terms, the 2015 Cayenne featured a cleaner but more pronounced 'egg crate' look for the front grills and more sculptured air dam. A new active aero feature was flaps that could close inside the grill to reduce drag when less cooling air was required. The tail was also redesigned to contribute to an overall wider, lower stance.

Although many Cayennes serve as dutiful family vehicles, it should be noted that a Turbo S broke the eight-minute barrier at the Nürburgring in 2015. This was close to the original 996 GT3 lap time at the 'Ring.

RIGHT: 2012 Cayenne Turbo in the snow. (Karl Noakes)

Diesels and Hybrids

Although the use of a diesel engine in a Porsche may have seemed contradictory to some, application in the Cayenne was logical, especially for the European market. That logic, and the development of Porsche-acceptable performance in the Audi-based V6, allowed Porsche to begin selling a diesel-powered Cayenne (Europe only) starting in 2009. The diesel Cayenne received good reviews from the press for its real-world performance and became a popular choice among European buyers where up to two-thirds of Cayennes sold were diesel-powered.

In 2011, Porsche introduced a lighter, more powerful V6 diesel with revised turbochargers. For 2012, an 'S' version of the diesel appeared with a V8 engine. As could be expected, a still higher performing 'Turbo S' diesel soon followed. For the 2013 model year, Porsche started US sales of the Cayenne diesel. Although the US market had a history of resistance to diesel passenger cars, the Cayenne started to build a following. However, at the end of 2015, the emissions scandal came to Porsche as a result of Porsche using the VW 'family' diesel. Large numbers of 2016 Cayenne diesels were left in limbo, voluntarily remaining unsold for a year or so while the emissions resolution was sorted out. Emissions issues aside, these terrific-performing SUVs are likely to remain sought-after for years to come.

In keeping with the close powertrain connection between the Panamera and Cayenne models, the Panamera received a diesel engine option for the 2012 model year. This also helped the Panamera match offerings from its competition in Europe. The single turbo, 250-horsepower diesel served the big sedan reasonably well despite a 300-pound weight penalty. It made for an excellent daily driver for executives who could drive their 911s on weekends. When driven carefully, the range exceeded 600 miles on a tankful.

As early as the mid-2000s, Porsche began exploring the concept of a hybrid drivetrain for the Cayenne. Although results were not encouraging at first, Porsche persisted given the need for a more environmentally responsible vehicle. Given the expense involved with developing hybrid drivetrains, much of the work and technology was shared with Volkswagen. Porsche first showed a hybrid Cayenne as a concept in 2007 at the Frankfurt show. Production Cayenne S Hybrids came to market in 2009, 2010 for the US. Sharing its 3.0-liter, direct injection V6 with Audi, this was the first supercharged Porsche. The supercharged engine and electric motor could work together (in 'parallel' hybrid fashion) to increase performance. Electric power could also be used to substantially increase fuel economy in the big SUV. This included the 'coasting' mode, where the combustion engine is switched off, and enhanced stop-start capability which prevents wasting fuel when the vehicle is stopped.

In the E22 generation, Porsche introduced the E-Hybrid plug-in Cayenne, a first in the premium SUV market segment. As a plug-in hybrid, the electric range increased significantly and, being a Porsche, overall performance improved as well. The battery pack was now lithium-ion and could store ten times the energy of the previous battery (nickel-metal-hydride).

The Panamera S Hybrid was introduced in 2011. The 47-horsepower electric motor enhanced performance and fuel economy in a manner similar to the E2 Cayenne. With a combined 380 horsepower, the first Panamera hybrid offered best-in-class fuel economy and emissions performance. It also offered a new feature, radar-based lane-change assistance. Press reception was mixed, with the hybrid driving experience being new in the Porsche line. At this time, before the diesel controversy, the weight and cost of the hybrid made the diesel option appear to be a better choice.

The Panamera E-Hybrid plug-in preceded the Cayenne version by one model year. This car was part of the program for the face-lifted Panamera that appeared in 2013 for the 2014 model

---

PORSCHE POINTS

» Porsche Engineering purchased the Nardò, Italy test track in 2012. The company now has engineering centers in five countries including one at Nardò.

» The 100,000th Panamera was an S E-Hybrid model made in 2013.

» In 2014, Porsche SE invested in INRIX of Kirkland, WA, a private company that provides real-time and historical traffic and other transportation-related analytics from over 140 countries around the world.

» 2015 was a banner year for Porsche sales with worldwide totals exceeding 200,000 for the first time and China passed the US as Porsche's largest single-country market (sales in Europe were still greater than in China).

Sapphire Blue Panamera Turbo S E-Hybrid Executive. (Karl Noakes)

year. In addition to an improved 'face', this latest Panamera went a long way toward resolving the awkward roofline and rear styling of the original. The Panamera was now available in super high-performance versions, like the GTS and Turbo S, or the longer wheelbase limo-style Executive edition, or the high-tech E-Hybrid. As with the E22 Cayenne, this was a much-improved car in the push toward electrification and was a first in its market segment. It offered very Porsche-like performance specs (5.5 seconds for zero-to-60, 168 mph top speed) with very high fuel economy ratings. Combined fuel economy was rated as high as 50 mpg with a maximum electric range above 15 miles from the 95-horsepower motor and lithium-ion batteries. Press reception was much more positive for this newest hybrid driving experience from Porsche.

Porsche 991

For the 2012 model year, Porsche introduced an almost all-new 911. Given the Type number '991' (with some controversy about being out of sequence, after 997), this car was entirely new except for the engine family which itself was a significant update of the 9A1.

The development of concepts and technologies for the 991 began in the mid-2000s. An interesting aspect of this long-term project was a procedure put in place to formally track good ideas that

Dieselgate

The Volkswagen diesel emissions scandal started in September 2015 when the US EPA discovered that VW TDI engines were emitting much greater pollution in regular driving compared to lab testing situations. This was due to software programming designed to fool the testing. Among other problems, this was a violation of the Clean Air Act.

The scandal led to the resignation of VW CEO Martin Winterkorn and several other senior executives. Countries around the globe sanctioned VW and there were prosecutions against the company and some individual employees. Although the direct financial cost to VW and its shareholders is hard to estimate, that tally is well in excess of $30 billion, including the damage to brand reputation. As a potential silver lining to the scandal, it did force Volkswagen, and to some extent the automotive industry, to focus their work on electrification. Numerous other manufacturers were subsequently caught cheating on emissions testing.

The effect on Porsche diesel sales in the US is noted above. There is an excellent Wikipedia summary on the scandal under the heading 'Volkswagen Emissions Scandal'.

did not make it into production. Never throwing anything away remains an important Porsche tradition.

The 991 featured an all-new body shell created after extensive studies of different materials. In the end, Porsche employed a blend of several different steel products in combination with aluminum and magnesium. They also used new techniques for bonding and fastening the various pieces together. The result was a structure of greater stiffness in every measure but also lighter. The finished 991 was lighter than the 997 despite being larger and having more interior room. Part of the challenge for this new structure was the car's longer wheelbase. In addition, Porsche Motorsport was involved throughout the process of design to ensure the basic platform could be competitive for racing.

In addition to the longer wheelbase, the 991 was designed for a wider track and the suspension components were redesigned to suit the new layout. A minor concern for traditionalists came with Porsche's new steering system for the 911. Being electric rather than hydraulic led to some differences of opinion about steering 'feel' even though Porsche conducted extensive testing and research with ZF on the motor unit controlling the steering. Also in the suspension and handling department, Porsche designed the 991 to be compatible with 20-inch wheels for an ever-greater contact patch. The full alphabet soup of chassis technology was available for the 991; PASM, PSM, PTV, PADM, PDCC and Sport Chrono, but in all cases the systems were upgraded or tailored to the needs of the new 911. This was especially true of PDCC which had originated in the Cayenne for off-road capability, but was now adapted to the 911 to add handling and ride options. PASM could now lower the ride height and change the aerodynamic action of the rear wing.

The 991 carried on with updated versions of the 9A1 engine now in 3.4 and 3.8-liter capacities. In traditional Porsche practice, the engineers worked for more power and better fuel economy. This was achieved through a combination of higher revs and better breathing, along with reduced weight for some of the components. The exhaust and catalytic converters formed a completely new system. Porsche also took the opportunity to improve the direct injection system for fuel delivery. Maximum horsepower in the smaller engine was up to 350 with an approximate 15% improvement in fuel efficiency. Although quieter overall, the 991 introduced the 'Symposer' concept which uses the rear bulkhead to conduct desirable sounds from the engine to the passenger compartment. The 991 also introduced the first-ever seven-speed manual transmission with an extra-high overdrive gear for fuel economy. Seven speeds were adopted for the PDK as well.

Styling and overall dimensions were similar to 997 but with short front and rear overhangs as a result of the longer wheelbase.

ABOVE AND OVERLEAF: 2012 promotional photographs show a new profile for the 991-generation 911.

981 Cayman GT4 in Racing Yellow.

The slightly lower 991's most distinctive feature was the new windshield with the leading edge moved further forward, giving a more aggressive-looking profile. A strong character line below the rear wing and much smaller taillights gave the tail a distinctly different look from the 997. Aerodynamic drag figures matched the 997 despite the larger front track and wheels.

Another minor controversy was caused by the 991's new sunroof. To help increase headroom, the design resulted in the lid deploying over the top of the roof. While it may not have been the most pleasing look aesthetically, 991 drivers would not really see how the car looks on the road with sunroof open. The action of the wider rear wing was adjustable to compensate for the altered airflow when the sunroof was open. The 991 interior was extensively redesigned, now featuring the sloping central console already seen in the Panamera. The seating position was revised for greater adjustability and headroom.

After well over a million test miles, the 991 took on the all-important Nürburgring lap time challenge. Porsche test driver Timo Kluck was able to better the 997-equivalent times by some 14 seconds. This was a credit to the great work of the 991 development team under Wolfgang Dürheimer and August Achleitner (who led 911 programs from 2001 until retiring in 2019). Just before the release of the 991, Dürheimer left Porsche to lead Bentley and Bugatti. The new head of engineering and development would be Wolfgang Hatz, returning to Porsche after having worked at Opel, Fiat, Audi and Volkswagen. Hatz was then caught up in the diesel emissions scandal and left Porsche in 2015.

In reasonably short order, Porsche began rolling out the expected squadron of 911 variants including four-wheel drive, Turbos and GT cars. A superbly stylish addition was the new 911 Targa with its complex folding top and retro-chic Targa bar. As with Targas of recent decades, the 991 was built only with four-wheel drive.

Porsche 981

Developed in parallel with the 991, the new Boxster and Cayman series began with a staggered launch for the 2013 model year. Also like the 991, the cars were substantially new aside from the evolution of the engines. This included lengthening the wheelbase and widening the track, especially at the front. In the now-established practice, the 981 shared many components with the 991, including the structure from the seats forward. The new structures were not only lighter but also increased in rigidity. The 981s featured striking, aggressive new exterior styling thanks to Matthias Kulla. As the contract with Valmet expired,

Porsche shifted 981 production both to Zuffenhausen and the ex-Karmann Osnabrück factory now owned by Volkswagen.

The new Boxsters appeared in mid-2012 with the Cayman sales beginning in March 2013. Porsche continued to offer two engine sizes, now 2706cc for the base model and remaining 3436cc for the 'S', in the revised configurations. Peak horsepower ratings increased slightly thanks to higher rpm capability. Fuel economy was tweaked with the addition of engine start-stop function, revised PDK programming and selectable driving modes (better fuel economy mapping when not driving in Sport mode). Similar to the 991, the 981s were converted to electric-assisted power steering and electric parking brake.

The new Boxster S passed the Nürburgring test with flying colors, twelve seconds faster than its predecessor and breaking the eight-minute barrier. It also had a faster-operating top, larger interior space and headlights inspired by the 918. Following previous practice and verging on 911 territory, Porsche brought out a variety of higher-performance 981s. These included GTS models, a new Boxster Spyder with a 3.8-liter engine and 375 horsepower, and even a 'GT4' Cayman, not to mention a track-only Clubsport GT4. The GT4 was developed by Porsche Motorsports under Andreas Preuninger as a 'junior' GT3 although there was nothing junior about its track performance capabilities.

Porsche Macan

In hindsight, the case for Porsche to enter the 'crossover' SUV market seems obvious. The case was started by Matthias Müller in his role as head of model strategy for VW. This was before he became CEO of Porsche, replacing Michael Macht in 2010. As usual, Porsche evaluated the concept carefully and considered

The Macan expanded Porsche's product line with entry into the crossover SUV segment.

the impact on brand image and whether they could bring a valid Porsche driving experience to the segment. Having the Audi A5 and Q5 platform as a starting point created an advantage, but in typical Porsche fashion, this new car was extensively changed and adapted to Porsche's requirements. At the early stages of development, the new vehicle was known as the 'Cajun', short for 'Cayenne Junior'. In 2012, Macan, an adaptation of an Indonesian word for tiger, was announced as the official name.

A significant aspect of the Macan project was the expansion of the Leipzig factory. Unlike the Cayenne and Panamera, whose bodies arrived from other factories, the Macan's body would be constructed and painted in a new extension of the Leipzig facility. The design process was managed by Oliver Laqua, as director of the SUV lines. Mitja Borkert was the main stylist and responsible for the overall look, including the Macan's curvaceous hips and character lines. The hips were appropriate given the wider track at the rear and Porsche's decision to employ different wheel sizes front to back. The Macan's unique, 'clamshell' hood, which encompasses the upper part of the front fenders and the headlight surrounds, was championed by none other than Ferdinand Piëch. It was a racing-inspired element unique to the Macan that challenged the manufacturing engineers to get right. Additional Porsche DNA came in the form of the taillights inspired by the 918. A bit of Ferrari DNA may have crept in at the front with the egg crate grill.

The mixed wheel widths and diameters helped put the sport in this SUV. The base Macan and diesel used suspension more directly derived from the Audi design. Climbing up the performance ladder, the Macan received Porsche-specific options such as PASM, air shocks, PTM and PTV (but not PDCC). Porsche judged the PDCC not needed, in part due to experience with the Cayenne, of which only about 2% saw off-road duty for customers. The four-wheel drive system was all Porsche, derived from the 911 and meant to offer maximum dynamic capability. Rear wheels were always driving but torque split to the front wheels could go up to 100% if the electronic sensors called for it. Steering was electromechanical not unlike the 991.

The entry-level, 2 liter (sold first in China) and diesel engines were based on Audi designs, however, they were altered and improved to suit the Porsche driving experience. The Macan S and Macan Turbo twin-turbo engines were also V6s but based on the Panamera's V8. Thus began a confusing trend where not all future Porsches powered by turbocharged engines were given 'Turbo' nomenclature. The high-end Turbo was good for 400 horsepower from 3.6 liters and zero-to-60 times under five seconds. Macans had 'Sport' mode as an option to alter the engine and PDK transmission mapping and further performance options could be purchased including Sport Plus, Sport Chrono and launch control. The Macan was the first Porsche never to be offered with a manual transmission.

The finished Macan only shared about 40% of its components with its Audi Q5 cousin. It certainly fulfilled the mission of being a fun-to-drive SUV with sufficient off-road capability. As with all Porsches, the performance limits were far beyond what most drivers would ever actually use, but owners like to have those reserves. Tellingly, the Macan made its world debut at the Los Angeles Auto Show in November 2013. Sales began in mid-2014 with the lowest-priced units just over $50,000 in the US. A highly-optioned Macan Turbo could run up over $150,000 on the sticker, although $100,000 would be more usual. Press reception was, no surprise, highly positive. Perhaps more important to Porsche was the huge percentage of Macan customers that were first-time Porsche buyers. In 2015, Porsche added the extra-sporty GTS option to the product line.

By 2016, the Macan became Porsche's best-selling vehicle and is currently the best-selling Porsche in the US.

Porsche 918

Generally recognized as Porsche's third 'supercar', after the 959 and Carrera GT, the 918 was Porsche's first hybrid sports car. It was first shown as a concept at the Geneva Salon in 2010 after a very short design and construction period. The concept and plan to actually produce the car provided a much-needed boost to morale after the wreckage of the 2009 Volkswagen takeover battle and financial crisis. The 918 also ushered in Porsche's 'Intelligent Performance' initiative along with the experimental flywheel 911 GT3 R Hybrid. Elements of electrification were

> **PORSCHE POINTS**
>
> » To test the concept of larger Macan wheel sizes at the rear compared to the front, Porsche tried it on an Audi Q5 and reduced the Nürburgring lap time by five seconds.
>
> » Porsche's hybrid 911 racer used the flywheel kinetic energy recovery system developed by Williams Grand Prix Engineering. The stored electrical energy drove motors for each front wheel. The car was an important experimental and learning platform for Porsche's engineers and nearly won the 24-hour race at the Nürburgring in 2010.

becoming a regular part of the discussion for new Porsche models. Naming the concept '918' was inspired by the slight resemblance to the 917 and the 917's 'game changer' status. It was also a nod to Ferdinand Piëch both as the father of the 917 and the original Type 918 design study from 1970 for a mid-engine, 908-based super sports car.

The 918 occupied a market strata now often referred to as a 'hypercar', something even beyond the old-school notion of a 'supercar'. The project certainly attracted a superb team of engineers and designers inside Porsche. Led initially by Wolfgang Dürheimer as head of R&D, the direct project lead was Frank-Steffen Walliser. Dürheimer was replaced by Wolfgang Hatz as head of R&D in the middle of the 918 project. Initial concept styling was credited to Hacan Saracoglu working under Michael Mauer. Born in Turkey, Saracoglu came to Porsche from California's Art Center College of Design and Ford. His work on the 987 Boxster and Cayman was good preparation for the mid-engine 918. In styling, the 918 differed from its hypercar competition, the La Ferrari and McLaren P1, taking a more subtle, classically refined Porsche form.

As with the 959, the 918 featured many new ideas. However, it also took inspiration from Porsche's most recent supercar, the Carrera GT with carbon fiber chassis sections, removable roof and motorsport-supplied engine. Although the 918's combustion V8 was derived from the RS Spyder, it was extensively redesigned for use in a street car. This included lengthening the stroke for a capacity of 4.6 liters and adopting a flat-plane crankshaft without the expected excess of vibration. In the design process, the engine was also made lighter. For packaging and heat management, the engine had 'hot inside' cylinder heads with the exhaust valves in the middle of the 'V'. This was combined with a unique exhaust system leading to upward-facing pipes. The exhaust configuration protected the Bosch-Samsung lithium-ion battery from damaging heat. The engine was carried in a separate carbon fiber structure bolted to the monocoque passenger compartment. The passenger seat was set slightly forward of the driver's seat to allow more room for the fuel tank. Improvements to the handling and molding of carbon fiber made construction of the 918 chassis much more efficient compared to the earlier Carrera GT.

911 GT3 R Hybrid shown at Rennsport Reunion VII in 2023. (Dean Holbrook)

918 announcement brochure.

The 918 offered a combined maximum of 887 horsepower with 944 foot-pounds of torque. This was the sum of its engine plus the Bosch-supplied electric motor/generator for the rear wheels and GKN Driveline front motor. Pure electric (E-Power) range was up to 20 miles (12 on the US EPA test cycle). The four hybrid driving modes allowed progressively higher performance. The top-end 'Hot Lap' performance map allowed up to 90% of battery energy to be applied by the driver's right foot. The seven-speed PDK was derived from the 911 Turbo, reconfigured and turned upside down for lower center of gravity and with ratios specific to the 918. All of this technology required software development of a high order to coordinate the engine, electric motors, battery, transmission, suspension and aerodynamic elements. The 918 was also a triumph of 'packaging', fitting all the new and complicated technology into a usefully compact sports car.

Adding to the long list of new ideas incorporated by Porsche in the 918, one must consider the following: plug-in battery recharging option for a sports car, active rear-wheel steering, active aerodynamics at the front in addition to the rear wing, LED headlights with daytime running lights, the stainless steel honeycomb engine cover and a new nine-layer paint process for carbon fiber body panels.

Although heavier than the Carrera GT, the 918 was significantly lighter than a Carrera GT would be with the added hybrid electric equipment. A further 90-pound weight saving was available to customers who ordered the Weissach package, which included some exposed carbon fiber body panels. The Pilot Sport Cup 2 tires were developed with Michelin for the 918 which ran ultra-light, center-lock magnesium wheels sized at 20-inch diameter front and 21-inch at the rear.

After building 24 prototypes, Porsche planned to build 918 examples with an announced price of $845,000. Deliveries began late in 2013 in Europe, mid-2014 in the US. Production finished by June 2015, with the US as the leading market representing almost 300 918 sales.

On September 4, 2013, the 918's Nürburgring lap time set a new record for a globally homologated road car at 6 minutes, 57 seconds, driven by Marc Lieb. This was almost 15 seconds faster than the previous record. Zero-to-60 in 2.2 seconds is hard to imagine, faster even than Porsche's 917 race cars. Yet the 918 was also capable of the equivalent of 67 mpg when run for maximum fuel economy. The top speed was listed at 214 mph for

> **PORSCHE POINT** — The 918 'RSR' was a hardtop, gullwing door, race-inspired version of the 918 that used the flywheel energy storage system.

THIS PAGE: 918 in black Porsche-Salzburg livery.

the brave driver who might find a long enough and clear enough stretch of autobahn.

Porsche 919

Related to the 918 by its hybrid concept, the 919 LMP1 racing car was developed at roughly the same time and of course, within the Porsche Motorsport organization. Creation of the 919 was prompted by a new set of rules for Le Mans and the World Endurance Championship. These rules allowed significant flexibility with the powertrain design and the principle regulation was the amount of energy to be used on a given lap. The energy was measured by 'joules' (a single joule being a measure of mechanical work relative to heat produced). Although Porsche was thought of as a Le Mans powerhouse, Audi had won 12 of 14 races at Le Mans since 2000 and Toyota was a committed player in the championship. Porsche essentially had to start a new team, in a new facility, to design and build a new car. This was done without any sharing of technology from VW family member, Audi.

The new formula was announced in 2012 after years of work by the FIA, ACO and the manufacturers. Porsche was already seeking a return to the top levels of motorsport. Interestingly, Ferdinand Piëch supported a Porsche return to Le Mans even though he had nurtured Audi dominance as Chairman of VW. Porsche would compete against Audi in the LMP1-H (hybrid) class which allowed for a sliding scale of energy recuperation versus fuel consumption per lap (monitored electronically by the officials). For the first season, Porsche chose the 6 megajoules formula for the 919 with battery storage for the hybrid system. Audi was committed to their diesel engine, so chose the lowest (2MJ) energy recovery option with flywheel storage. Toyota also chose the 6MJ category with a non-turbo combustion engine, but with supercapacitor storage for electrical energy.

Porsche's new racing team was led by Fritz Enzinger under Wolfgang Hatz. Ex-BMW man Enzinger recruited talent including expertise from the F1 ranks. Alexander Hitzinger came from Red Bull to lead design of the 919. For the combustion engine, Porsche designed a unique V4 with a single turbo. The 2-liter V4 was loosely based on the V8 engine from the RS Spyder. Energy recovery came in part from an exhaust-driven generator. The engine could produce nearly 500 horsepower but suffered from severe vibration problems at first. Resolution of the vibration issues was a lengthy development exercise. Only the front wheels were powered electrically, using a GKN motor/

919 wins at Le Mans in 2015.

919 takes the checkered flag at Le Mans in 2016.

generator similar to that of the 918. The combined hybrid drive system could produce a maximum of 750 horsepower. The chassis was a two-piece carbon fiber structure with the engine as a stressed member. After 2014, the chassis was changed to a single-piece carbon fiber tub. Aerodynamic design was critical to performance, in part because of the cooling needs of the battery and hybrid systems. Most fans would not have judged the 919 to be the prettiest Porsche ever, but some ungainly features, such as the center 'spine' or fin, were mandated by the rules.

The 919 first ran at Weissach in June of 2013, driven by Timo Bernhard, starting a schedule that allowed for plenty of testing ahead of the 2014 season. That first racing season produced mixed results but was generally positive for a new team with a new car. The 919 led for a time at Le Mans in the late stages of the race but neither car finished. It also showed good speed with some podium finishes in the WEC races. In the last six-hour race of the season, at Interlagos in Brazil, the 919 notched up its first win, with drivers Marc Lieb, Romain Dumas and Neel Jani. Audi won at Le Mans and Toyota took the season championship with Porsche third on points.

For 2015 Porsche designed a nearly all-new 919, now with a one-piece carbon monocoque. The suspension, braking system and combustion engine were extensively redesigned. Engine capacity was slightly larger and maximum horsepower up above 500. The turbo was new and the hybrid drive and energy recovery system was also redesigned including a new, more powerful battery pack that was also lighter. The 919 would now run in the 8MJ category for energy recovery per lap. Total maximum horsepower was now close to 1000. Reduced overall weight was a key feature of the new car along with revised aero packages to better suit different race tracks.

At Le Mans, the 919 not only sat on the pole but won the race overall thanks to a steady drive by the third team crew of Earl

Bamber, Nick Tandy and F1 driver Nico Hülkenberg. The other team cars placed second and fifth, with Audis third and fourth. The WEC season also produced strong results for Porsche. They won the manufacturer's championship and the driver's title for the team of Timo Bernhard, Brendon Hartley and Mark Webber. These were the first such season championships for Porsche since 1986. Although Porsche's business was in excellent shape, the return to winning at the highest levels of sports car racing was a welcome boost to morale for employees and fans alike.

For the 2016 season, Audi and Toyota moved closer to Porsche, using lithium-ion battery storage for their hybrid systems and Toyota switched to a smaller, turbocharged combustion engine. Porsche and Toyota opted for the 10MJ category, with Audi moving up to 6MJ in combination with their large diesel engines. Rule changes required all of the teams to use less fuel per lap in their engines. The 919s were not completely new but were extensively developed under team principal Andreas Seidl who also took on the technical director role when Alexander Hitzinger left for Apple's electric car project.

2016 is best remembered for a thrilling Le Mans race and heartbreak for Toyota on the last lap. The Japanese giant was still trying to win Le Mans, having first tried in 1985. At the end of the race, Toyota's Kazuki Nakajima led with Neel Jani in the 919 on the same lap. With only five minutes remaining, the Toyota lost power when the connection between the turbo and its intercooler failed. Nakajima stopped at the start/finish line with no drive at all. Neel Jani took the checkered flag for Porsche and his teammates Marc Lieb and Romain Dumas. It was a lucky win for the persistent, reliable Porsche team but nearly everyone felt for the Gazoo guys from Toyota. Porsche again won the manufacturer and drivers' titles for 2016, the Le Mans-winning team also won the season championship. Audi was second on manufacturer points and elected to withdraw from top-level sports car racing after 18 very successful seasons.

2017 would be the last season for the 919 and it was a straight, head-to-head competition between Porsche and Toyota in the LMP1 category, both opting for the 8MJ power strategy.

The V4 combustion engine was updated with a new technology. Plasma-jet injection was used in F1 and Porsche elected to adopt it to the 919. The high temperature, supersonic ionized gas entering the combustion chamber from a separate port causes more complete and powerful combustion of the main fuel/air mixture. The latest 919 also featured improved battery cooling with cockpit air conditioning. Attention was paid to the two aero packages allowed by rule, with an eye on improving top speed at Le Mans. Another new idea was a hydraulic linkage between

919 Evo at Rennsport Reunion VI, 2018, Laguna Seca. (Sherwin Eng)

front and rear suspension to maintain optimal pitch of the car for aerodynamic performance. Andreas Seidl estimated the cars were 60 to 70% new for 2017.

Winning a third consecutive Le Mans race in 2017 took a classic measure of the racer's 'never get up' persistence. Toyota took first and second in qualifying and had a third car to Porsche's two 919s. The Toyotas led early and apparent disaster struck on Saturday evening when the race number 2 Porsche was stopped for over an hour to replace the front drive motor. However, all three Toyotas dropped out with mechanical problems, leaving the other 919 in the lead on Sunday morning. Disaster struck Porsche again when the lead car's engine suffered connecting rod failure while running 13 laps ahead of the field. Given the reliability of modern racing cars, it was something of a shock to see an LMP2 car leading Le Mans. Meanwhile, the remaining 919 had been driven to the limits since Saturday night in an effort to catch up from 54th place, 19 laps down. By 1 PM, Timo Bernhard was on the same lap with the leader. The 919 passed for the lead on lap 347 and pulled away gradually to win by one lap. The performance by Bernhard, Earl Bamber, Brendon Hartley and the Porsche team was fittingly heroic. It was also reminiscent of Porsche's and Jacky Ickx' 1977 performance with the 936. Forty years earlier, Ickx won the race with Barth and Haywood after being down in 41st place. 2017 was Porsche's 19th overall win at Le Mans.

Passing of Butzi Porsche and Ferdinand Piëch

These influential cousins within the Porsche family both passed away during the decade. F.A. 'Butzi' Porsche, the 911 stylist, had advanced Porsche's design language and even after leaving day-to-day involvement in 1972, he remained a loyal and enthusiastic guide for the company. He started his own Porsche Design business in the 1970s and this was eventually folded back into the main company. He passed away in 2012.

Butzi's ambitious, hard-driving cousin Ferdinand Piëch passed away in 2019. Father of the 917 and Audi Quattro, Piëch had risen to the heights of the industry as Volkswagen CEO and Chairman. Known for his imperious style, his immense accomplishments were tainted by the management failures leading to the Volkswagen emissions scandal. Although sometimes at odds with his cousins, Piëch lived to see the Porsche family become controlling shareholders of the VW empire that he built in the 1990s and 2000s.

Porsche Customs

The 2010s saw a blossoming of Porsche customization. Musician Rob Dickinson's Singer Design in Los Angeles tapped in to demand for restored and modernized ('re-imagined'), high-performance and highly personalized air-cooled 911s. Using 964 tubs, these stylish 911s cost upwards of a half-million dollars. Various other 911 customizers entered the market during the decade including Bruce Canepa with his company's modernized 959SC package.

The 356 'Outlaw' movement also flourished in the recent decade. Led by Emory Motorsport, these cars offer stylish, often higher-performance takes on classic 356s. Although not 'correct' in the traditional sense, these customs have an appeal different from purist restorations. Fortunately, the car hobby remains friendly to all manner of Porsches including ultra-original cars and 'barn finds' which soared in value during the 2010s.

Bernhard, Bamber and Hartley won the drivers' title for the 2017 WEC and Porsche won the manufacturer championship although Toyota won five races to Porsche's four.

Although most of the 919s were retired to the museum, the program wasn't quite finished. Porsche cobbled together a small budget to create and run the 919 'Evo' in 2018. This was meant as a demonstration effort to show what the car would be capable of without various rules restricting performance. Endurance-related and night-driving equipment was removed and the aero package was revised. The new package featured higher downforce balanced with an F1-like drag reduction system built

> **PORSCHE POINTS**
>
> » In 2014, Patrick Long and Howie Idelson founded the Luftgekühlt event series celebrating air-cooled Porsches. The events have grown into major 'happenings' within the Porsche fan culture.
>
> » At Le Mans in 2017, the LMP2 team that nearly won the race was partly owned by film star, Jackie Chan.
>
> » The last demonstration runs for the 919 Evo were at Laguna Seca during Rennsport Reunion VI, in September 2018. Chassis 1704 was then retired to the Porsche Museum collection.

into the wing and rear diffuser. Michelin provided a special tire designed for the record runs that Porsche had in mind. Without restrictions on fuel use and electric power, the Evo had a combined 1160 horsepower to put on the road.

At Spa, Neel Jani was the driver and he broke Lewis Hamilton's F1 track record by less than a second. This record was then re-broken by Sebastian Vettel in his F1 Ferrari later in 2018. As always, the true test came on the Nürburgring Nordschleife. Timo Bernhard was the driver for the record attempts, targeting Stefan Bellof's 1983 qualifying time set in a 956 at 6:11.130. After much work in the simulator, June 29, 2018 was the date to try it for real. Bernhard's first lap shattered the record and he was then able to lower it significantly on two further laps. The record now stands at 5:19.546, an incredible 51 seconds faster than the 956, at an average speed of 145.3 mph. This other-world technical and driver performance will live on for everyone to see on YouTube.

Some at Porsche were inspired by Mark Donohue's closed course record project with the 917/30. The 919 Evo was perhaps a spiritual successor in the 1,000-plus horsepower club and it certainly fulfilled the mission to show the absolute capabilities of an unlimited hybrid sports racer.

Porsche 718 Boxster and Cayman

In 2016, Porsche introduced a new generation of mid-engine sports cars. Named '718' in tribute to the sports racing cars of the 1950s, they are also referred to as the 982 generation. In terms of styling, the change was evolutionary. Under the skin, however, major changes were made under Porsche's sports car leader, August Achleitner. As could be expected, Porsche engineered for still greater rigidity along with weight reduction. This generation of mid-engine cars made much greater use of aluminum in the structure, with far less steel, to achieve the conflicting goals. The new 718 platforms were developed alongside the future 992 911, the platform being referred to as the MMB.

In the engine room, the new cars also had to balance the conflicting goals of greater power and better fuel efficiency. Continuing a Porsche trend, this was achieved with the use of turbocharging for all 718s. That turbocharging was applied to flat four-cylinder engines created a perception challenge for the new cars. The engines clearly met the challenge of producing greater peak horsepower and torque than their predecessors with better fuel mileage. The sound of the new fours, particularly at low revs, played into a time-honored tradition within the Porsche culture, for the curmudgeonly fans to resist change. It's hard to believe there was a time when 911-resistant 356 owners believed that the only 'real' Porsches had four-cylinder engines.

The new Boxster and Cayman received attention in every area including revised suspension and steering, improved engine mounting, updated electronics and new interiors. As is now customary, the 2.0 and 2.5-liter base and 'S' models were joined before long by a GTS option, then the new 'T' (a lower-priced, base 'touring' model but with Sport Chrono, PASM and larger

PORSCHE POINTS

» During the 2010s, Porsche expanded their Experience Center facilities, including the 2018 Shanghai opening, the first in Asia. The centers offer an optimal setting for test-driving current Porsche models.

» Michael Mauer is only the fifth man to head the styling function for Porsche. He follows Erwin Komenda, Butzi Porsche, Tony Lapine and Harm Lagaaij.

BELOW: In 2016, Porsche introduced the '718' generation Boxster.

wheels) plus the GT4 Clubsport variant (sneaking a flat-six back in, for a track-focused machine). The 718 GTS proved its mettle by lapping the Nürburgring 13 seconds faster than the equivalent 981.

Four-Door Fortitude

In the second half of the 2010s, Porsche introduced new versions of the Cayenne, Macan and Panamera. While some may have been concerned about a loss of emphasis on sports cars, Porsche forged ahead, intent on producing the sportiest four doors in their given market segments. In partnership with fellow VW brands, this was accomplished efficiently through platform sharing as well as the adaptation of family engines.

The third-generation Cayenne would share its basic 'MSB' platform with SUVs from VW, Audi, Bentley and Lamborghini. As the reader might expect, this newest Cayenne made much greater use of aluminum for its structure, achieving both greater strength and moderate weight saving. It came with either V6 or V8 engines from the VW family and a new 8-speed Tiptronic transmission from ZF. Following another trend, the Cayenne was designed for different size wheel and tire combinations front to back. The restyling by Michael Mauer's team was evolutionary, making use of upper and lower spoilers at the rear. The cars also featured new interiors with a larger central display screen and other new technology options such as PSCB, for Porsche Surface-Coated Brakes. All new-generation Cayennes would be built at the VW plant in Bratislava, Slovakia.

Press reception was generally favorable with some concerns expressed about improved comfort and ride qualities at the expense of the sporting character shown by earlier Cayennes. In 2018, Porsche introduced the E-Hybrid plug-in Cayenne offering a combined 462 horsepower and over 500 foot-pounds of torque. In 2019, Porsche further expanded the range with the Coupé version of the Cayenne. Intended for a youthful, sportier audience, the new version featured a lowered roofline with a sloping, fastback shape at the rear. The new roof could be had in either panoramic glass or lightweight, aerodynamic carbon fiber.

The updated Macan appeared in 2018, again with only evolutionary styling changes, but with significant changes under the hood. A new family of turbocharged engines designed by Audi featured the 'hot V' concept, with the turbocharger(s) positioned within the 'V' of the engine. This layout provides fast and direct flow of exhaust to spin the turbo(s). Porsche also introduced a new dashboard and display for the Macan with updated suspension and engine mounts for yet better handling character. The sporty, medium-size crossover SUV has been a stunningly successful product for Porsche in terms of sales numbers, reaching a whole new audience of potential buyers. In 2019, Porsche announced the intention that the next generation of Macan will be an EV, although combustion engine Macans may continue to be produced during a transition period.

The styling of the newest Panamera was a stunning success. Presenting a fully resolved sport-coupé sedan design, the Porsche no longer gave anything away to its rivals, including the Tesla Model S, in the beauty contest. The all-new 2017 Panamera was built on the VW Group MSB platform shared with Audi and Bentley. Similar to the Macan, the Panamera used family engines based on Audi designs in either V6 or V8 form. The V8 Panamera Turbo was the first Porsche to feature cylinder deactivation for the benefit of fuel economy (an idea that had been explored as far back as the 928). The new car was also the first with an eight-speed PDK, again mainly to help with fuel economy from the overdrive 7th and 8th gears. Ride comfort was enhanced by the new suspension and new air spring option while handling got an assist from rear-wheel steering. New software in the form of 4D-Chassis Control coordinated all of the various sensors and systems to better manage any driving situation.

The new Panamera range was extended with the expected E-Hybrid version, plus an Executive 'limo' version with a longer wheelbase (intended to be chauffeur-driven). The range-topper was the Turbo S E-Hybrid. In 2017, Porsche added the Sport Turismo 'shooting brake' version with its unique rear roof treatment, and also a sportier, GTS option. This latest generation of Porsche sedans sold very well from the start although suffering from some complaints reminiscent of the 928: 'too good' or 'not enough like a Porsche' in driving character.

All of the latest generation four-door Porsches shared the common goals to be lighter, stronger and faster. They also gained in terms of comfort with the Cayenne and Panamera helping to point the way toward an electric future.

OPPOSITE: Mamba Green Macan at Mt. Rainier. (Norbert Kremsner)

Special 911s

Inspired by the ultra-lightweight 911 racing prototype from 1967, Porsche created the 991 911 R in 2016. By returning a manual transmission to a GT3 engine, it made for a purist driver's car. The six-speed 'sport' manual had 500 horsepower to work with and the car also borrowed rear-wheel steering from the GT3. The R was a lighter-weight package although nothing like as light as its 1967 predecessor. It had a subtle, standard body and no fixed wing (just a movable spoiler) with period livery options. Porsche only made 991 of these sublime 911s which were quickly gobbled up by insiders and collectors.

On May 11, 2017, Porsche built the one millionth 911, an Irish Green Carrera S. 911 production had started on September 14, 1964. It was a proud moment for Wolfgang Porsche to celebrate the success of Porsche's signature model. Porsche sent the car on a world tour before putting it in their museum collection.

Perhaps the ultimate example of a generation-ending 'special' was the 2019 991.2 '935'. Built as a tribute to the 'Moby Dick' 935/78 version of the 935 racing car, this was a 700-horsepower, seven-speed PDK GT2 RS with revised bodywork, 919 taillights, RSR mirrors and 908-style titanium exhaust pipes. It was also equipped for track duty with a roll cage, racing seat, air jack system and fire suppression system. Porsche offered only 77 of these non-road-legal 991s.

ABOVE: 991 911R at the Porsche Experience Center in California, 2018. (Jay Gillotti)

BELOW: One millionth 911 shown with Wolfgang Porsche.

Wolfgang Porsche

Born in 1943, Wolfgang is the youngest son of Ferry Porsche. He studied business at the University of World Trade in Vienna and had his own business importing Yamaha motorcycles into Austria. He worked for Daimler-Benz for five years, starting in 1976 and was appointed to the Supervisory Board of Porsche AG in 1978. In 2007 he became Chairman of Porsche AG and in 2008 joined the board of Volkswagen AG.

He is currently the Chairman of Porsche SE, the holding company that manages the Porsche/Piëch family investments including its interest in Volkswagen AG and Porsche AG. Wolfgang has an organic farm in Austria and his own collection of rare Porsches. He can be considered the 'head of the family' and often appears at important Porsche events or occasions. A genial figure, Wolfgang remains a top ambassador for the Porsche brand and a direct connection to his father and grandfather.

Porsche 992

Following a familiar pattern, Porsche produced a second-generation '.2' 991 for the 2017 model year. The biggest change was a new family of smaller displacement, all turbo engines. A minor point of confusion came with base models having turbo engines but not the 'Turbo' designation, which remained reserved for the higher-performance models. The goal was to improve fuel efficiency while maintaining or improving performance across the line. The 991.2 met the goals and was produced in all the expected variations during the final development period for an all-new 911, the 992.

In typical Porsche fashion, models like the Turbo S were faster and better. In this case, the Nürburgring lap time was seven seconds faster. The uncompromising, 700-horsepower GT2 lapped the 'Ring at 6:47, 31 seconds faster than its luxurious, Turbo S cousin. A new variation appeared as the 911 'T'. The T was a base model but with shorter gearing, Sport PASM suspension, optional rear-wheel steering and a weight-savings package. It was a driver-oriented package that could rival the

Ultra high-performance 911 for the street – the 991.2 GT3 RS.

PORSCHE POINTS

» During 2018, England's Lee Maxted-Page Porsche specialist company completed the restoration of the 911 RSR (Chassis 0588) that won the 1973 Targa Florio. Known to aficionados by its internal designation 'R6', it may be the most valuable of all 911s.

» In 2018, Porsche created the Ferry Porsche Foundation to support projects in education, social issues and youth development. This was part of Porsche's 70th Anniversary celebration and the initial funding was 20 million Euros.

» In 2019, the Ferry Porsche Foundation endowed a professorship at Stuttgart University for corporate history. This was partly inspired by Professor Wolfram Pyta's work studying the history of Volkswagen and Porsche during the Nazi era.

» In 2018, Porsche took a 10% ownership stake in Rimac (Ree-mats) of Zagreb, Croatia, to partner in development of high-performance EV technology.

Carrera S. Following another tradition, Porsche ended the 991 run with a Speedster model, building 1,948 copies to celebrate Porsche's 70th year as a manufacturer. This Speedster was harder-edged and more performance-focused than earlier 911 Speedsters, in this case using a GT3-based engine and PDK.

The 992 was the last 911 from the era of August Achleitner's successful stewardship of the 911 and the sports car range. It was first shown in November 2018 and production began with the 2020 model year. The 992 introduced an all-new platform, the MMB (for modular mid-engine componentry), that was intended to strengthen the production efficiencies between the 911 and Boxster and Cayman in the future. As with all previous water-cooled 911s, the 992 improved in structural strength and crash safety even though the use of aluminum increased dramatically (with more than a 50% reduction in the use of steel). Exterior body panels were now all aluminum.

From a styling perspective, the work of Peter Varga and team was classic 911, but with significant changes front and rear. At the front, the full-width opening for ducting broke with the tradition of disguising the 911's transition to water cooling for

992 in Carmine Red. (Ben Przekop)

	991 2012-2019	992 2020-Present
ENGINES	Improved 9A1 engines, 991.2 transition to all turbo engines, 350+ horsepower	Improved engines from 991.2 family, 385+ horsepower
CHASSIS	Nearly 100% new compared to 997, longer wheelbase, wider track, larger wheels, electric steering, seven-speed manual transmission	All-new platform, all-aluminum body, improved rear-wheel steering, enhanced electronic driving modes
STYLING	Lower profile, shorter overhangs, new windshield positioning, wider rear spoiler	Lower-look tail, more openly-ducted air dam, all versions with wide-body styling
OTHER	Larger interior, new seating position, new sunroof operation, all-new folding-roof Targa	Automatic accident avoidance and lane-change assist, almost fully digital controls

the engine. At the rear, functional and design elements worked to make the car look lower to the ground. The shape was defined by a narrow red lighting band below which the Porsche lettering was fixed between narrow taillight clusters. The rear spoiler was wider and longer. Front cooling flaps were designed to be active depending on operating temperatures but also to coordinate with speed and driving mode. The overall package helped the 992 achieve low drag and improved fuel economy. Another significant decision was to discontinue offering a narrow-body 911. All 992s would share the same basic, wider-body lines.

The 992 carried over the basic engine family from the 991.2 but with the expected and significant improvements to performance. Larger turbos, now counter-rotating, contributed to an overall redesign of the airflow and increased power. The intercoolers were also larger and placed directly behind the engine and above the exhaust, rather than tucked into the rear fender area. Porsche revised the direct injection system and designed a new variable valve control where each inlet valve could open at different levels within the same cylinder, to improve combustion mixture. Carrera S horsepower for the 992 was 450 (approaching 959-level power). To aid in overall stability and rigidity, the cylinder heads were linked to the chassis.

Owners could now select between a seven-speed manual or eight-speed PDK with further fuel-saving overdrive ratios. Transmission/differential packaging was redesigned to make room for an electric motor in a future hybrid 911. The design of the floor was also set to receive battery packaging if needed. The 992 was the first 911 designed to run on larger diameter wheels at the rear across all versions of the model line. This enhances traction and handling by enlarging the rear contact patch. Rear-wheel steering could now move the rear wheels with greater range either to the left or right depending on speed and the direction of turning. No surprise, the 992 was faster at the Nürburgring compared to the equivalent 991.2, by five seconds.

The 992 offered other new features such as automatic braking for accident avoidance and lane-change assistance. If the car senses an impending collision, it activates the seat belt tensioners and closes the windows and sunroof. A further safety feature is the wet road sensors that adjust driving dynamics to wet conditions while in Normal mode. Further, the 992 was granted a Full Wet Mode which coordinates all of the driver aids, suspension and spoiler to give the driver optimal traction and control in the rain. Other driving modes were Normal, Sport, Sport Plus and the temporary Sport Response option allowing 20 seconds of even higher-performance levels above Sport Plus. In the interior,

the 992 retained a traditional five-gauge cluster ahead of the driver, however, all but the tachometer were digital displays and the gauges no longer overlap. A larger center display helped the driver with programming and managing the car's systems including new options like Night Vision Assist.

Speeding into the 2020s, the 911 remains alive and well, familiar but always newly-developed. It continues to defy logic, with the engine in the 'wrong' place. It is not only Porsche's flagship but also a car that Ferry Porsche would instantly recognize and enjoy.

Porsche Taycan

At the Frankfurt Auto Show in September 2015, Porsche showed the Mission E concept vehicle, a pure electric car. Smaller than a Panamera, it was a four-door, four-passenger car pointing the way to future EV development. A key feature of the concept was its fast-charging, 300 kilowatt/800 volt technology. The goal was to be able to charge to 80% of battery capacity in 15 minutes. Maximum range was pegged at something over 300 miles, depending on weather conditions and driving style. Porsche-designed motors powered both the front and rear wheels. The structure of the concept was an innovative mix of steel, aluminum and a large amount of carbon-reinforced polymer. Mission E designers, under Michael Mauer and Mitja Borkert, were relative newcomers to Porsche, Emiel Burki and Fabien Schmöltz. Betraying a desire to project sportiness, the Mission E was lower at the roofline than a 356 or early 911. It was clear that any future electric Porsche had to be a Porsche in every sense.

By the end of 2015, Porsche had committed to putting an EV into production, roughly by the end of the decade. The 'Taycan'

> **PORSCHE POINTS**
>
> » Porsche considered 600 different naming ideas for the production version of the Mission E, before deciding on 'Taycan'.
>
> » VW and Porsche CEO Oliver Blume was the first German student to earn a doctorate at the Institute for Automotive Engineering at Tongji University in Shanghai (2001). He has worked at Audi and SEAT in addition to Porsche and Volkswagen AG.
>
> » The TAG Heuer Porsche Formula E team had its first race at the opening event of the 2019-2020 season, at the Ad Diriyah e-Prix in Saudi Arabia.

(loosely translated as 'soul of a spirited young horse' from Turkish) was ready for production by 2019. Its styling resembled the Mission E with some changes for practicality, including conventional door opening. The look of the car blended cues from the 911, 718, 919 and the Panamera to present a distinctly Porsche form. The interior was a fresh design with all-digital instruments. Ahead of the driver, these were arranged in a traditional grouping of five displays, with those displays changing based on driving mode. A center console allowed for control of systems for climate, navigation, entertainment, etc., while the front passenger also had his or her own display.

Although the Taycan platform was originally meant to be exclusive, it was eventually decided to share with Audi, within the VW family. With the transition from Matthias Müller to Oliver Blume as Porsche CEO, the decision was made to make a massive investment in Porsche's Stuttgart facilities to produce the Taycan (and components such as the electric motors). Those motors would be of the permanently excited synchronous machine type, a brushless AC motor with magnets embedded in the rotor, giving efficient, fast and dynamic performance. Feeding the motors were underfloor lithium-ion batteries contributing to a low center of gravity (always good for handling and even lower than a 911). EPA ratings put the Taycan's range at approximately 200 miles, however, as with all EVs, this depends on weather and other driving conditions plus driving style. The 800-volt technology is compatible with 400-volt charging systems if needed.

The Taycan now comes in ten different performance and body style versions. The highest performance models are named 'Turbo' and 'Turbo S' following a Porsche tradition although hard to take for traditionalists given that the cars have no turbochargers. Even before the Taycan was released for sale, Porsche had shown a Cross Turismo, 'shooting brake' concept. Taller than a standard Taycan, it was Panamera-height. The Cross Turismo is now available in five versions.

Press reception has been generally favorable and owners that the author has spoken to seem to agree that the Taycan fulfills the mission of being the Porsche of EVs. Porsche's goal is to achieve 50% of the model range as EVs by 2025. The Taycan expansion at Zuffenhausen is meant to produce a zero emission vehicle in a net-zero emission facility. At the 120-year mark of Porsche vehicle design, propulsion methods had come full circle.

The Taycan Turbo S powered Porsche into the EV market.

2020s

To celebrate Porsche's 75th anniversary as a manufacturer, the Mission X hypercar concept was revealed as part of the celebrations in Stuttgart, on June 8, 2023…

2020s

Future historians will have to assess the current decade in its full scope. It is clear that the first four years have been eventful and consequential for Porsche. The company entered 2020 with more than 35,000 employees and a solid product line including a new 911 and the EV Taycan. Below in bullet point form are just some of the events and activities that Porsche historians will be assessing and writing about in the future.

2020

- The global Covid-19 pandemic affected business and automobile production across the board. The disruption to global supply chains limited Porsche's ability to produce enough cars to meet the demand which recovered quickly in the second half of 2020. 2020 Porsche vehicle sales came to 272,162, down only 3% year-over-year.

- The Panamera Turbo set the 'Executive Vehicle' record at the Nürburgring, 7.29.81, in August 2020.

- Porsche built the one millionth Cayenne in December 2020.

- In late 2020, Porsche released the 'Porsche Unseen' design studies and prototypes from 2005 through 2019. There was a book and museum exhibit for these unique vehicles including a 'street' 919 and 917 'Living Legend' hypercar design.

2021

- In March 2021, Porsche announced a goal for their value chain (including suppliers) to be net carbon-neutral by 2030.

- In the spring of 2021, Porsche introduced the Taycan Cross Turismo shooting brake, with the Sport Turismo arriving early in 2022.

- In 2021, Porsche showed that combustion engines were still alive and well by introducing a flat-six, 4.0-liter Cayman GTS. Reviews were glowing for the engine and chassis combination.

911 RSRs racing in the 24 Hours of Daytona, 2020. (Sherwin Eng)

- Also in 2021, Volkswagen's Bugatti brand came under Porsche management and was spun off to a joint venture with Rimac. The plan was to produce the hypercar Bugatti Chiron and EV Rimac Nevera.

- Porsche Engineering began discussing its work on ADAS (Advanced Driver Assistance System) and HAD (Highly Automated Driving) systems. The challenges in development were to reach 100% safety and reliability while taking into account driving dynamics and the autonomous driving infrastructure.

- In 2021, Porsche auctioned an NFT (non-fungible token) digital art piece for the first time. The sketch of the Taycan Cross Turismo by Head of Exterior Design, Peter Varga, sold for the equivalent of $92,000 in crypto currency. The physical, paper sketch was included in the sale. Proceeds were donated to Viva con Agua, a non-profit which works to expand access to clean drinking water around the world.

- In 2021, Porsche formed a joint venture with Custom Cells GmbH for battery development and production. In 2022, they announced an investment in Group 14 (Woodinville, WA), further seeking technology to improve battery materials and performance along with Cellforce, another company in which Porsche has invested.

- In June 2021, Porsche set the Nürburgring road-legal lap record with a 911 GT2 RS equipped with the Manthey performance kit. Lars Kern was the driver and the time was 6:38.835. The record was then broken in October 2022 by a Mercedes-AMG One. Also in 2021, the Cayman GT4 RS put down a blistering 7:09.3 Nürburgring lap in the hands of development driver and Porsche racer Jörg Bergmeister.

- In August 2021, the Pebble Beach Concours d'Elegance celebrated the 50th anniversary of the Porsche 917 and Porsche's first two overall wins at the 24 Hours of Le Mans.

THESE TWO PAGES: Pebble Beach Concours d'Elegance celebrated the 50th anniversary of the 917 in 2021. (Kurt Oblinger)

(Dean Holbrook)

Mission R – Porsche's concept for an EV sports/racing car.

- Another 2021 introduction was the Mission 'R' EV racing car. It presented Porsche's vision for pure electric 'customer' sports car racing in a sustainable future.

- In late 2021, PCNA announced a further expansion at the Atlanta HQ and Experience Center campus, with a larger, more versatile test track and increased sustainability by use of solar power and green walls.

- In 2021, Porsche delivered 301,915 vehicles in spite of microchip shortages and other pandemic issues. It was the first time Porsche exceeded 300,000 vehicles sold. China was the largest market at approximately 88,000 sales through 136 outlets in China. Also in 2021 the Taycan outsold the 911 for the first time, 41,296 to 38,464.

2022

- During 2022, the Porsche Museum completed a mechanical restoration on the 1922 Austro-Daimler 'Sascha'. This allowed the car to be in running condition, 100 years after it was built.

- Porsche's first win in open-wheel Formula E racing came in the Mexico E-Prix on February 12, 2022. The winning driver was Pascal Wehrlein.

- Russia's invasion of Ukraine in February 2022 created further supply chain issues for Porsche and other brands sourcing components (mainly wiring harnesses) from factories in Ukraine.

- Also in February 2022, the automobile cargo ship, *Felicity Ace*, caught fire on the Atlantic Ocean. It sank on March 1 in 9,800 feet of water after having been stabilized for towing. Porsche lost 1,117 cars among many other vehicles from Volkswagen Group brands. The exact cause of the fire and sinking will likely remain unknown. It was a significant misfortune at a time when parts shortages were already hampering production.

- In March of 2022, the first Porsche Cayenne assembly in Malaysia was completed. The factory at Kulim is Porsche's first outside of Europe. Bodies in white are shipping from Slovakia to Malaysia for final assembly.

- The 911 GT3 won Motor Trend's Performance Vehicle of the Year in 2022.

- In 2022, Porsche announced the current 718 lineup will be replaced by EVs in 2025. EV Boxster and Cayman prototypes were spotted later in the year. The delayed EV Macan was also on the way, now expected for the 2024 model year. The SSP, Scalable Systems Platform, was announced for an electric SUV above the Cayenne, along with the next generation Panamera, Taycan, and multiple other VW Group vehicles.

- Another 2022 product announcement was the 718 GT4 RS, a 4.0-liter, track-focused car that may be the last combustion engine, ultra-high performance Cayman.

- At Le Mans in 2022, a 911 RSR won the final GTE Pro class race. Porsche's tally stood at 107 class wins at Le Mans since 1951.

- At the 2022 Goodwood Festival of Speed, the Porsche 963 LMDh made its debut. The new rules for top-level sports car racing allow either Le Mans Daytona Hybrid or Le Mans Hypercar builds. The LMDh choice by Porsche involves a budget limit, rear-wheel drive only, and use of a spec Multimatic chassis (one of four options from chassis makers). The first race for the 963 was the 24 Hours of Daytona in 2023 where the 963s showed good speed as one of the Penske-run cars qualified second. In the race, the 963s ran into mechanical problems and the surviving car finished 14th overall.

- In August 2022, Porsche's one-off 'Sally Special' 992 inspired by Sally Carrera from *Cars* movie sold for $3.6m at RM Sotheby's Monterey auction to benefit Girls Inc. and the UN Refugee Agency.

THIS PAGE: Porsche fans come in all shapes and sizes, at the Porsche Club Werks Reunion Monterey, 2021. (John Ficarra)

Porsche's new LMPh Hypercar, the 963, debuted at the 24 Hours of Daytona in 2023. (Sherwin Eng)

- Discussion of a possible IPO of Porsche AG from Volkswagen began around 2021 with speculation on the value of Porsche between $55 and $110 billion. The Porsche/Piëch family supported the IPO in concept and planned to keep their VW stake at 53% of the voting shares and own 25% of Porsche AG voting shares under Porsche SE. Porsche SE owns approximately 31% of total VW shares. VW sold 12.5% (total of 911 million shares) of Porsche AG in the public offering and shortly after, Porsche valuation was greater than the VW brand. Porsche SE also purchased 12.5% of Porsche AG directly from VW. The IPO was completed on September 29, 2022, valuing Porsche at approximately 75 billion Euros ($73 billion). The European stock symbol for Porsche AG is P911.

- The Porsche IPO coincided with Oliver Blume becoming CEO of Volkswagen AG while retaining the CEO position at Porsche AG. Blume replaced Herbert Diess at VW. Diess had success in guiding VW to recovery after Dieselgate but ran into major issues with delays in software development, a critical area for current and future EVs.

- The Porsche IPO documents disclosed an ongoing lawsuit by the daughter of Erwin Komenda under Germany's "best seller" law. The law allows an original creator of a work to receive a share of profits if the work is more successful than expected. The suit contends that Komenda was responsible for the original 911 styling, not Butzi Porsche. The claim against Porsche had been rejected in court twice previously, as had Ms. Steineck's claim against VW for the new Beetle.

- Yet another 2022 announcement was Porsche's increase to $100m total investment in Highly Innovative Fuels (HIF Global) based in Haru Oni, Chile. The company creates sustainable e-fuels made from hydrogen and carbon dioxide. The announcement explained: "Electricity-based fuels, or e-fuels, are clean, carbon-neutral fuels produced from renewable, green hydrogen and carbon dioxide taken from the atmosphere. They can act like gasoline, allowing owners of current and classic vehicles a more environmentally friendly way to drive." Synthetic, carbon-neutral fuels could be a solution for combustion engines for decades into the future.

- During 2022, a major topic of discussion was Porsche's apparent plan to buy 50% of the Red Bull F1 team. However, the deal fell thru and any Porsche F1 plans for the new formula in 2026 remain uncertain as of this writing.

- 992 development continued apace with the new GT3 RS, perhaps the most track-oriented Porsche GT ever. It doubled the downforce of the previous GT3, and featured active aerodynamics, DRS, 9,000 rpm rev limit and 525 horsepower. A GT3 RS with the Weissach package set a Nürburgring lap time at 6:49.328 in the hands of Jörg Bergmeister. Another 911 development was the introduction of the seven-speed manual transmission 992 Carrera T. Porsche announced that Frank Moser would replace Frank-Steffen Walliser as head of sports cars, Walliser moving to VP of Vehicle Architecture.

- In July 2022, Porsche announced a partnership with Los Angeles-based UP.Labs to fund innovative start-up companies in businesses such as predictive maintenance, supply chain transparency and digital retail.

- In November 2022, Porsche announced the 911 Dakar edition, offering a new car option for fans of the 911 'safari' trend. Only 2,500 will be built.

- For 2022, Porsche set a US sales record at 70,065 cars, beating 2021 by 40 vehicles. Full year global sales for Porsche were just under 310,000.

2023

- On April 15, 2023, the Porsche 963 got its first victory. At the Long Beach, CA IMSA race the winning drivers were Nick Tandy and Mathieu Jaminet. Ricky Taylor's Acura crashed trying to pass for the lead with two laps left in the race.

- Also in April 2023, Porsche announced an extensively redesigned Cayenne for the 2024 model year. Significant changes to the styling, chassis and drivetrain are joined by a new 'digital-first' approach to the interior.

- Porsche's Q1 2023 global sales increased by 18% year-over-year, helped by strong performance in China. US sales shattered the old record with a 38% increase year-over-year.

- To celebrate Porsche's 75th anniversary as a manufacturer, the Mission X hypercar concept was revealed as part of the celebrations in Stuttgart, on June 8, 2023. The pure EV hypercar could be a successor to the 959, Carrera GT and 918.

2023 reveal of the Mission X.

Afterword

Brief thoughts on buying and owning a new, used, vintage or collectible Porsche

Having spent most of my life around cars and car enthusiasts, I think the best advice I have heard is: 'buy what you like'. This advice argues against focusing on cars as an investment vehicle (excuse the pun). Making money with cars can be a fine and rewarding profession, but I feel it is best left to those professionals. The follow-on to 'buy what you like' is, in my opinion, the most important consideration when thinking about buying a Porsche (or any automobile that is intended for fun and enjoyment beyond basic transportation).

That follow-up question is: 'How will you use it?' A mismatch between the car you purchase and its intended use will very likely lead to dissatisfaction and the potential costs involved with selling out and buying something else again. Will your Porsche be a daily commuter or just a fun weekend driver? Will you plan long trips and tours? What about track driving or autocross? Will you do car shows and Concours events? Are you building a collection of historically significant Porsches? Do you need four seats or only two? The answers to these questions and more should guide your decision on what to buy. Of course, it is possible that one car could satisfy many uses. A new, daily driver Porsche could also be perfectly at home on the track and you can proudly detail it for a local Concours event. Versatility is a time-honored Porsche trait, however, using a vintage Porsche for daily driving or long-distance touring could be challenging.

There are numerous dimensions to consider within your intended use. Do you want a super-fast car that will exceed 150 mph on the straight at your local track (and how will that car be to drive in city traffic)? Or, are you more of a slow-car-fast driver? Do you want modern ride and handling or the decidedly vintage experience of a torsion bar 356 or 911? Do you like an open car or closed, or something in between like a Targa?

For those contemplating a vintage Porsche (which I consider to be about 20 years old, or older), be sure to understand your local service providers. As an example, owning an early 928 might offer a really cool, somewhat retro experience with a very capable and stylish machine. However, having at least one local shop that is well-versed in the foibles and complexities of the 928 is critical. Even the newest 928s are now close to 30 years old. In addition, it is important to connect with the local community of owners for vintage Porsches. This can be done through the network of Porsche clubs (and other organizations). Porsche owners generally love to share their experiences, recommendations and knowledge. Note also that Porsche and its dealers have continued to expand the parts and service offerings under Porsche Classic.

Color is a popular topic of discussion among enthusiasts and Porsche owners are no exception. Porsche has always worked to provide an interesting color palate, although the standard offerings have changed with fashion over the decades. Many interesting Porsches were created by the application of custom 'paint-to-sample', special order or just unusual colors (perhaps standard offerings that sold in tiny numbers). In the past couple of decades, vintage Porsche buyers have tended to seek out bold, bright and unusual colors, sometimes paying a hefty premium. So, as a vintage Porsche buyer it is important to know how your color preferences play in the market. If you are chasing a 3.2 911 Carrera and like the basic, dealer stock colors (mainly white, black and red), you'll have an easier time than the person whose

The author driving his Prussian Blue 1985 911. (Christian Bouchez)

heart is set on a Prussian Blue or Granite Green. The Porsche Club of America's color database, at Rennbow.org, is a great resource, cataloging colors and relative rarity with sample photos. It's also important to think about interior color since the overall combination of interior color and exterior paint can change the personality of a car.

This book should convey a sense that all owners experience some aspect of Porsche history. There is a through-line of Porsche personality, quality and style that dates back 75 years (and more). Porsche design and development, with a dedication to continuous improvement, can be directly experienced. Whether it be sitting in a 356, 911, Cayenne or Taycan, the driver will know he or she can only be in a Porsche. When you paddle the gear changes in a new PDK-equipped 911, you are following in the footsteps (fingertips?) of Derek Bell, Hans Stuck and the others who raced with the first PDKs in the 1980s. If you are driving a Boxster on a curvy road and notice the front fenders help you place the front wheels precisely in a turn, you are following quite literally in Ferry Porsche's wheel tracks.

In the Porsche Club of America, we often say '*You come for the cars, stay for the people…*' It is undoubtedly true that membership in the Porsche Club has changed my life in many wonderful ways. I have met countless fascinating, knowledgeable and just plain nice people whom I never would have met were it not for joining and participating in the Porsche community. It has been my privilege to sit down and talk history with hands-on Porsche racing legends like Brian Redman, Derek Bell, Vic Elford, Hurley Haywood, John Horsman, Patrick Long and others. Although I have 'stayed for the people', I think I stay for the cars as well. I can find no end to the fascinating story of Porsche cars and the people who made them. The real authors of this book are those dedicated people who designed, tested, built, raced and sold Porsches over the years. My great hope is that readers will experience the joy of owning or driving a Porsche but also that they will find their own pathways to discovering the depth of the Porsche story.

Jay Gillotti
February 2024

Appendix 1 *Porsche Project Numbers*

- 1 to 6 — unused
- 7 — Medium-size, six-cylinder car design for Wanderer
- 8 — Prototype for larger, four-door, straight-8 Wanderer
- 9 — Wanderer fastback, supercharged coupe prototype, retained as Dr. Porsche's personal car
- 12 — Small car design for Zündapp, with three or five-cylinder radial engine
- 22 — Auto Union Grand Prix racing car

Auto Union Grand Prix car driven by Phil Hill at the 1982 Monterey Historic Races. (Allan Caldwell)

- 32 — Small car design for NSU
- 50 — Torsion bar spring front suspension for Triumph (England)
- 52 — Auto Union 'Supercar' design for road-going, V16 mid-engine, three-seater
- 55 — 1,000 horsepower aircraft engine for Suddeutsche Bremsen
- 60 — Volkswagen Type 1 'Beetle' (KdF-Wagen)
- 62 — First version of the Kübelwagen
- 64 — Berlin-to-Rome streamlined Type 60 (60K10)
- 70 — 32-cylinder aircraft engine for DVL
- 80 — Land speed record car for Mercedes-Benz
- 82 — Kübelwagen and its various configurations
- 100 — Leopard tank design
- 101 to 103 — Tiger tank design proposals, engines and configurations (later became the 'Ferdinand' tank)
- 110 — Small tractor *Volkspflug* 'people's plow'
- 114 — 1.5-liter V10, mid-engine, two-seat sports car
- 128 — Schwimmwagen, first Kübelwagen-based version
- 166 — Improved, production version of the Schwimmwagen
- 199 to 204 — unused
- 205 — Maus tank
- 287 — Volkswagen four-wheel drive Command car
- 312 — Gasoline engine tractor for Brazil
- 313 — Diesel two-cylinder engine for Type 312
- 315 — Rope tow ski lift with VW engine for Santner
- 332 — Diesel tractor for Cisitalia
- 335 — Cable winch for mining, Carinthian government
- 352 — Pre-356 sports car design for the Swiss Rupprecht von Senger
- 356 — Porsche sports car and its variations
- 360 — Cisitalia Grand Prix car
- 370 & 372 — Sports car design studies for Cisitalia
- 411 to 424 — unused
- 425 — Diesel tractor for South America
- 426 to 499 — unused
- 502 — 1.5-liter, 55 horsepower engine for 356
- 514 — 356 SL for 1951 24 Hours of Le Mans
- 519 — Synchromesh four-speed transmission for 356
- 530 — Four-seat Porsche with 2400 mm wheelbase
- 534 — Small sports car design for Volkswagen

Type 530 prototype.

- ▶▶ 535 to 538 — One- through four-cylinder diesel tractor designs for Allgaier
- ▶▶ 540 — 356 America roadster
- ▶▶ 542 — Studebaker engine designs
- ▶▶ 547 — Four-cam, Fuhrmann-designed racing engine for the 550; note sub-Types such as 547/3, 547/4
- ▶▶ 550 — Two-seat, spyder sports racing car
- ▶▶ 597 — Off-road, jeep-style 'hunting' vehicle (*Jagdwagen*) for West German military
- ▶▶ 611 — Synchromesh transmission for Ferrari Grand Prix car

(Numerous other Type numbers from this period refer to synchromesh transmission designs for racing cars, passenger cars and commercial vehicles for a range of manufacturers.)

- ▶▶ 616 — 1.6-liter engine for the 356
- ▶▶ 621 to 624 — Additional Type numbers for Allgaier-manufactured tractors
- ▶▶ 647 — Synchromesh transmission for Ferrari 250 GT
- ▶▶ 633 — Studebaker sedan design
- ▶▶ 695 — Design for 356 successor
- ▶▶ 700 — Large-volume car study for Volkswagen
- ▶▶ 702 — One-man helicopter for Gyrodyne
- ▶▶ 718 — Spyder RSK racing car, sub-types for center-seat and Formula Two cars
- ▶▶ 745 — 2.0-liter, six-cylinder engine design led to 821
- ▶▶ 753 — Eight-cylinder engine for Formula One
- ▶▶ 754 — Sports car design (T9), partial forerunner to the 911

718 GTR coupe shown at Werk I in May 1962.

▶▶ 756	Abarth-Carrera GTL	
▶▶ 771	2.0-liter, eight-cylinder racing engine	
▶▶ 804	Eight-cylinder Formula One racing car	
▶▶ 821	2.0-liter, six-cylinder engine for the 911 (from Type 745)	
▶▶ 823 to 900	unused	
▶▶ 901	Original Type number for the 911 (1963)	
▶▶ 902	Type number for the 912, four-cylinder version of the 911	
▶▶ 904	1964 *Carrera GTS*, mid-engine, fiberglass semi-monocoque racing or street car	
▶▶ 905	Sportomatic transmission designs	
▶▶ 906	1966 *Carrera 6*, mid-engine, tubular spaceframe, plastic body sports racing car	
▶▶ 907	1967 and 1968 sports racing car, six or eight-cylinder, right-hand drive	
▶▶ 908	1968 eight-cylinder sports racing car, short or long tail coupe, or spyder (01)	
▶▶ 908/02	1969 spyder version of the 908, 'Flounder' body	
▶▶ 908/03	1970 and 1971 spyder version of the 908	
▶▶ 909	1968 ultra-lightweight hill climb racing car	
▶▶ 910	1967 development of the 906, six and eight-cylinder racing and hill climb cars	
▶▶ 911	Model name also used for numerous sub-Types such as engines and transmissions	
▶▶ 912	12-cylinder engine for the 917 racing cars	
▶▶ 914	VW/Porsche mid-engine sports car	
▶▶ 915	Four-seat 911 study; number also used for a series of 911 transmissions	
▶▶ 916	First used for twin-cam 911 six-cylinder engine. Also used for five-speed transmission for 908, and then for the fixed-roof version of the 914/6.	
▶▶ 917	12-cylinder racing car, 1969 to 1971, K (short) or LH (long) tail coupe, or spyder	
▶▶ 917/00	& 917/01 16-cylinder engines for the 917 (not raced)	
▶▶ 917/10	1971/1972 Can-Am style 917 spyder, turbocharging introduced for 1972	
▶▶ 917/20	1971 Le Mans special designed by SERA, aka 'Pink Pig'	
▶▶ 917/30	1973 turbocharged 917 spyder	
▶▶ 918	Study for mid-engine, 908-based sports car, number repurposed for 2010s hybrid supercar	
▶▶ 919	PDK prototype transmission, number repurposed for hybrid endurance racer 2014 through 2017	

▶▶	920	Four and five-speed transmissions for the 917	▶▶ 924	Sports car design for VW (EA 425) eventually built and sold as Porsche's entry-level car
▶▶	921	Design study for turbine-powered 4wd racing car concept similar to the 917 in layout and appearance	▶▶ 925	Various versions of the Sportomatic transmission
▶▶	922	Four-valve engine design for 917, number also used for three-speed automatic for the 928	▶▶ 926	Study for off-road vehicle (also initial designation for the 936)
▶▶	923	Engines and transmissions for the 912, 912E and 916	▶▶ 927	Reserved for four-valve 917 engine

Hurley Haywood in his Brumos 917/10 at Rennsport Reunion VI. (Dean Holbrook)

▶▶ 928	Design for eight-cylinder 911 engine, also used for 908 engine with water-cooled, four-valve cylinder heads, number repurposed for production front-engine V8 sports car
▶▶ 929	911 turbo study
▶▶ 930	911 with turbocharged engine, dozens of sub-types for engines and transmissions
▶▶ 931	924 Turbo
▶▶ 932	924 Turbo with right-hand steering
▶▶ 933	Study for SCCA 924 racer, number also used for 3.2-liter aircraft engine
▶▶ 934	Group 4 racing version of the 930
▶▶ 935	Group 5 racing version of the 930 with numerous sub-types for racing engines and iterations of the 935 racing cars
▶▶ 936	Group 6 sports and endurance racer, 1976 through 1981
▶▶ 937	924 Carrera GT
▶▶ 938	924 Carrera GT with right-hand steering
▶▶ 939	924 Carrera GTP Le Mans entry
▶▶ 940	Interscope/Parnelli Indianapolis racing car
▶▶ 941	924 Targa study
▶▶ 942	924 Targa study with right-hand steering
▶▶ 943	Tiptronic transmission for 964

Factory Porsche 956 team at Le Mans scrutineering, 1982.

▶▶ 944	Production sports car, update to the 924 with Porsche-designed engine
▶▶ 945	944 with right-hand steering
▶▶ 946	924S production car with 944 engine
▶▶ 947	924S with right-hand steering
▶▶ 948	Study for alloy-body 928, number also used for Cayenne V8 engine
▶▶ 949	924 GTP engine and transaxle with 944-based block for 1981 Le Mans race
▶▶ 950	Getrag-built G50 gearbox for 1987 911s replacing the Type 915
▶▶ 951	944 Turbo left-hand drive
▶▶ 952	944 Turbo right-hand drive
▶▶ 953	911-based desert racer for the Paris-Dakar Rally
▶▶ 954	911 SC/RS, lightweight, 290 horsepower homologation car for Group B
▶▶ 955	Used for Cayenne model line
▶▶ 956	Monocoque, ground-effect sports and endurance racing car first appearing in 1982
▶▶ 957	944 Turbo GT study (957 and 958 also used for later generation Cayennes)
▶▶ 958	944 Turbo GT study with right-hand steering
▶▶ 959	911-based supercar originally intended for Group B in the 1980s
▶▶ 960	Experimental car based on the 928, also used for 928 four-speed automatic transmission
▶▶ 961	Circuit racing version of the 959
▶▶ 962	Longer wheelbase version of the 956 sports and endurance racing car first appearing in 1984
▶▶ 963	Five-speed version of 911 Sportomatic transmission, number repurposed for current LMPh endurance racer
▶▶ 964	Extensively redesigned 911 for 1989 through 1994 model years
▶▶ 965	High-performance 911 potentially with a V8 engine (based on the 959)
▶▶ 966	Development of 911 Speedster
▶▶ 968	Final version of the 944, model years 1992 through 1995

Porsche RS Spyder, Type 9R6.

▶▶	969	Reserved as marketing designation for 965 (never entered production)	▶▶ 995	Study for 1990s-era future 928
▶▶	970	Panamera four-door sedan	▶▶ 996	911 intended for 1996 production, model years 1999 to 2004
▶▶	971	Second-generation Panamera	▶▶ 997	911 starting with 2005 model year
▶▶	980	LM 2000 racing spyder, program canceled before the car ever raced, V10 engine adapted for the Carrera GT supercar, number repurposed for Carrera GT	▶▶ 1866	Design for VW Beetle successor with under-floor, mid-engine design (VW project EA266)
▶▶	981	Third-generation Boxster and Cayman starting in model year 2013	▶▶ 1983	Racing transaxle for Le Mans-winning Matra MS670 series (1972-1974)
▶▶	982	Fourth-generation Boxster and Cayman starting in model year 2017 using 718 marketing designation	▶▶ 1906	Leopard Tank (1970)
			▶▶ 2564	Air-cooled V engines for Harley-Davidson
			▶▶ 2590	Study for electric vehicle (1980)
▶▶	984	Study for Porsche 'junior' sports car (mid-1980s)	▶▶ 2602	Aircraft cockpit for Airbus (1980s)
			▶▶ 2612	PDK transmission for the 962C
▶▶	986	First production version of the Boxster	▶▶ 2623	F1 V8 engine for TAG-McLaren
▶▶	987	Second-generation Boxster, first-generation Cayman	▶▶ 2708	CART/IndyCar design (9M0 engine)
			▶▶ 2800	Study for 1990s WSC racing car using Type 3512 engine
▶▶	989	V8, four-door sedan design from 1989 (never entered production)	▶▶ 2889	Study for 1990s China family car (C88)
▶▶	991	911 starting with 2012 model year	▶▶ 3512	V12 race engine for Footwork/Arrows F1 (aka 9M1)
▶▶	992	911 starting with 2020 model year		
▶▶	993	911 intended for 1993 production, model years 1995 to 1998		
▶▶	994	Development for 964 Turbo		

Note that there are other small gaps in the early Type number sequence beyond the larger gaps listed above. Wikipedia has an easy-to-access listing of Porsche Type numbers.

Appendix 2 *Porsche in the Movies*

1960s

Harper Paul Newman's portrayal of a hard-boiled, Bogart-style California private detective includes a battered 356 Speedster as the protagonist's car. It is an early example of the California-cool ethos of air-cooled Porsches.

Bullitt Steve McQueen famously played San Francisco police lieutenant Frank Bullitt, solidifying his reputation as the 'king of cool'. The film features a seminal large-scale, long-form car chase. Bullitt's girlfriend is played by the equally attractive Jacqueline Bisset and her character's 356C Cabriolet plays a role in an important scene toward the end of the film. It is further evidence of California-cool Porsche style.

Downhill Racer Not unlike *Bullitt*, *Downhill Racer* makes use of a Porsche owned by the protagonist's girlfriend. Robert Redford challenges Steve McQueen in the charisma department, playing a racer on the US Ski Team. His love interest, played with authentic European style by Camilla Sparv, drives a Bahama Yellow 911T Sportomatic. The scene featuring Redford driving the 911 makes for a very effective Porsche advertisement and advances the relationship between the characters. Redford became known as a Porsche enthusiast and owned a green 904 among others.

THIS PAGE: Steve McQueen and Jacqueline Bisset in *Bullitt*. (Alamy)

1970s

Le Mans The ultimate expression of Porsche in cinema, *Le Mans* became a cult favorite and beloved by racing fans for capturing the essence of a classic period in sports car racing. It is a must-see for its place in Porsche history and beautiful, documentary-style cinematography. As fate would have it, the filming coincided with Porsche's first overall win at Le Mans.

Actor and racer Steve McQueen hoped to drive a 917 with Jackie Stewart at Le Mans in 1970 as part of filming *Le Mans*. He prepared by winning two SCCA regional events and finishing second with Peter Revson at the 12 Hours of Sebring in a 908/02 'Flounder'. McQueen drove with a broken foot resulting from an accident in a motorcycle race. A minor controversy occurred when Porsche produced a victory poster using McQueen's likeness without his permission. The posters were mostly withdrawn but those that remain are highly valuable to collectors. Conversely, Gulf Oil printed some eight million of their *Le Mans* movie posters to be given away for free at Gulf gasoline stations.

The Gumball Rally Numerous cool cars appear in this, one of two 1976 films based on the Cannonball Baker Sea-To-Shining-Sea Memorial Trophy Dash. The Porsche 911E Targa in the film is driven by two sisters, portrayed by Susan Flannery and Joanne Nail. The Porsche finishes third in the illegal cross-country race.

ABOVE: Gulf gasoline station poster for *Le Mans*. (Alamy)

LEFT: Susan Flannery and Joanne Nail in *The Gumball Rally*. (Alamy)

1980s

Arthur An SCCA D-Production Porsche 924 makes a brief cameo appearance with Dudley Moore and Sir John Gielgud in this classic comedy. The scene was filmed in Connecticut at the old Danbury Fair Racearena (in the author's hometown).

Condorman In this spoofy spy movie, a group of KGB assassins is portrayed driving black Porsche 911s. The lead assassin drives a 935-looking, slant nose 911.

48 Hours A dusty Porsche Speedster owned by Eddie Murphy's character is portrayed as the hiding place for stolen money in the 1982 film that launched the actor to movie stardom.

Risky Business Arguably the film that put Tom Cruise on the road to stardom, *Risky Business* includes a Porsche 928 prominently in the storyline. The movie car sold at auction in 2021 for $1.98 million. The film displays an awareness of Porsche culture including Cruise's delivery of the advertising slogan: '*Porsche, there is no substitute.*' In case the reader has not seen the film, the author won't spoil the funniest line ever delivered by a Porsche service advisor.

Tom Cruise and the 928 in *Risky Business*. (Alamy)

Scarface Al Pacino's ambitious and violent Cuban gangster character is shown purchasing a Porsche 928 as a symbol of his drug-dealing success on the way up.

Against All Odds In this 1984 noir-style LA story, Jeff Bridges' character is a football player who drives a 911 SC Cabriolet. The car participates in a street racing scene with its natural competitor of the era, a Ferrari 308 GTS QV.

16 Candles In John Hughes' debut as the master director of 1980s teen comedy-drama, the object of Molly Ringwald's character's affections drives a Porsche 944.

Top Gun Once again, an open Porsche (Speedster) plays a role as the girlfriend's car. Kelly McGillis did the honors as the love interest to Tom Cruise's character as the film rocketed Cruise to superstar status.

No Man's Land Charlie Sheen plays a dedicated Porsche thief in this 1987 crime drama. The author would classify it as a standard 1980s 'B' movie, but with lots of Porsches. The film was written by Dick Wolf of *Law & Order* fame.

1990s

True Romance (1993) A white 1988 911 Turbo Flachbau Cabriolet is driven by Bronson Pinchot's Elliot Blitzer character, in this film written by Quentin Tarantino early in his career. Although not a success at the box office, it was acclaimed by the critics and features an all-star cast. The Porsche scene involves questionable and stereotypical behavior for the era.

Bad Boys In this 1995 buddy-cop vehicle for Will Smith and Martin Lawrence, the stars drive a 1994 911 Turbo (the movie car was owned by the director, Michael Bay).

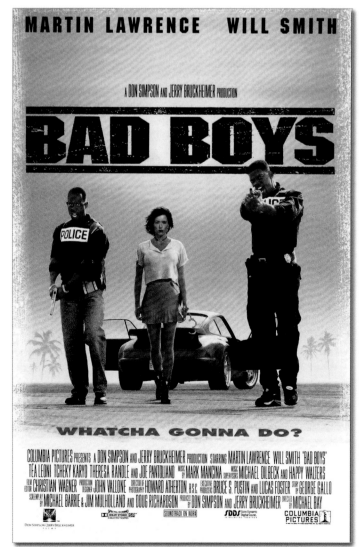

The English Patient Austrian László Almásy, was a World War I pilot who later worked for Steyr and did some car racing. He also became fascinated with the North African deserts which he explored extensively. After considerable intrigue during and after World War II, he returned to Egypt and worked at several jobs including the sale of European cars in Cairo. He sold three early Porsches there including a 356/2 Beutler Cabriolet, sold to a nephew of King Farouk. Almásy knew Prince Abdel Moneim from his earlier exploration trips to Egypt in the 1930s. *The English Patient* is a fictionalized story loosely based on Almásy's life. Author Karl Ludvigsen presents an excellent sidebar on the Almásy story in his book *Origin of the Species*.

Office Space The 911 takes on a negative connotation (in its reserved parking spot) as owned by the corporate vice president character, Bill Lumbergh, in Mike Judge's classic 1990s comedy-satire on the world of work.

2000s

Gone In 60 Seconds The window-smashing 996-era 911 was actually a lightweight 911 SC modified to look like a 996. The lighter-weight vehicle was needed to perform the stunt safely.

Spy Game Porsche enthusiast Robert Redford starred in this film and his CIA agent character is portrayed driving a vintage Irish Green 912 (the film is set in 1991).

Cars In Pixar's 2006 animated film, the Sally Carrera character was based on a 996-generation 911. Actor Bonnie Hunt provided Sally's voice.

996-based Sally Carrera from *Cars*. (Alamy)

2010s

Red 2 In another CIA-related action film, Bruce Willis' character is portrayed driving a 2012 997 GTS (although a Lotus Exige steals the spotlight in a big chase scene).

Atomic Blonde In this 2017 release, a period, cold war action story, the MI6 agent portrayed by actor James McAvoy is shown driving a 964-era 911 (with incorrect wheels for the period in which the story is set).

2020s

Top Gun, Maverick Vintage Porsches are still cool at the movies and Tom Cruise's character still goes for the ladies who drive them. In the sequel to the 1986 original, Jennifer Connolly's character is portrayed as owning and driving a silver 1973 911S.

Transformers: Rise of the Beasts For the 2023 *Transformers* installment, a 964-era 911 RS 'transforms' into the character known as Mirage. Given the rarity of real 964 3.8 Carrera RSs, five replicas were built to be used in filming.

Transformers display at Rennsport Reunion VII. (Dean Holbrook)

Bibliography and References

Books

Ian Bamsey – *Auto Union V16 Supercharged – A Technical Appraisal* (Foulis/Haynes, 1990)

Author's Note: This book is an excellent, in-depth summary of the technical details and development of the Auto Union Grand Prix cars through 1937. There are also brief but valuable summaries of the individual competition entries. Recommended for those who desire an increased understanding of the 1934 to 1937 period and the cars themselves.

Jürgen Barth & Gustav Busing – *The Porsche Book, The Complete History of Types and Models* (David Bull Publishing, 2009)

Author's Note: An excellent resource for in-depth research and model-specific information. Enjoyable for details such as the description of all the 928 prototypes and how they were used.

Neal Bascomb – *Faster* (Houghton Mifflin Harcourt, 2020)

Author's Note: A highly readable story book on the 1930s Grand Prix and sports car racing scene, featuring the partnership between Lucy Schell and René Dreyfus.

Mitch Bishop & Mark Raffauf – *IMSA 1969-1989* (Octane Press, 2019)

Colin Burnham & Paul Jeffries – *Porsche 356, Rear Engine Jewel* (Osprey, 1993)

Markus Caspers – *Designing Motion* (Birkhäuser, 2016)

RM Clarke (compilation) – *Le Mans, The Porsche & Jaguar Years, 1983-1991* (Brooklands Books, Ltd.)

David De Jong – *Nazi Billionaires* (HarperCollins, 2022)

Author's Note – This is a work of narrative nonfiction that details several industrial families and their involvement with the Nazi regime and German war production in World War II.

Mark Donohue with Paul Van Valkenburgh – *The Unfair Advantage* (Bentley Publishers, 1975)

Vic Elford – *Reflections on a Golden Era in Motorsports* (David Bull Publishing, 2006)

Richard von Frankenberg – *Porsche: the Man and his Cars* (Robert Bentley, 1969 version with translation by Charles Meisl)

Author's note: Von Frankenberg was the first editor of Christophorus magazine and knew members of the Porsche family well. He also had numerous successes as a driver, racing and record-setting in Porsches during the 1950s. The book is highly readable even as a translation from German. It is often the small anecdotes that are most entertaining.

Paul Frère – *The Racing Porsches* (Arco, 1973)

Paul Frère – *Porsche Racing Cars of the 70s* (Arco, 1980)

Author's Note – These two volumes do have some overlap, however, when viewed together they form a trove of highly technical information on the 'plastic' Porsche era, from the 904 through to the 936 and including the 911-based 934 and 935.

Paul Frère – *Porsche 911 Story* (Patrick Stephens Ltd., 1997)

Jay Gillotti – *Gulf 917* (Dalton Watson Fine Books, 2018)

Hurley Haywood & Sean Cridland – *Hurley From The Beginning* (Visions of Power Press, 2018)

Andrea Hiott – *Thinking Small – The Long, Strange Trip of the Volkswagen Beetle* (Ballantine Books, 2012)

Chris Jones – *Road Race* (David McKay Company, 1977)

Jung, Frank et al – *Porsche 75 – Driven By Dreams*, Edition Porsche Museum published by Porsche AG and Delius Klasing & Co., 2023)

Brian Long – *Porsche 914 914-6* (Veloce Publishing, 1997)

Author's note: A good general overview of the 914 story and reference for year-to-year changes, options and technical details.

Randy Leffingwell – *Porsche 75th Anniversary – Expect the Unexpected* (Motorbooks/Quarto Publishing, 2022)

Karl Ludvigsen – *Battle for the Beetle* (Bentley Publishers, 2000)

Karl Ludvigsen – *Excellence Was Expected* (Bentley Publishers, 2019 Revised Edition)

Author's Note – Revised several times over a forty year period, this is Ludvigsen's seminal work on the history of Porsche from 1948 to 2020. Considered the bible for post-war Porsche history, the final edition is four hefty volumes and over 2,400 pages. A must-have for serious Porsche historians and fans. Still available new from Bentley Publishers.

Karl Ludvigsen – *Ferdinand Porsche, Genesis of Genius* (Bentley Publishers, 2007)

> *Author's Note: This is a superb Ludvigsen deep dive into Ferdinand Porsche's early life, automotive and aviation work up to 1933. The book includes such gems as Ferdinand Porsche's horoscope and translations of some of his writing. Notable for cooperation from Ernst Piëch, eldest son of Anton Piëch and Louise Porsche, and an enthusiastic curator of his grandfather's legacy. Still available new from Bentley Publishers.*

Karl Ludvigsen – *The Mercedes-Benz Racing Cars* (Bond/Parkhurst Books, 1971)

Karl Ludvigsen – *Porsche, Origin of the Species* (Bentley Publishers, 2012)

Karl Ludvigsen – *Professor Porsche's Wars* (Pen & Sword Books, 2014)

> *Author's note: Highly recommended for anyone with an interest in Ferdinand Porsche's career. This book covers military projects in great detail but also addresses connections to automobiles and racing where appropriate. The book is comprehensive and thoroughly illustrated in the manner of Ludvigsen's other Porsche-related books.*

Laurence Meredith – *Original Porsche 356* (Bay View Books Ltd., 1995)

Laurence Meredith – *Ferdinand Porsche and the Legacy of Genius* (Bank House Books, 2010)

Christian Moity – *The Le Mans 24-Hour Race, 1949-1973* (Edita SA/Chilton)

Walter Näher – *Porsche 917, Archive and Works Catalog* (Delius Klasing, Edition Porsche Museum, 2014)

W. Robert Nitske – *The Amazing Porsche and Volkswagen Story* (Comet Press Books, 1958)

> *Author's Note: An interesting early attempt at the history of Porsche and Volkswagen. Some well-known anecdotes are present in the story even then. Note that some facts or interpretations may have changed over the decades. It is notable that the author studiously avoided using the word 'Nazi' at any point in the book.*

J.J. O'Malley – *Daytona 24 Hours* (David Bull Publishing, 2009)

Bill Oursler – *Porsche Prototype Era, 1964-1973 in Photographs* (David Bull Publishing, 2005)

> *Author's Note: This book provides a visual tour of the 'plastic Porsche' era from 1964 to 1973. There are many excellent and rare photos. The author presents a good, basic overview of the development sequence from the 904 to the 906 and 910.*

David Owen – *Targa Florio – Seventy Epic Years of Motor Racing* (Haynes, 1979)

Dominique Pascal – *Porsche Au Mans* (Editions Presse Audiovisuel, 1983)

Ferry Porsche with John Bentley – *We At Porsche* (Doubleday & Co., 1976)

> *Author's note: This autobiography has been criticized in recent years for its handling of the relationship between Porsche and Adolf Rosenberger. The Rosenberger story was adjusted slightly in the later book, Cars Are My Life. The books have also been shown to misrepresent the correct history of Ferry Porsche's SS membership in the 1930s. Porsche experts may find other detail inaccuracies in these period books. However, they remain valuable as statements directly from the father of the modern Porsche company.*

Ferry Porsche with Günther Molter – *Cars Are My Life* (Patrick Stephens Ltd., 1989)

Porsche Museum – *Ferry Porsche, 100 Years* (Edition Porsche Museum, 2009)

Porsche Museum – *Ferdinand Porsche, Hybrid Automobile Pioneer* (DuMont, 2010)

Lee Raskin – *James Dean on the Road to Salinas* (StanceAndSpeed.com, 2013)

Norbert Singer with Michael Cotton – *24:16* (Coterie Press Ltd., 2006)

> *Author's Note: This book presents a good overview of Norbert Singer's Porsche career with a focus on the Le Mans victories. It presents many interesting photos, technical details and anecdotes, including the story of Dr. Fuhrmann threatening not to serve dinner to his team or guests at his home unless he was satisfied with the results of a meeting!*

Norbert Singer with Wilfried Müller – *My Racing Life With Porsche* (Sportfahrer Verlag, Edition Porsche Museum, 2020)

> *Author's Note: An excellent expansion on 24:16 with more technical information and additional story details.*

Quentin Spurring – *Le Mans, 1970-79* (Haynes Publishing, 2011)

Paul Schilperoord – *The Extraordinary Life of Josef Ganz* (RVP Publishers, 2012)

Schneider/Slonigers – *Porsche Museum*, (catalog) published by Porsche AG, 1985)

> *Author's note: Beautiful look at the cars in the Porsche Museum collection as of the mid-1980s along with a significant amount of historical information and period photos.*

Peter Schutz – *The Driving Force* (Entrepreneur Press, 2003)

> *Author's Note: Highly recommended business book and autobiography, with a generous helping of insight and detail on Schutz' time as Porsche CEO (with a few minor historical inaccuracies). Especially useful for Schutz' explanation of the differences between the management board structures of German companies compared to US companies.*

Strache/Schrader/Slonigers – *100 Years of Porsche Mirrored In Contemporary History* (Porsche AG, 1975)

> *Author's Note: An interesting book published on the centennial of Dr. Porsche's birth. It features excellent historical photos combining Porsche history with world history events, year-by-year, from 1875 through 1975.*

James Taylor – *Supercharged Mercedes in Detail 1923-1942* (Herridge & Sons, 2013)

Alexander Ulmann – *Mercedes, Pioneer of an Industry* (Carroll Press, 1968)

Julius Weitmann – *Porsche Story* (Arco Publishing, 1968 with English translation by Charles Meisl.

> *Author's Note: This is an early overview of Porsche's motor sport history covering the early 1950s to 1968.*

Peter Zimmermann – *The Used 911 Story* (PMZ Publications, 1995

Automobile Quarterly

Volume 3, Number 2 – "The SS Revived" by Mervyn Kaufman

Volume 3, Number 3 – "Marcus Mysterious" by Carl Wagner

Volume 5, Number 3 – "Mercedes, the Man Behind the Name" by Alec Ulmann

Volume 6, Number 3 – "Grand Prix Cars of the Thirties" by Karl Ludvigsen

Volume 7, Number 1 – "I Remember Rudi" by Alfred Neubauer

Volume 8, Number 1 – "The Auto Union Grand Prix Car" by Karl Ludvigsen

> *Author's Note: This is a deep technical analysis of Auto Union Grand Prix cars through 1939 and includes a complete listing of competition results.*

Volume 8, Number 2 – "Cisitalia" by Stanley Nowak

Volume 9, Number 2 – "Porsche: The Man" and "The Cars" by David Owen

Volume 9, Number 2 – "The Origins of Supercharging" by Karl Ludvigsen

Volume 9, Number 2 – "Porsche: The Cars To Beat, John Wyer's Gulf-Sponsored 917s"

> *Author's Note: Automobile Quarterly 9/2 is highly recommend for anyone with an interest in Porsche history. In addition to the highly informative articles, the photos and Walter Gotschke paintings are wonderful to have in a library.*

Volume 10, Number 2 – "The Baron of Park Avenue" by Karl Ludvigsen

> *Author's Note: An in-depth article on Max Hoffman*

Volume 11, Number 4 – Series of six related articles on the Targa Florio by David Owen and Cullen Thomas

Volume 12, Number 2 – "Wilhelm and Karl: A Tale of Two Maybachs" by Carl Owen

Volume 16, Number 3 – "A Siren Song from Stuttgart – The History of the Supercharged Mercedes-Benz K and S" by Griffith Borgeson

> *Author's Note: This article is a very detailed examination of the "S" Mercedes series with significant and fascinating technical information including Borgeson's firsthand experience as an owner.*

Volume 16, Number 4 – "The All-Steel World of Edward Budd" by Stan Grayson

Volume 18, Number 4 – "In the Name of the People, Origins of the VW Beetle" by Griffith Borgeson

> *Author's Note: A very interesting and important article on the origins of the 'people's car' concept that includes images of notes written by Hitler during his first meeting with Dr. Porsche to discuss the small car project. It also covers background information on Rumpler, Ledwinka, Ganz and quotes from Rosenberger. Perhaps most interesting is a deep dive on the period political and contextual meaning behind the term 'volkswagen'.*

Volume 19, Number 2 – "The Remarkable Career of Robert Eberan von Eberhorst" by Jerry Sloniger

> *Author's Note: A superb overview of the career of an engineer closely associated with Auto Union as well as the Cisitalia and Porsche 356 projects. Written with direct interview material and quotes while Eberan-Eberhorst was still alive.*

Volume 19, Number 4 – "Mark Donohue and the Porsche 917/30" by Burge Hulett

> *Author's Note – A good period overview of the then-recent Porsche Can-Am program, focused on Donohue's involvement. Also includes a detailed account of the closed course speed record attempts in 1975 and an excellent look at 917/30 Chassis 005 which was then newly-built by Porsche for Gerry Sutterfield.*

Magazine & Newspaper Articles

Autoweek, January 2010 – "Action Express Pulls off a Stunning Win at the Rolex 24 Hours of Daytona" by Steven Cole Smith

Christophorus 1959 356B Special Edition

Christophorus #396 – "The Mind Is Free" by Thomas Ammann

Christophorus #398 – "How It All Began" by Manfred Schweigert

Christophorus #400 – "Between Heaven and Earth" by Donald von Frankenberg

Christophorus #401 – "Crypto Art" by Heike Hientzsch

Christophorus Editorial 2018 – "Is Porsche A She?" by Josef Arweck

Classic Porsche #82 – "Beauty in Speed" by Karl Ludvigsen

Classic Porsche #82 – "War Chest" by Shane O'Donoghue

Classic Porsche #82 – "Four Play" by Karl Ludvigsen

Classic Porsche #97 – Active at the Creation (Ernst Piëch) by Karl Ludvigsen

Classic Porsche #97 – "Those Pesky Poopers" by Karl Ludvigsen

Classic Porsche #97 – "Crest of a Wave" by Karl Ludvigsen

Excellence #211 – "Helmuth Bott" by Kieron Fennelly

Excellence #229 – "The Private Life of 356-001" by Karl Ludvigsen

Excellence #248 – "Profile: Harald Wagner" by Dr. Susanne Roeder

Excellence #271 – "The Sascha, Ferdinand Porsche's Early Open-Wheeler"

Excellence #271 – "Porsches, Argentine Style" by Karl Ludvigsen

Excellence #279 – "Porsche's Volkstraktor" by Karl Ludvigsen

Motor Sport, March 1928 – "Another Grand Prix Abandoned"

Motor Sport, July 1959 – "43rd Targa Florio" by Denis Jenkinson

Motor Sport, May 2018 – "Star & Stripes" by Andrew Frankel

Motor Trend, December 2022 – "RS Evolution" by Aaron Gold

Octane, November 2019 – "Recaro Seats" by Delwyn Mallett

Octane, November 2019 – "Ferdinand Piëch, 1937-2019" by Delwyn Mallett

Octane, July 2022 – "The Future of Fuel" by Andrew English

Panorama #749, August 2019 – "Ground Zero (Type 64)" by Johan Dillen

Panorama #761, January 2021 – "550 spyder, Prototypical Porsche Racer" by Randy Leffingwell

Panorama #780, March 2022 – "Porsche Makes Bicycles?" by Jim Hemig

Panorama #782, May 2022 – "The Colors of Summer" by David Mathews

Panorama #782, May 2022 – "Take Two" by Johan Dillen

Panorama #783, June 2022 – "A Hit By Any Measure" by Bruce Sweetman

Panorama #788, November 2022 – "Street Talk" by Bob Rassa and Doug Lloyd

PCA/PNWR Spiel, August 2019 – "914 50th Anniversary" by Allan Caldwell

PCA/PNWR Spiel, October 2021 – "Porsche Electronic Driving Aids" by Allan Caldwell

Road & Track, March/April 2020 – "Weather Helm (America Roadster)" by Zach Bowman

Road & Track, 2021 Vol. 3 - "King of America" (959SC) by John Pearley Huffman

Sports Car Market, March 2021, "1993 Porsche 968 Club Sport" by Prescott Kelly

Sports Car Market, July 2021, "2016 Porsche 911 GT3 RS" by Prescott Kelly

Sports Car Market, March 2022, "1987 Porsche 959 Komfort" by Prescott Kelly

Sports Car Market, April 2022, "1989 Porsche 944 Turbo" by Prescott Kelly

Sports Car Market, May 2022, "2004 Porsche Carrera GT" by Prescott Kelly

Total 911 #175 – "Rolf Sprenger" by Kieron Fennelly

Total 911 #182 – "The 20 Biggest 911 Developments"

Total 911 #184 – "Remembering Ferdinand Piëch" by Kieron Fennelly

Total 911 #185 – "The Weissach Axle" by Kieron Fennelly

Total 911 #185 – "The Genius of Ferdinand Porsche" by Jack Williams

Total 911 #186 – "The Story of Max Hoffman" by Jack Williams

Total 911 #186 – "Manfred Jantke" by Kieron Fennelly

Total 911 #186 – "911 Hero Tony Lapine" by Chris Randall

Total 911 #189 – "Tackling Africa" by Chris Randall

Total 911 #189 – "911 Hero Huschke von Hanstein" by Jack Williams

Total 911 #214 – "The History of PDK" by Ben Barry

Total 911 #214 – "Valentin Schäffer" Interview by Glen Smale

Total 911 #215 – "Sportomatic – An Automatic Choice?" by Kyle Fortune

Total 911 #215 – "The Big Interview: Grant Larson"

Total 911 #215 – "Story of the 964 RS America" by Kieron Fennelly

Total 911 #216 – "Carrera from the Start" by Glen Smale

Total 911 #219 – "Carrera Cup USA" by Kieron Fennelly

Total 911 #220 – Louise Piëch – The Porsche Powerhouse– by Kieron Fennelly

Wall Street Journal 3/29/19 – "The Real Reason Porsche Ignitions Are Left of the Wheel" by Dan Neal

Internet Articles & References

www.1000sel.com – History on 1980s customizers of Porsches and other cars

Butzi Squared – "A Short History of the Fuchs Wheels" by Holger Wilcks, 8/21/13

CNBC.com – "Porsche Ups its Investment in eFuels…" by Michael Wayland, 4/6/22

Collier Auto Media – "The Story of the Nürburgring" by Alexander Davidas

Collier Auto Media – "The Rennmezger – Excellence Was Achieved" by Karl Ludvigsen

Collier Auto Media – "Mercedes-Benz, An Appreciation" by Philip Richter

Collier Auto Media – "Porsche 935 - The Old Warhorse" by Wouter Melissen

Collier Auto Media – "Porsche TWR WSC95" by Wouter Melissen

Collier Auto Media – "Porsche's Serious Sports Car for Cisitalia" by Karl Ludvigsen

Ferdinand Magazine – "Butzi Porsche Helped Design the Cayenne" by John Glynn

FlatSixes.com – "The First Porsche at Indianapolis" by Bradley Brownell

Hagerty.com – "The Haunting Story of James Dean's Little Bastard" by Nik Berg

Hagerty.com – "Janis Joplin's 1964 Porsche 356C" by Jeff Peek

Hagerty.com – "Galvanization Sensation" by Benjamin Hunting

Hagerty.com – "C88 – Porsche's Ill-Fated Chinese Eco-Box" by Ronan Glon

Hagerty.com – "Legends of Motorsport: Bernd Rosemeyer" by Don Sherman

LoveForPorsche.com – "Ferdinand Alexander 'Butzi' Porsche" by Tim Havermans

MotorAuthority.com – "Stillborn Supercar: 1934 Auto Union Type 52" by Thomas Bey

PCA.org – "Model Guide, 356: The Simple Porsche" by David Seeland – 6/27/2017

PCA.org – "Model Guide, The First Porsche 911s" by Ed Mayo – 2/20/2018

PCA.org – "Model Guide, Type 993 – The Last Air-Cooled Porsche" by Tony Callas and Tom Prine 10/9/2018

PCA – Rennbow.org (Porsche Club of America exterior colors reference website)

PorscheAviation.com – Information about the Mooney program and numerous other aircraft with Porsche engines. See also: Porsche Fails to Outpace Suit Over Plane Engines | Courthouse News Service

Porsche.com – "Game Changers – The Remarkable Story of the Porsche 911 Carrera RS"

PorscheHolding.com – History of Porsche Konstruktionen (Porsche-Salzburg)

Porsche Newsroom – "The Porsche Code" – 12/21/2015

Porsche Newsroom – "R = Racing: The Historical Roots of the 911 R" – 3/2/2016

Porsche Newsroom – "The Philosophy of F.A. Porsche" – 3/5/2017

Porsche Newsroom – "The History of the Porsche Engineering Office – 10/10/2017

Porsche Newsroom – "Wolfgang Porsche" – 5/4/2018

Porsche Newsroom – "The Art of Speed" (Erich Strenger) – 7/23/2018

Porsche Newsroom – "Made In Austria" – 11/6/2019

Porsche Newsroom – "There Can Only Be Two" (993 Speedsters) – 12/28/2019

Porsche Newsroom – "Porsche Indy – Porsche's History at the Indianapolis Speedway" – 5/20/2020

Porsche Newsroom – "Porsche Mourns the Death of Hans Mezger" – 6/11/2020

Porsche Newsroom – "20 Year Anniversary of Porsche in Mainland China" – 4/29/2021

Porsche Newsroom – "Weissach Development Center: the Track of Legends" – 10/4/2021

Porsche Newsroom – "Porsche and UP.Labs Plan to Build Tailormade Start-Ups" – 6/7/2022

Porsche Newsroom – "Porsche and Rosenberger gGmbH Commission Joint Research Project" – 10/31/2022

Porsche Newsroom – "The Ducktail: 4.5 Kilometers Per Hour For All Eternity" 11/7/2022

Porsche Newsroom – "The History of the Austro-Daimler ADS-R" – 11/10/2022

Porsche Road and Race – "Ernst Fuhrmann – The Inspiration Behind the 911 Turbo" by Kieron Fennelly

Porsche Road and Race – "Weissach – Centre of Porsche Excellence" by Kieron Fennelly

Porsche Road and Race – "Porsche 956/962 Remembered" by Glen Smale

Porsche Road and Race – "The Sonderwunsch Man – Rolf Sprenger" by Kieron Fennelly

(Note: Porsche Road and Race content has transferred to Stuttcars.com)

RacingSportsCars.com (General reference for sports car race entries and results)

Rennlist – "What Makes Porsche's PDK Transmission So Great" by Brett Foote – 9/28/22

Sports Car Digest – "Porsche 550 Spyder: Giant Killer" by Djordje Sugaris

Sports Car Digest – "Discover The Incredible History of the GP Ice Race"

Sports Car Digest – "The Strange Tale of the Ital Design Porsche Tapiro" by Matt Stone

Stuttcars – "Porsche Type 64 (1939-1940)" by Nick Dellis

Stuttcars – "Peter Falk – Porsche's Enduring Engineer" by Kieron Fennelly

Stuttcars – "Arno Bohn – The Porsche Outsider" by Kieron Fennelly

Stuttcars – "Porsche 984 Concept Car (1984-1987)" by Nick Dellis

Stuttcars – "Porsche 928 – The Story" by Nick Dellis

Stuttcars – "Porsche C88 (1994)" by Nick Dellis

Stuttcars – "Porsche 9R3 'LMP 2000' (1999)" by Nick Dellis

Stuttcars – "From the Scrap Heap to Winning Le Mans: The TWR-Porsche WSC" by Martin Raffauf

Stuttcars – "Porsche's Most Enigmatic Model – The America Roadster" by Karl Ludvigsen

The Drive – "901 Cabriolet on the Market for the First Time" by Bradley Brownell, 11/15/16

The Drive – "Long Forgotten 911 HLS" by Peter Holderith, 5/14/2020

The Drive – "Porsche IPO Docs Reveal It's Still Fighting Lawsuit Over Who Designed The 911" by Stef Schrader, 2/21/22

The Truth About Cars – "Porsche's Forgotten Man, Adolf Rosenberger" by Ronnie Schreiber

https://www.thetruthaboutcars.com/2016/04/porsches-third-forgotten-man-adolf-rosenbergerdr-porsches-jewish-partner-part-one/

> Author's note: Highly recommended two-part article about Rosenberger's life, his role in founding the Porsche design office and the controversial resolution of his post-WWII restitution claims. It also includes the conflicting views on Porsche's involvement in getting Rosenberger released from a concentration camp in 1935.

www.type550.com – Andrew Hosking's website dedicated to 550 histories.

YouTube Video Suggestions

"Supercharged Grand Prix Cars 1924-1939"

"1937 Vanderbilt Cup Race" Newsreel

"Great Cars – Silver Arrows"

"Mercedes-Benz T80 'Blackbird'" – The Reich's Land Speed Record Challenger

"Burning Passion: The Story of Otto Mathé and His Fetzenflieger"

"Was the Porsche that Killed James Dean Cursed?"

"Porsche 356 – Made By Hand – Documentary"

"Mark Donohue Talladega Speed Record – Porsche 917/30"

"East African Safari Rally 1978"

"Mercedes-Benz C111-IV (Experimental 1979)

'Paris-Dakar Rally 1984' and 'Paris-Dakar Rally 1986'

"Porsche 911 Secrets: V8 Concept"

"Dr. Ferdinand Porsche Commercial, Time"

"Ferdinand Porsche: Driven to Perfection" (Biography series)

"Different Flat & Boxer Engine Configurations Explained"

"The 919 Tribute Tour: On-Board Record Lap, Nordschleife"

"RS – Fresh Brewed & Air Cooled Deep Tracks with Grant Larson/S2E1"

"Porsche Holding Corporate Film 2023" (Porsche-Salzburg history)

"Discover Mirage, the Porsche 911 from Transformers: Rise of the Beasts"

The Porsche Club of America YouTube channel provides dozens of informative videos covering technical topics and model-specific information: *Porsche Club of America - YouTube*

Suggested Wikipedia searches,
for more detail on these people and entities:

Emil Jellinek

Count Alexander Kolowrat

Camillo Castiglioni

Alfred Neubauer

Robert Ley

Fritz Todt

Albert Speer

Robert Eberan von Eberhorst

Heinrich Nordhoff

Piero Dusio

László Almásy

Porsche SE

Volkswagen AG

Volkswagen Emissions Scandal

Index

PORSCHE MODELS

356 39, 40, 42, 44, 60, 62, 75, 78, 79, 80, 81, 84, 85, 86, 87, 89, 90, 92, 93, 94, 95, 96, 97, 99, 102, 104, 105, 106, 107, 108, 109, 110, 114, 115, 116, 118, 119, 120, 122, 123, 124, 125, 126, 128, 129, 131, 142, 143, 144, 158, 168, 169, 178, 188, 199, 206, 277, 305, 307, 313, 324, 325, 326, 327, 332, 335

356/1 62, 78, 79

356/2 62, 79, 81, 335

356A 84, 108, 109, 116, 117

356B 84, 116, 117, 119, 123

356C 124, 131, 144, 332

356 America Roadster 93, 106, 2262

356 Carrera 109, 168

356 Carrera 2 123, 225, 226

356 Continental 59, 108

356 SL 87, 89, 326

356 Speedster 78, 106, 107, 277, 332

550 43, 84, 88, 99, 101, 102, 103, 104, 105, 106, 109, 111, 204, 242, 327

550A 84, 110, 111, 112, 113, 114, 270

645 111

718/2 115, 121

718 GTR 121, 327

718 RS60 120, 121, 242, 268, 270

718 RS61 120, 121

718 RSK 114, 115, 121, 242

804 121, 122, 131, 328

901 118, 124, 125, 126, 128, 129, 131, 133, 137, 138, 143, 154, 158, 197, 206, 215, 328

902 129, 147, 328

904 118, 126, 130, 131, 132, 133, 152, 270, 328, 332

906 131, 132, 133, 135, 136, 137, 139, 141, 146, 147, 179, 328

907 97, 137, 138, 139, 140, 259, 328

908 25, 133, 138, 139, 140, 141, 147, 150, 153, 154, 155, 157, 165, 180, 260, 263, 264, 272, 298, 310, 328, 330, 333

908/02 140, 157, 328, 333

908/03 133, 140, 150, 157, 165, 180, 260, 263, 264, 272, 328

909 133, 140, 157, 328

910 133, 136, 137, 138, 140, 328

911 31, 42, 47, 80, 102, 118, 119, 123, 124, 125, 126, 128, 129, 131, 133, 137, 138, 139, 141, 142, 143, 144, 145, 146, 147, 150, 151, 152, 153, 154, 157, 158, 159, 161, 162, 163, 164, 165, 166, 168, 169, 171, 172, 173, 176, 177, 178, 179, 183, 185, 186, 188, 189, 190, 191, 193, 195, 196, 197, 198, 199, 205, 207, 208, 209, 215, 218, 220, 221, 222, 224, 225, 226, 227, 228, 229, 232, 233, 234, 236, 237, 238, 239, 241, 243, 246, 247, 248, 249, 250, 251, 255, 261, 262, 263, 264, 265, 266, 270, 272, 274, 275, 276, 277, 279, 283, 284, 285, 286, 287, 288, 291, 292, 293, 296, 297, 298, 300, 305, 306, 307, 310, 311, 312, 313, 315, 317, 318, 320, 321, 322, 323, 324, 325, 327, 328, 330, 331, 332, 334, 335

911 3.2 Carrera 195, 196, 197, 199, 205, 226, 232

911 Dakar 161, **3**23

911E 333

911 GT1 222, 239, 246, 247, 248, 262

911 GT2 237, 263, 264, 277, 283, 284, 285, 310, 311, 318

911 GT3 8, 145, 158, 251, 255, 261, 262, 263. 264, 277, 283, 284, 285, 288, 296, 297, 298, 310, 311, 312, 320, 323

911 HLS 143

911R 145, 146, 310

911 RS 150, 154, 161, 164, 264, 277, 335

911 RSR 31, 163, 164, 165, 179, 311, 321

911 RSR Turbo 179

911S 128, 142, 144, 145, 153, 154, 165, 335

911 SC 150, 171, 188, 189, 195, 196, 197, 330, 334, 335

911 SC/RS 197, 330

911T 146, 152, 154, 332

911 Turbo 166, 168, 169, 171, 172, 188, 196, 197, 199, 209, 226, 227, 261, 262, 263, 276, 279, 283, 284, 293, 300, 311, 330, 334

912 124, 126, 129, 144, 147, 151, 159, 169, 328, 329, 335

912E 169, 329

914 131, 139, 144, 146, 150, 151, 152, 153, 154, 169, 177, 203, 242, 270, 328

914/6 152, 153, 270, 328

914/8 153

916 138, 153, 270, 328, 329

917 10, 56, 118, 141, 147, 148, 150, 151, 154, 155, 156, 157, 158, 159, 160, 161, 162, 164, 168, 172, 177, 179, 180, 195, 200, 201, 213, 215, 223, 264, 266, 272, 298, 300, 305, 306, 317, 318, 328, 329, 333

917/10 159, 160, 161, 179, 201, 215, 328, 329

917/20 156, 328

917/30 160, 161, 164, 201, 306, 328, 339, 342

918 139, 282, 283, 296, 297, 298, 299, 300, 301, 302, 303, 323, 328

919 33, 282, 283, 302, 303, 304, 305, 306, 310, 315, 317, 328

924 150, 168, 169, 177, 178, 179, 183, 184, 188, 194, 195, 198, 203, 204, 215, 227, 228, 257, 329, 330, 334

924 Carrera GT 194, 203, 330

924 Carrera GTR 194

924 Carrera GTS 194

924 GTP 203, 330

924S 190, 204, 330

924 Turbo 178, 179, 194, 203, 330

928 150, 168, 177, 179, 183, 184, 185, 186, 187, 188, 190, 193, 195, 203, 204, 215, 217, 218, 220, 222, 225, 227, 228, 229, 234, 251, 258, 309, 311, 324, 329, 330, 331, 334

928 GTS 228, 229

928S 215, 216, 217, 218

928 S4 204, 217, 218

930 and 911 Turbo 150, 166, 168, 171, 172, 188, 196, 197, 199, 204, 209, 226, 261, 262, 263, 276, 279, 284, 300, 334

934 150, 172, 195, 330

935 150, 165, 172, 173, 174, 176, 177, 180, 181, 182, 191, 195, 208, 212, 215, 221, 310, 330, 334

935/78 176, 182, 310

936 150, 172, 179, 180, 181, 183, 195, 200, 305, 329, 330

944 190, 191, 193, 194, 203, 204, 205, 207, 215, 226, 227, 228, 257, 300, 330, 334

944S 204

944 S2 204, 228

944 Turbo 190, 204, 205, 207, 330

953 208, 225, 330

956 33, 190, 191, 199, 200, 201, 202, 203, 204, 207, 208, 212, 215, 221, 306, 330

959 190, 206, 207, 208, 209, 211, 212, 225, 226, 233, 234, 237, 262, 264, 284, 297, 298, 313, 323, 330

959 SC 211

961 209, 330

962 161, 190, 191, 203, 212, 213, 215, 221, 233, 239, 262, 273, 330

963 114, 263, 321, 322, 323, 330

964 168, 199, 222, 224, 225, 226, 227, 232, 234, 236, 237, 258, 305, 330, 331, 335

965 209, 226, 234, 330, 331

968 207, 222, 227, 228, 242, 330

981 293, 296, 309, 331

987 270, 271, 272, 274, 276, 283, 285, 286, 287, 298, 331

989 190, 220, 234, 277, 331

991 265, 282, 291, 292, 293, 296, 297, 310, 312, 313, 331

992 282, 306, 311, 312, 313, 321, 323, 331

993 185, 199, 215, 220, 222, 227, 232, 233, 234, 236, 237, 238, 241, 242, 243, 247, 250, 251, 258, 262, 263, 264, 274, 331

996 220, 222, 232, 233, 236, 241, 242, 249, 250, 251, 253, 255, 256, 257, 258, 261, 262, 263, 264, 270, 274, 275, 284, 288, 331, 335

997 256, 270, 274, 275, 276, 277, 278, 283, 284, 285, 287, 291, 292, 293, 312, 331, 335

2708 218, 221, 331

Abarth-Carrera GTL 119, 328

Boxster 215, 222, 237, 241, 242, 243, 246, 249, 250, 251, 255, 256, 258, 263, 266, 267, 268, 270, 272, 274, 282, 283, 285, 286, 287, 293, 296, 298, 306, 307, 312, 320, 325, 331

C88 222, 239, 331

Carrera GT 194, 203, 256, 264, 265, 266, 272, 279, 285, 297, 298, 300, 323, 330, 331

Cayenne 131, 249, 256, 257, 258, 259, 261, 264, 265, 275, 277, 278, 279, 282, 283, 284, 286, 287, 288, 290, 291, 292, 297, 309, 317, 320, 323, 325, 330

Cayman 247, 249, 256, 267, 270, 271, 272, 274, 282, 283, 285, 286, 287, 293, 296, 298, 306, 307, 312, 317, 318, 320, 331

Cayman GT4 293, 296, 309, 318, 320

LM 2000 264, 265, 331

Macan 282, 283, 286, 288, 296, 297, 309, 320

Mission X 316, 323

Ollon-Villars Spyder 133

Panamera 139, 249, 256, 265, 277, 278, 279, 283, 287, 288, 290, 291, 293, 297, 309, 313, 315, 317, 320, 331

Panamericana 199, 234, 237

RS Spyder 256, 257, 272, 273, 274, 298, 302, 331

Taycan 18, 282, 283, 313, 315, 317, 318, 320, 325

WSC 95 Spyder 148, 238, 239, 240, 241, 247

NON-PORSCHE MODELS

Acura NSX 223

Alfa Romeo 12C 52

Alfa Romeo 33TT12 163

Alfa Romeo 8C 52

Alfa Romeo P2 31, 257

Amphicar 64

Audi 100 183

Audi 80 177, 232

Audi Avant RS2 Sportwagen 232

Audi Q5 297

Audi Quattro 207, 305

Audi R10 273, 274

Austro-Daimler 22/86 'Prince Henry' 22

Austro-Daimler 28/32 19

Austro-Daimler ADS-R 'Sascha' 25

Austro-Daimler Maja 19

Auto Union Grand Prix 38, 44, 45, 46, 49, 50, 53, 54, 74, 78, 326

Bentley Old Number 1 148

Benz Tropfenwagen 44

BMW 2002 Turbo 169

BMW 328 95

BMW 507 90, 125

BMW CSL 173

Bugatti Chiron 71, 318

Chaparral 137

Cisitalia 202 71, 78

Cisitalia Grand Prix 69, 71, 72, 74, 75, 102, 326

Cooper Mark VIIIR 106

Daimler DZVR 33

Daimler-Mercedes 35 HP 19

Egger-Lohner C2 14, 16

Elva-Porsche 131

Ferrari 206 SP 136

Ferrari 275 GTB 177

Ferrari 275P 148

Ferrari 288 GTO 207

Ferrari 312P 140

Ferrari 312PB 163

Ferrari 330 P3 136

Ferrari 360 Modena GT 263

Ferrari 365 Daytona 177

Ferrari 512S 155

Ferrari 860 Monza 112

Fetzenflieger 101

Ford GT40 138, 140, 141, 148

Ford RS200 207

Gulf-Mirage 163, 165

Jaguar C-Type 88
Jaguar XJR-14 148, 238, 241
Justicialista Sport 96
Kremer K8 239
La Ferrari 298
Lada Samara 193
Lancia 037 207
Lohner La Toujours Contente 16, 17
Lohner Model 27 16
Lohner Semper Vivus 16, 17
Lohner Type J 16, 17
Lohner Voiturette 16
Lola T380 180
Lola T70 138
Lola T92/10 247
March 90P 223
March/Porsche Indycar 190, 191, 218, 219, 221, 223, 227, 331
Marcus Strassenwagen 13
Maserati 300S 112
McLaren F1 59, 204, 237, 239, 247
McLaren M8D 159
McLaren MP4/2 204
McLaren P1 298
Mercedes 130 H 41
Mercedes 170 H 41
Mercedes 190 SL 90
Mercedes 300 SL 64, 90
Mercedes 300 SLR 99
Mercedes C-111 194
Mercedes SL 183
Mercedes T80 38, 58
Mercedes W125 Grand Prix 50
Mercedes-Benz 500 E 232
Mercedes-Benz 630K 24, 35
Mercedes-Benz 630S 33
Mercedes-Benz 680S 24, 35
Mercedes-Benz G-Class 37, 257
Mercedes-Benz SS 35, 36, 65
Mercedes-Benz SSK 36
Mercedes-Benz SSKL 36, 40
Mercedes-Electric 19
MG Metro 6R4 207
Miersch 356 97
Mini Cooper S 132
Nissan 300 ZX 223
Renault A442B 182
Rimac Nevera 318
Ruf CTR2 236

Sauber C7 201
Shadow 161
Steyr Austria 37
Steyr Type 20 37
Steyr Type 30 36, 37
Teram Puntero 96
Toyota Supra 223
Volkswagen 'Beetle' (KdF-Wagen) 38, 54, 57, 69, 326
Volkswagen 60K10 38, 59, 60, 61, 78, 80, 101, 137, 326
VW Golf/Rabbit 145, 177
VW Passat/Dasher 177
VW Scirocco 177, 178
VW Super Beetle 177
Wanderer W21/22 40
Zunder 1500 97

VEHICLES/AIRCRAFT

Austro-Daimler M17 23
Austro-Daimler Pferd 22, 23
Ferdinand Tank 66, 67
Kübelwagen 62, 63, 64, 65, 78, 97, 108, 326
Leopard Tank 118, 130, 331
Maus Tank 66, 67, 68, 75, 326
Mooney M20 198
Pinzgauer 37
Porsche Jagdwagen 108, 257, 327
Porsche Tractor Models 193
Porsche Wiesel 150, 177
Russian T-34 Tank 67
Schwimmwagen 62, 65, 326
Tiger Tank 66, 326

COMPANIES/ORGANIZATIONS/INSTITUTES

Action Express 284
Adler 42, 44
Airbus 193, 331
AiResearch 159
Aisin 275
Alfa Romeo 27, 30, 33, 42, 44, 53, 69, 75, 88, 90, 99, 130, 137, 138, 157, 163
Allgaier 108, 110, 327
Alpine-Renault 147
AMAG 75, 78, 81
AMC 125, 177
American Sunroof Company 204
Apple 304

Ardie 42
Arrows 32, 223, 224, 331
Art Center College of Design 298
Aston Martin 114
Audi 43, 86, 88, 108, 141, 143, 146, 152, 177, 179, 183, 198, 201, 202, 203, 207, 209, 220, 222, 225, 228, 232, 233, 241, 243, 250, 274, 277, 290, 293, 297, 302, 303, 304, 305, 309, 313, 315
Austro-Daimler 12, 18, 19, 20, 21, 22, 23, 24, 25, 26, 27, 30, 36, 37, 39, 41, 42, 54, 75, 320
Autoar 71, 96
Autocar 186
Auto Union 33, 38, 40, 42, 43, 44, 45, 46, 47, 49, 50, 51, 52, 53, 54, 57, 58, 59, 68, 69, 74, 78, 143, 213, 326
Ballot 30
B&B 221
Bentley 35, 88, 141, 148, 293, 309
Bertone 119
BMC 147
BMW 23, 50, 56, 90, 95, 99, 123, 125, 161, 169, 173, 201, 220, 227, 228, 246, 248, 250, 277, 284, 302
Boeing 154
BorgWarner 97, 198, 283
Brembo 243
Bridgestone 208
BRM 122
Brumos 120, 154, 163, 165, 227, 263, 272, 284, 329
Budd 51, 63
Bugatti 44, 64, 78, 88, 141, 293, 318
Burmester 279
CART 190, 191, 218, 221, 223, 331
Caterpillar 193
Cellforce 318
Cessna 198
Chaparral 137
Chevrolet 44, 96, 97, 159
Chip Ganassi Racing 284
Chrysler 13, 51, 68, 99, 193
Cincinnati Milling 51
Cisitalia 62, 69, 71, 72, 74, 75, 78, 96, 102, 326
Citroen 42
Continental 59, 108
Coventry-Climax 122
CTS 263, 275
Cummins 193
Custom Cells 318
DAF 56, 57
de Dion-Bouton 17
Delugan Meissl 281
DKW 43, 108
Dodge 125, 237

DP 221
Drauz 110
Dunlop 155, 163, 208
Elva 131
Emory Motorsport 305
Fabcar 272
Facel Vega 90
Ferrari 27, 52, 53, 88, 99, 112, 113, 114, 115, 116, 120, 121, 132, 136, 137, 140, 148, 155, 163, 165, 177, 184, 207, 221, 238, 239, 248, 263, 284, 297, 298, 306, 327, 334
FIA 88, 95, 104, 120, 121, 132, 133, 140, 147, 148, 156, 159, 163, 164, 172, 180, 199, 207, 213, 227, 233, 242, 247, 272, 302
Fisher Body 51
Fletcher 95
Flying Lizard Motorsports 284
Footwork 223, 224, 264, 331
Ford 51, 57, 88, 104, 108, 125, 136, 137, 138, 140, 144, 148, 161, 169, 193, 207, 298
Frua 153
Fuchs 142, 144, 152, 168, 224
GE 223
Gemballa 221
General Motors 168
Getrag 99, 228, 250, 275, 330
Ginzkey 13, 15
GKN Driveline 300
Glöckler 78, 99, 101, 105
Glyco 237
Goetze 102
Grant 241, 242, 243, 264, 270, 274, 277
Greyp 223
Group 14 318
Gulf 10, 137, 140, 141, 148, 151, 155, 157, 163, 165, 195, 333
Gunther Werks 221
Gyrodyne 95, 327
Hanomag 42
Harley-Davidson 263, 331
Henschel & Sohn 66
HIF Global 322
Hirth 92, 93, 101
Hispano-Suiza 30
Hofherr Schrantz 110
Holden 42
Honda 204, 273
Horch 33, 40, 43, 45, 86
Hueliez 153
IAME 96
IMSA 150, 154, 161, 164, 165, 172, 176, 194, 195, 212, 213, 215, 227, 239, 247, 273, 323
INRIX 290

Interscope 191, 330
Ital Design 152
Jaguar 88, 90, 148, 161, 213, 214, 238, 241
Judd 238
Junkers 63
JW Automotive Engineering 148
Kaizen Institute 238
Karmann 96, 123, 152, 296
Karosseriewerke Weinsberg 205
KdF 38, 54, 57, 60, 63, 69, 326
KHD 193
Koenig 221, 233
Kremer Brothers 176, 221
Kugelfischer 121
Lada 193
Lamborghini 141, 286, 309
Lancia 90, 147, 163, 164, 200, 207
Land Rover 257
Larbre Competition 226
Lee Maxted-Page Porsche 311
Lincoln 108
Lockheed 45, 54, 80, 86
Lohner 12, 14, 15, 16, 17, 18, 19, 22, 30, 86
Lola 122, 131, 138, 180, 221, 247
Lotus 121, 122, 133, 200, 335
Magna International 275
Mahle 86, 92, 116, 123, 154, 161
Mannesmann 110
Martini Racing 154
Martini & Rossi 160, 163
Matra 163, 165, 331
Maybach 19, 66
Mazda 207
Mazdaspeed 238
McLaren 59, 158, 159, 204, 237, 239, 247, 248, 298, 331
Megaline 272
Michelin 300, 306
Mid-American Research 95
Mitsubishi 203
Mod Works 198
Momo 137
Mooney 198
Morris 2, 11, 42
Motor-Kritik 42, 61
Motor Trend 279, 320
Multimatic 321
NASA 154
NASCAR 161
Nazi Party 58, 63

Nissan 99, 161, 207, 223, 248, 284
NSU 41, 42, 43, 143, 177, 326
ONS 60, 95
OPEC 165, 177, 183
Opel 19, 39, 64, 97, 125, 145, 168, 185, 241, 293
OSCA 112
Packard 96
Pebble Beach Concours 318
Peugeot 30, 69, 71, 124, 203
Pininfarina 60
Porsche Automobil Holding SE 158, 280
Porsche Club of America 10, 105, 154, 276, 325
Porsche-Salzburg 79, 96, 108, 155, 157, 280, 301
Racemark 161
RDA 43, 54, 55
Recaro 118, 123
Renault 69, 147, 180, 181, 182, 203
Reutter 41, 60, 81, 85, 86, 92, 97, 118, 123, 136, 207
Richard Lloyd Racing 202
Riley 272, 284
Rimac 311, 318
Rinspeed 221
Risi Competizioni 263
RM Sotheby's 60, 321
Rolls-Royce 241
Rotwild 223
Rudge Whitworth 45
Scania 86, 241
Scania Valbis 86
SCCA 121, 132, 144, 160, 161, 164, 172, 194, 227, 237, 330, 333, 334
SEAT 215, 241, 313
Shadow 161
Shin-Gijutsu 238
Singer 10, 150, 161, 164, 165, 172, 179, 199, 201, 206, 221, 226, 233, 239, 247, 256, 272, 305
Skoda 22, 39
Solex 101, 109, 126
Sonauto 92, 152
Standard Superior 42
Stark 242
Steyr 21, 24, 36, 37, 39, 42, 63, 78, 94, 123, 171, 335
Storck 223
Strosek 221
Studebaker 96, 97, 327
Talbot 35
Tata 95
Tatra 42
Teram 96
Teves 54, 124

Tongji University 313
Toyota 99, 161, 169, 223, 233, 247, 248, 302, 303, 304, 305
Valmet 246, 247, 272, 293
VEAG/Béla Egger 16, 17
Vespa 95
Vienna University of Technology 15
Volkswagen 38, 39, 42, 43, 44, 51, 54, 55, 56, 57, 58, 59, 60, 61, 62, 63, 65, 68, 69, 70, 71, 75, 78, 79, 80, 81, 82, 85, 86, 90, 92, 95, 96, 97, 99, 105, 108, 109, 113, 118, 123, 124, 125, 141, 143, 146, 148, 151, 152, 158, 177, 178, 183, 193, 197, 203, 204, 206, 241, 256, 257, 261, 263, 276, 280, 286, 290, 291, 293, 296, 297, 305, 311, 313, 318, 320, 322, 326, 327
Volvo 42, 203, 206, 227, 241
Votec 223
Wanderer 40, 41, 43, 326
Webasto 237
Weber 74, 75, 101, 102, 111, 133, 142
Weidenhausen 105
Wendler 105
Williams 204, 297
Xtrack 272
Yamaha 241, 311
Zagato 120
Zenith 109
ZF 47, 64, 128, 131, 185, 225, 258, 279, 292, 309
Zündapp 41, 42, 326

PEOPLE

Abarth, Carlo 69, 75, 119, 120, 130, 133
Achleitner, August 274, 293, 306, 312
Ahrens, Kurt 148
Aiello, Laurent 248
Akin, Bob 212
Alboreto, Michele 239
Almásy, László 335
Alzen, Uwe 248
Ampferer, Herbert 171, 177, 246, 264
Andretti, Mario 155
Andretti, Michael 212
Andrews, Major 69
Arkus-Duntov, Zora 102
Ascari, Alberto 30
Ascari, Antonio 30, 31
Attwood, Richard 148, 156
Baker, Wayne 195
Baldi, Mauro 233
Bamber, Earl 303, 305
Bamsey, Ian 52, 53
Bandini, Lorenzo 136

Bantle, Manfred 208
Barbosa, João 284
Barbour, Dick 195
Barényi, Béla 42
Barilla, Paulo 202
Barnard, John 204
Barth, Edgar 113, 115, 120, 130
Barth, Jürgen 181, 204
Bäumer, Walter 95
Bay, Michael 334
Beddor, Steve & David 236
Behra, Jean 113, 114, 115
Beinhorn, Ellie 52
Bell, Derek 156, 181, 195, 201, 202, 213, 214, 325
Bellof, Stefan 33, 176, 202, 213, 214, 306
Bergmeister, Jörg 263, 318, 323
Berkheim, Count 'Tin' 86
Bernhard, Timo 33, 263, 272, 303, 304, 305, 306
Bez, Ulrich 220, 223, 224, 226, 232, 236, 237, 241
Bianchi, Lucien 140
Binder, Robert 171
Bischoff, Klaus 97
Bishop, John 161
Bisset, Jacqueline 332
Blank, Bernhard 80, 81
Blume, Oliver 313, 315, 322
Bohn, Arno 223, 224, 249
Böhringer, Eugen 132
Bonnier, Jo 115, 120, 121, 122
Borcheller, Terry 284
Borkert, Mitja 286, 287, 297, 313
Bott, Helmuth 99, 103, 115, 124, 130, 132, 138, 159, 190, 193, 201, 206, 207, 212, 213, 220, 223
Boutsen, Thierry 247
Braess, Hans-Hermann 185
Branitzki, Heinz 159, 223
Bridges, Jeff 334
Briscoe, Ryan 273
Brodbeck, Tilman 164
Brooks, Tony 115
Buchler, Kevin 263
Burki, Emiel 313
Burst, Hermann 164, 208
Busby, Jim 212
Buzzetta, Joe 121, 137
Cabianca, Giulio 112
Campbell, Malcolm 58
Canepa, Bruce 11, 209, 305
Caracciola, Rudolf 31, 32, 33, 35, 36, 40, 45, 47, 50, 52, 53, 86

Carlson, Bob 257, 260
Castellotti, Eugenio 112
Castiglioni, Camillo 23, 27, 37
Chiron, Louis 71
Clark, Jim 122
Cobb, John 59
Collard, Emmanuel 274
Connolly, Jennifer 335
Cooper, Jacques 153
Cooper, John 106
Cruise, Tom 334, 335
Daimler, Gottlieb 19, 115
Daimler, Paul 19, 27, 31, 33
Dalmas, Yannick 233
Dalziel, Ryan 284
Dauer, Jochen 233
Davis, Colin 132
de Adamich, Andrea 163
de Beaufort, Carel 113, 121
de Cadenet, Alain 180
Dean, James 104
Dennis, Ron 204
Depailler, Patrick 181
Dickinson, Rob 305
Diess, Herbert 322
Domnick, Dr. 94
Donohue, Mark 159, 160, 161, 163, 164, 306
Dowe, Tony 238, 239
Dubonnet, André 30
Dumas, Romain 272, 273, 274, 303, 304
Dünninger, Michael 97
Durand, Georges 88
Dürheimer, Wolfgang 266, 278, 293, 298
Dusio, Piero 62, 69, 71, 74, 75, 96
Dyson, Rob 212
Elford, Vic 137, 138, 139, 144, 145, 146, 148, 155, 156, 157, 226, 325
Enzinger, Fritz 302
Eyb, Wolfgang 183
Eyston, George 58, 59
Fabi, Teo 219, 221
Facetti, Carlo 163
Falk, Peter 97, 138, 148, 183, 190, 201, 206, 215, 217, 232, 237
Farina, Guiseppe 52, 53
Farouk, King of Egypt 335
Faroux, Charles 71, 88, 89
Fennelly, Kieron 206
Ferdinand, Franz (Archduke) 17
Ferrari, Enzo 14, 27

Flannery, Susan 333
Flegl, Helmut 148, 164, 179, 183, 208, 215
Florio, Ignazio 31
Florio, Vincenzo 31, 112
Follmer, George 160, 163, 172
Ford, Henry 51, 57
Forstner, Egon 111
Foyt, AJ 137, 182
France, Bill 161
Franzen, Major 68
Freisinger, Manfred 143
Frère, Paul 101, 113, 201
Fröhlich, Karl 39, 55
Fuhrmann, Ernst 25, 84, 89, 92, 99, 100, 102, 109, 111, 121, 150, 158, 159, 161, 164, 165, 169, 179, 183, 188, 191, 193, 327
Ganz, Josef 42
Garretson, Bob 195
Gartner, Jo 214
Gendebien, Olivier 112, 113, 115, 120
Giacosa, Dante 71
Gielgud, John 334
Ginther, Richie 105
Giugiaro, Giorgetto 152, 153, 178
Giunti, Ignazio 138
Glemser, Dieter 146
Glöckler, Walter 99, 101
Glotzbach, Dieter 163
Goering, Hermann 63
Goldinger, Josef 22, 55
Gorissen, Wolfhelm 183
Graves, John 164
Gregg, Peter 163, 165
Gregory, Masten 121
Gurney, Dan 115, 120, 121, 122, 137
Haas, Carl 131
Haberl, Fritz 236
Hahn, Carl 146, 148
Haldi, Claude 164
Hamilton, Lewis 306
Hart, E.W. 16, 17
Härter, Holger 280
Hartley, Brendon 304, 305
Hasse, Rudolf 53
Hatter, Tony 234, 247, 264
Hatz, Wolfgang 293, 298, 302
Hawkins, Paul 137
Haywood, Hurley 154, 163, 164, 165, 172, 181, 182, 195, 202, 227, 233, 284, 305, 325, 329
Heinkel, Ernst 57

Heintz, Josh 78
Helmick, Dave 163, 164
Henn, Preston 212
Hensler, Paul 177, 203
Hering, Manfred 143
Herrarte, José 102
Herrmann, Hans 101, 102, 103, 105, 112, 113, 114, 115, 120, 121, 130, 136, 137, 138, 141, 146, 148, 156
Hetmann, Richard 215
Heuer, Eric 93
Hezemans, Toine 172
Hickman, Bill 104
Hill, Graham 120, 121
Hill, Phil 105, 120, 326
Hirst, Major Ivan 71
Hitler, Adolf 32, 39, 42, 43, 44, 50, 51, 54, 55, 56, 57, 59, 63, 64, 65, 66, 67, 68
Hitzinger, Alexander 302, 304
Hoffman, Max 71, 84, 90, 94, 96, 105, 108, 146
Holbert, Al 201, 212, 213, 218, 221
Holbert, Bob 120, 121
Horch, August 86
Horsman, John 148, 325
Hruska, Rudolf 69, 75
Hughes, John 334
Huidekoper, Weit 247
Hülkenberg, Nico 304
Hulme, Denny 159
Hunt, Bonnie 335
Ickx, Jacky 140, 141, 148, 163, 172, 173, 176, 177, 181, 182, 195, 201, 202, 209, 239, 305
Idelson, Howie 305
Jabouille, Jean-Pierre 181
Jaminet, Mathieu 323
Jani, Neel 303, 304, 306
Jantke, Manfred 150, 176, 199
Jarier, Jean-Pierre 226
Jaussaud, Jean-Pierre 182
Jellinek, Emil 12, 19
Jellinek, Maja 19
Jellinek, Mercedes 19
Jenatzy, Camille 19
Joest, Reinhold 156, 180, 202, 239
Johansson, Stefan 239
Jones, Davy 239
Joplin, Janis 144
Joplin, Laura 144
Judge, Mike 335
Juhan, Jaroslav 102, 105

Kaes, Ghislaine 39, 51, 54, 83, 86, 130
Kahnau, Bernd 237, 251, 274
Kaiser, Peter 78
Kales, Josef 39, 55
Kampik, Hans 198
Kauhsen, Willi 146, 156, 215
Kautz, Christian 53
Kelleners, Ralf 247
Kern, Hans 130, 159
Kern, Lars 318
Keyser, Michael 163
Kinnunen, Leo 151, 157, 164
Kirchdorffer, Gerhard 203
Kissel, Wilhelm 36
Klass, Günter 144
Klett, Arnulf 82, 173
Klie, Heinrich 125, 144, 152
Kluck, Timo 293
Koehler, Otto 54
Kolb, Eugen 136
Kolowrat, Alexander 'Sascha' 25, 27
Komenda, Erwin 39, 55, 60, 62, 76, 78, 79, 80, 85, 86, 125, 131, 152, 307, 322
Koskull, Cecilia 86
Kraus, Johannes 87
Kristensen, Tom 239, 241
Kuhn, Fritz 25, 27
Kulla, Matthias 293
Lafferentz, Bodo 51, 60
Laffite, Jacques 181, 226
Lagaaij, Harm 164, 178, 220, 222, 228, 234, 242, 257, 274, 307
Lai, Pinky 250, 270
Lang, Hermann 50, 53
Lange, Karlheinz 227
Lapine, Anatole 'Tony' 96, 150, 164, 168, 208, 307
Lapper, Herrmann 94
Laqua, Oliver 297
Larrousse, Gérard 141, 146, 156, 165
Larson, Grant 241, 242, 243, 264, 270, 274, 277
Lauda, Nikki 33, 204
Lautenschlager, Christian 30, 31
Lawrence, Martin 334
Ledwinka, Hans 42
Leven, Bruce 212, 213
Lieb, Marc 300, 303, 304
Lindner, Arno 97
Linge, Herbert 81, 102, 104, 106, 113, 116, 119, 130, 136, 146, 215
Lipe, Tippy 106
Lohner, Ludwig 15, 18

Long, Patrick 274, 305, 325
Loos, Georg 172
Lotz, Kurt 151
Lovely, Pete 106
Lövstad, Anton 42
Lozano Brothers, 284
Ludvigsen, Karl 11, 17, 18, 20, 30, 42, 55, 64, 65, 86, 87, 102, 106, 124, 125, 132, 148, 159, 188, 195, 270, 335
Ludwig, Klaus 202, 214
Luhr, Lucas 272, 273, 274
Lurani, Giovanni 69
Lutz, Bob 169, 193
Maassen, Sascha 272
Macht, Michael 280, 286, 296
Mackenzie, Angus 279
Maglioli, Umberto 105, 112, 138
Mahle, Eberhard 116
Mairesse, Willy 136
Mäkinen, Timo 132
Marchart, Horst 223, 241, 243, 251, 266
Marcus, Siegfried 13
Marko, Helmut 157
Masetti, Giulio 30
Mass, Jochen 172, 173, 176, 177, 195, 214
Mathé, Otto 60, 101
Mauer, Michael 274, 277, 298, 307, 309, 313
May, Michael 111, 168
Maybach, Wilhelm 19
Mays, Rex 52, 53
McAvoy, James 335
McGillis, Kelly 334
McNish, Allan 248
McQueen, Steve 155, 332, 333
Merckle, Adolf 280
Merz, Otto 33, 35, 36
Messerschmitt, Willy 57
Metge, René 209
Mezger, Anne 60
Mezger, Hans 124, 126, 131, 132, 147, 150, 158, 159, 171, 204, 221, 223, 250
Mickl, Josef 39, 55, 59, 66, 71, 102, 111
Miersch, Hans 97
Miles, Ken 106
Millanta, Corrado 69
Mitter, Gerhard 133, 137, 140
Möbius, Wolfgang 168, 185, 208
Moneim, Prince Abdel 335
Moore, Dudley 334
Moss, Stirling 121
Motto, Rocco 120

Mouche, Edmond 92
Mullen, Jim 195
Müller, Herbert 136, 163, 164, 165
Müller, Jörg 248
Müller, Peter 242
Müller, Petermax 99
Murkett, Steve 257
Murphy, Eddie 334
Nail, Joanne 333
Nakajima, Kazuki 304
Nauck, Alfred 69
Neal, Dan 97
Neerpasch, Jochen 137, 138, 146
Neubauer, Alfred 21, 25, 26, 27, 30, 31, 32, 35
Neumeyer, Fritz 41, 42
Newman, Paul 177, 195, 332
Nibel, Hans 36
Nierop, Kees 195
Noppen, Rudi 223
Nordhoff, Elisabeth 108
Nordhoff, Heinz 78, 86, 87, 97, 105, 108, 148, 151
Nurse, Dave 154
Nuvolari, Tazio 33, 52, 53, 69, 71, 74
Oliver, Jackie 141, 156, 161, 224
Ongais, Danny 191
Ortelli, Stéphane 248
Pacino, Al 334
Penske, Roger 114, 120, 159
Perón, Juan 71, 75
Pescarolo, Henri 165, 181, 202
Peugeot, Jean-Pierre 69, 71
Peyron, Christina 86
Piëch, Anton 21, 30, 40, 57, 63, 69, 71, 96, 141
Piëch, Ernst 108
Piëch, Ferdinand 10, 83, 96, 118, 122, 125, 126, 132, 139, 141, 147, 148, 156, 157, 159, 177, 183, 191, 206, 215, 220, 257, 280, 282, 297, 298, 302, 305
Pinchot, Bronson 334
Pironi, Didier 181, 182
Pöcher, Lambert 25
Polak, Vasek 144, 172
Popp, Franz Josef 56
Porsche, Anna (Ferdinand's mother) 13
Porsche, Anna (Ferdinand's sister) 15
Porsche, Anton 13, 14
Porsche, Dorothea Reitz 21, 46, 47, 96, 249
Porsche, F.A. 'Butzi' 10, 47, 115, 118, 125, 130, 131, 152, 164, 223, 236, 251, 257, 280, 282, 305, 307, 322
Porsche, Ferdinand 10, 12, 13, 14, 15, 16, 17, 18, 19, 20, 22, 23, 26, 27, 30, 31, 32, 33, 36, 39, 40, 51, 54, 56, 76, 85, 87, 158, 247, 249

Porsche, Ferry Multiple
Porsche, Gerhard 57
Porsche, Hans-Peter 63
Porsche, Hedwig 15
Porsche, Oskar 15
Porsche, Wolfgang 16, 68, 109, 280, 282, 310, 311
Porsche-Piëch, Louise 17, 21, 32, 80, 84, 96, 125, 141, 171, 249
Preuninger, Andreas 284, 296
Prince Henry, 12, 19, 20, 22
Prinzing, Albert 82, 86, 110
Prost, Alain 204
Pruett, Scott 284
Pucci, Antonio 132
Pyta, Wolfram 311
Rabe, Karl 23, 27, 39, 55, 74, 78, 96
Rahal, Bobby 195
Redford, Robert 332, 335
Redman, Brian 11, 118, 140, 148, 151, 157, 161, 163, 195, 257, 260, 325
Rees, Alan 224
Reeves, Colonel 68
Regazzoni, Clay 163
Reimspiess, Franz Xaver 55, 94, 131
Reitter, Horst 221, 247
Reuter, Manuel 239
Revson, Peter 333
Richards, Dave 144
Ringwald, Molly 334
Rockenfeller, Mike 284
Rodriguez, Pedro 140, 151, 156, 157
Rohr, Jochen 247
Röhrl, Walter 209, 264, 284
Rommel, Manfred 173
Rondeau, Jean 195
Rose, Mauri 53
Rosemeyer, Bernd 47, 50, 51, 52, 53
Rosenberger, Adolf 31, 32, 39, 40, 43, 60
Ruf Jr., Alois 124, 221
Rumpler, Edmund 38, 42, 44
Saracoglu, Hacan 298
Sauber, Peter 213
Sauerwein, Rudolph 89
Scaglione, Franco 119
Scarfiotti, Ludovico 137
Schäffer, Valentin 159
Schechter, Roy 120
Schmid, Leopold 75, 97, 123
Schmöltz, Fabien 313
Schröder, Gerhard 125

Schulthess, Hermann 78
Schultz, Michael 198
Schuppan, Vern 195, 202, 233
Schutz, Peter 190, 191, 193, 195, 198, 199, 201, 203, 206, 218, 223, 224, 249
Schütz, Udo 137, 140
Schwab, Fred 226
Seaman, Dick 50, 52, 53
Seidel, Wolfgang 116
Seidl, Andreas 304, 305
Seinfeld, Jerry 236, 237
Senden, Hans-Leo 143
Senna, Ayrton 214
Shaw, Wilbur 53
Sheen, Charlie 334
Sholar, Bill 105
Siffert, Jo 137, 138, 139, 140, 146, 148, 151, 155, 156, 157
Singer, Norbert 10, 150, 161, 164, 165, 172, 179, 199, 201, 206, 221, 226, 233, 239, 247, 256, 272, 305
Sloniger, Jerry 201
Smith, Will 334
Snyder, Jimmy 53
Söderberg, Dick 168, 178, 208, 234
Sommer, Raymond 53, 71
Sparv, Camilla 332
Speer, Albert 66, 68, 69
Spielberger, Walter 117
Spoerry, Dieter 146
Sprenger, Rolf 217
Springer, Alwin 238
Srock, Rainer 251
Stein, Al 137
Steinemann, Rico 146
Steiner, Michael 278
Stewart, Jackie 33, 333
Stommelen, Rolf 137, 148, 163, 174, 180, 195, 259
Stone, David 146
Stout, Pete 287
Strähle, Paul Ernst 116, 119
Strenger, Erich 118, 147
Strom, Michael 263
Stuck, Hans (Junior) 50, 213, 214, 227, 247, 277, 325
Stuck, Hans (Senior) 44, 45, 46, 47, 49, 50, 53, 58, 59, 158
Szodfridt, Imre 215
Tambay, Patrick 181
Tandy, Nick 304, 323
Tarantino, Quentin 334
Taruffi, Piero 112, 113
Taylor, Ricky 323
Thirion, Gilberte 89

Tito, Josip Broz 22
Titus, Jerry 144
Todt, Fritz 66
Toivonen, Pauli 147
Tomala, Hans 123, 126, 132
Trippel, Hans 64
Trostmann, Bruno 100
Tschudi, Jolanda 75
Uhlenhaut, Rudolf 50
Vaccarella, Nino 136, 163
van Lennep, Gijs 154, 157, 163, 164, 165, 180, 181
Varga, Peter 277, 312, 318
Vettel, Sebastian 306
Veuillet, Auguste 87, 92, 152
Vögele, Charles 146
von Brauchitsch, Manfred 50
von Braun, Wernher 68
von Delius, Ernst 50
von Eberhorst, Eberan 52, 53, 74, 80
von Frankenberg, Richard 27, 47, 57, 86, 95, 101, 111, 114, 147
von Goertz, Albrecht 125, 153
von Guilleaume, Paul 89
von Hanstein, Huschke 95, 112, 133, 137, 144, 176
von Neumann, John 104, 106
von Oertzen, Klaus 43, 44
von Rücker, Klaus 102, 123
von Senger, Rupprecht 81, 326
von Trips, Wolfgang 111, 112, 115, 120, 121
von Veyder-Malberg, Hans 40
Vorwig, Wilhelm 55
Wagner, Harald 142
Walb, Willi 35
Waldegård, Björn 147, 154
Walker, Rob 121
Walkinshaw, Tom 213, 238
Walliser, Frank-Steffen 298, 323
Walter, Hans 119
Webber, Mark 304
Weber, Friedrich 75
Wehrlein, Pascal 320
Werlin, Jakob 42, 51
Werner, Christian 30, 31, 35, 36
Wiedeking, Wendelin 222, 223, 237, 238, 239, 241, 242, 249, 250, 256, 257, 264, 277, 278, 280
Willis, Bruce 335
Winkelhock, Manfred 214
Winter, John 202
Wolf, Dick 334
Wolf Family 57

Wollek, Bob 214, 226, 247, 248, 265
Woolfe, John 148
Wurz, Alex 239
Wüst, Rainer 277
Wütherich, Rolf 104, 132
Wyer, John 140, 148, 155, 157
Zadnick, Otto 68
Zahradnik, Josef 39, 55
Zborowski, Count 31

RACE/TRACK/EVENT

Ad Diriyah, Saudi Arabia 313
Alpine Rally 60
Avus, Germany 30, 31, 32, 45, 111
Berlin-to-Rome 60, 326
Brands Hatch, England 140
Can-Am 114, 130, 131, 137, 150, 157, 159, 160, 169, 180, 201, 272, 328
Cannonball Run Trophy Dash 333
Carrera Panamericana, Mexico 86, 101, 102, 103, 104
Dakar Rally, North Africa 208, 209, 330
Daytona, Florida 131, 136, 137, 138, 148, 151, 155, 157, 163, 164, 165, 177, 195, 212, 239, 260, 263, 284, 317, 321, 322
Feisberg Hill Climb, Germany 46
Gabelbach Hill Climb, Germany 36
German Grand Prix 31, 32, 33, 35, 45, 47, 49, 103, 121, 173
Goodwood, England 321
Gordon Bennett Cup 19, 31
Indianapolis 500 30, 31, 32, 137, 191, 221
Interlagos, Brazil 303
IROC 164
Kartellfahrt Rally, Austria 32
Kyalami, South Africa 215
Laguna Seca, California 160, 221, 260, 272, 273, 304, 305
Le Mans 10, 31, 50, 59, 71, 84, 87, 88, 89, 90, 92, 97, 99, 101, 103, 111, 112, 113, 114, 115, 119, 121, 130, 132, 136, 137, 138, 139, 140, 141, 142, 144, 148, 150, 152, 154, 155, 156, 157, 160, 161, 164, 165, 168, 172, 173, 176, 177, 180, 181, 182, 183, 190, 191, 193, 194, 195, 200, 201, 202, 203, 206, 209, 213, 214, 215, 222, 226, 227, 233, 238, 239, 240, 241, 246, 247, 248, 262, 263, 264, 270, 272, 273, 274, 283, 284, 302, 303, 304, 305, 318, 321, 326, 328, 330, 331, 333
Liège-Rome-Liège Rally 95, 109, 146
Lime Rock, Connecticut 260, 274
Marathon de la Route 145, 146, 152
Meadowlands, New Jersey 219, 221
Midnight Sun Rally, Sweden 86
Mid-Ohio, Lexington, Ohio 161, 213, 221, 273, 274
Mille Miglia 31, 40, 71, 95, 99, 102
Miller Motorsports Park, Utah 274
Mitholz-Kandersteg Hill Climb, Switzerland 78

Monte Carlo Rally 132, 146, 154, 215

Monterey Historic Races/Monterey Motorsport Reunion 111, 248, 326

Monza, Italy 27, 31, 121, 139, 140, 146, 215

Mugello, Italy 172

Nardò, Italy 161, 209, 218, 241, 290

Nelson Ledges racetrack, Ohio 204, 205

Norisring, Germany 173

Nürburgring 24, 33, 35, 45, 47, 49, 50, 103, 112, 115, 119, 120, 121, 122, 136, 137, 139, 140, 146, 148, 155, 157, 174, 180, 202, 251, 259, 263, 264, 266, 275, 284, 285, 288, 293, 296, 297, 300, 306, 309, 311, 313, 317, 318, 323

Ollon-Villars Hill Climb, Switzerland 133

Ontario Motor Speedway, California 191

Paris-Dakar Rally 208, 209, 330

Paul Ricard, France 164, 180, 200

Professor Porsche Memorial Races 101

Reims, France 121

Rennsport Reunion 256, 257, 260, 298, 304, 305, 329, 335

Riverside Raceway, California 213

Road America track, Wisconsin 274

Road Atlanta, Georgia 160

Rossfeld Hill Climb, Germany 137

Rouen, France 122

Safari Rally, Kenya 150, 161, 189

Sebring, Florida 120, 131, 136, 138, 140, 155, 156, 163, 178, 195, 274, 284, 333

Semmering Hill Climb, Austria 16, 17, 19, 27, 33, 36

Silverstone, England 176, 200, 213

Snetterton, England 218

Spa-Francorchamps, Belgium 53, 140, 156

Targa Florio 10, 24, 25, 26, 27, 30, 31, 32, 33, 112, 114, 119, 120, 121, 130, 132, 135, 136, 137, 138, 142, 155, 157, 163, 311

Tour de Course Rally, Corsica 144, 146

Tour de France 146

Trans-Siberia Rally 261

Vanderbilt Cup 50

Watkins Glen, New York 121, 140, 160, 173, 260

Zandvoort, Holland 122

Zeltweg, Austia 121, 133, 140, 148

Zhuhai, China 226

PLACES

Austria 10, 13, 17, 19, 21, 22, 23, 25, 36, 37, 39, 54, 58, 60, 62, 65, 66, 68, 69, 71, 78, 80, 81, 85, 90, 96, 108, 110, 148, 311

Austria-Hungary 13, 19, 23, 54

Bohemia 13, 272

Bonneville 59, 218

Bratislava 309

Gmünd 68, 69, 71, 74, 75, 76, 79, 80, 81, 82, 86, 89, 96, 97, 102, 106, 109, 141

Hemmingen 257

Hungary 13, 17, 19, 23, 54

Kulim 320

Leipzig 257, 259, 265, 266, 277, 297

Maffersdorf 13, 17, 168

Malaysia 320

Neckarsulm 177, 203, 205, 207

Paris Salon 37, 86, 87, 94

Paris World Exposition 16

Porsche Museum 15, 16, 26, 40, 60, 75, 78, 97, 114, 125, 159, 176, 194, 232, 239, 241, 256, 276, 281, 305, 320

Reichenberg 14, 15, 64

Sicily 27, 31, 33, 120, 132, 137, 140, 142, 157, 163

Solitude Palace 86

Stuttgart 10, 19, 20, 21, 25, 27, 31, 39, 41, 45, 54, 57, 59, 60, 63, 64, 68, 81, 82, 84, 85, 86, 87, 94, 96, 97, 105, 108, 109, 122, 123, 130, 136, 146, 152, 158, 169, 173, 177, 193, 206, 255, 257, 266, 281, 311, 315, 316, 323

Swabia 21

Vienna 13, 14, 15, 17, 21, 23, 90, 102, 281, 311

Weissach 118, 123, 130, 159, 183, 184, 185, 192, 200, 201, 204, 206, 209, 213, 215, 217, 225, 234, 239, 241, 257, 273, 300, 303, 323

Wolfsburg 57, 69, 71, 81, 86, 105

Zell am See 58, 68, 69, 70, 87, 101

Design	Jodi Ellis Graphics
Printer	Interpress Ltd.
Page Size	230 mm x 280 mm
Text paper	150 gsm Magno Gloss
End paper	170 gsm Woodfree Offset
Dust jacket	150 gsm Glossy Artpaper
Casing	Foil stamping on front and spine, on Surbalin Rubinrot Honan Prague, over 3 mm board
Chapter Heads	36pt. Eurostile Bold
Subheads	15pt. Eurostile Bold
Main Body Text	11pt. Baskerville Regular
Captions	10pt. Univers 55